Oceanography of the Southeastern U.S. Continental Shelf

Coastal and Estuarine Sciences

A series devoted to advancing our knowledge of physical, chemical, and biological processes in coastal and estuarine regimes and their relevance to societal concerns.

1. Coastal Upwelling
2. Oceanography of the Southeastern U.S. Continental Shelf

Coastal and Estuarine Sciences 2

Oceanography of the Southeastern U.S. Continental Shelf

L. P. Atkinson
D. W. Menzel
K. A. Bush
Editors

American Geophysical Union
Washington, D. C.
1985

Oceanography of the Southeastern U. S. Continental Shelf

Library of Congress Cataloging in Publication Data

Library of Congress Cataloging-in-Publication Data

Main entry under title:

Oceanography of the southeastern U.S. continental shelf.

(Coastal and estuarine sciences, ISSN 0733-9569; 2)
Includes bibliographies.
1. Oceanography—Atlantic Coast (U.S.)
2. Continental shelf—Southeastern States. 3. Marine
biology—Atlantic Coast (U.S.) I. Atkinson, L. P.
II. Menzel, D. W. III. Bush, K. A. IV. American
Geophysical Union. V. Series.
GC511.027 1985 551.46'148 85-9130
ISBN 0-87590-251-0
ISSN 0733-9569

Printed in the United States of America.

CONTENTS

PREFACE

The AGU Monograph Series on Coastal and Estuarine Regimes is intended to provide timely summaries and reviews of major process and regional studies, both observational and theoretical, and of theoretical and numerical models. It grew out of an IAPSO/SCOR/ECOR working group initiative several years ago designed to enhance scientific communications on this topic. The series' authors and editors are drawn from the international community. The ultimate goal is to stimulate bringing the theory and observations of coastal and estuarine regimes together on the global scale.

The study of coastal and estuarine regimes is important scientifically because they are where the "oceans meet the continents." In other words, it is through the estuaries and coastal oceans that materials are exchanged between the oceans and continents. From a geophysical fluid dynamics perspective, estuarine and coastal waters present a rich array of challenging phenomena due to the extreme ranges of density stratification and topographic variation found there. Especially in the coastal oceans, the effects of the earth's rotation (Coriolis force) are dominant, too. The coastal oceans are notable for their intense variability, due in part to their extraordinary responsiveness to the passage of atmospheric weather systems. Another great source of variability in addition to river runoff and atmospheric forcing has only been recently appreciated: meandering boundary currents and synoptic/mesoscale eddies of the open ocean impinge upon the continental margins and entrain waters from the coastal oceans. Conversely, some of the eddies of the open ocean may originate from unstable flows in the coastal ocean. Closely associated with the intense physical variability of coastal and estuarine regimes is the rich and varied structure of coastal ecosystems and their well-known high concentrations and productivity at all trophic levels. How the physical and biological aspects of the coastal and estuarine systems are connected is yet to be fully understood. However, the present volume reports a study which takes significant steps in this direction.

The first volume in the series was entitled *Coastal Upwelling,* which summarized the state of multidisciplinary knowledge, on an international basis, of the coastal upwelling process as of 1980.

The present volume on the South Atlantic Bight, the second in the series, summarizes the state of multidisciplinary knowledge on an entire continental shelf regime, providing the first comprehensive synthesis of the major new understanding acquired through a decade-long scientific campaign on a particular shelf regime. This campaign is notable for the variety of physical, chemical, and biological studies undertaken in a sustained and adaptive fashion as the scientific investigators pursued mounting evidence for the role of Gulf Stream meanders in driving the coastal ecosystem of interest through eddy fluxes of nutrients from offshore, a striking new finding which reversed conventional wisdom. The efficacy of well-placed biological samples based upon adequately resolved representations of the synoptic state of physical and chemical fields is amply demonstrated, which provides a proven sampling strategy that could be adopted in other regions.

Other forthcoming volumes in the series will treat dynamical and modeling topics, as well as further process and regional studies. Scientists interested in proposing volumes for this series should contact the series editor or AGU headquarters for further information.

Christopher N. K. Mooers
Series Editor

Introduction: Oceanography of the Southeast United States Continental Shelf

L. P. ATKINSON AND D. W. MENZEL

Skidaway Institute of Oceanography, Savannah, Georgia

The continental shelf area between West Palm Beach, Florida, and Cape Hatteras, North Carolina, known as the South Atlantic Bight (SAB), is a shallow shelf geomorphologically isolated from adjacent continental shelves by Cape Hatteras to the north and the very narrow south Florida shelf to the south. The width of the shelf varies from a few kilometers off Florida to nearly 200 km off Georgia. The SAB extends from approximately 27° to 35°N latitude, with climatic variations ranging from subtropical to temperate. The large latitudinal range results in significant latitudinal variation in weather-related physical processes. The cross-shelf variation in physical, chemical, and biological processes is also significant because of interaction with the estuaries and marshes along the coast and with the Gulf Stream at the shelf break. The interaction with coastal marshes is strongest in the central SAB where tidal range is highest. Interaction with the Gulf Stream also varies latitudinally in a more complex manner. The papers in this book describe the distributions of physical, chemical, and biological variables in the SAB and the processes that control them.

Introduction

The southeastern United States continental shelf, commonly called the South Atlantic Bight (SAB), is distinguished by the proximity of the Gulf Stream to the shelf break and by the dominating effect of the stream on adjacent shelf waters. The influence of the Gulf Stream, combined with the unusually broad and shallow SAB shelf, results in a complex interplay of oceanic and atmospheric forces that affect the physics, chemistry, and biology of the shelf waters in diverse and often unusual ways. Large contrasts in the distribution of properties, the strength of oceanic and atmospheric forces, and the high frequency (4–12 days) at which these forces vary have created a unique natural laboratory in which a variety of oceanic processes may be studied. Since 1975 the authors whose contributions are included in this book have investigated the oceanography and meteorology of the SAB, using sampling strategies and techniques designed to study specific oceanic events as synoptically as possible. These observations have been supplemented with long-term records from moored instruments to determine the frequency and spatial correlation of oceanic processes. In the following papers, various aspects of the oceanography of the region are discussed. This paper provides a brief description of the SAB and reviews the history of oceanographic research in the area.

The South Atlantic Bight

The South Atlantic Bight, as defined here, extends from West Palm Beach, Florida, where the narrow south Florida shelf begins to broaden, to Cape Hatteras, North Carolina, where the shelf narrows again (Figure 1). Maximum shelf widths in the central SAB are nearly 200 km. Major geomorphic features include several prominent capes which significantly affect shelf circulation. Water depth at the shelf break is about 50 m, in contrast to that near the Florida Keys (10–20 m) and the Middle Atlantic Bight/Georges Bank area (150 m) [*Emery and Uchupi,* 1972]. The mean axis of the Gulf Stream (Plate 1) follows the shelf break from the Florida Straits to about 32°N, where it veers offshore owing to the influence of a submarine ridge, forming a semipermanent meander [*Pashinski and Maul,* 1973; *Legeckis,* 1979; *Bane and Brooks,* 1979]. North of the ridge the Gulf Stream again tends to follow the shelf break, although here meanders of the current have larger spatial scales than south of 32°N. Thus the effect of the Gulf Stream on the shelf is expected to vary latitudinally. Cape Hatteras and the adjacent Gulf Stream tend to form an oceanic barrier to southerly flow from the Middle Atlantic Bight, except during intense southward wind events [*Stefansson et al.,* 1971].

Nearshore processes in the SAB also vary latitudinally. Fresh water enters the Bight from many small rivers between the St. Johns River to the south and the Cape Fear River to the north. Most of the river discharge is rather evenly distributed in the central SAB, so that runoff can be considered a line source of fresh water. The extreme northern and southern portions of the SAB receive little runoff from rivers (Figure 2). River discharge is usually

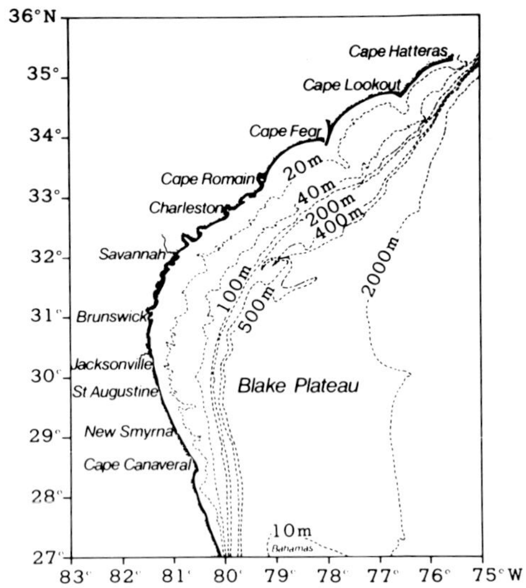

Fig. 1. Bathymetric chart of SAB showing capes, general bathymetry, and cities.

maximum in March when monthly flows may reach 7.7 km^3. During low runoff, flow is about 3 km^3/month.

The coastline of the SAB can be described as a perforated barrier, especially in the central two thirds where extensive salt marshes and sounds are connected to the sea by tidal inlets spaced about 10–20 km apart (Plate 2). Because tidal ranges in the central SAB are almost 3 m, most inshore waters are in daily contact with salt marshes that surround the tidal inlets, with important physical, chemical, and biological implications. Table 1 summarizes some of the physical features of the SAB.

In this book the terms inner, middle, and outer shelf are often used to denote geographic and physical process domains. The inner shelf is dominated by tidal currents, river inflow, winds, and interaction with coastal sounds. The middle shelf, which is dominated by winds, has less interaction with the coastal environment but is influenced by the Gulf Stream. The outer shelf is dominated by the Gulf Stream. In general, the depth divisions for the inner, middle, and outer shelf are 0–20, 20–40, and 40–60 m, respectively, although to the north and south Gulf Stream effects are felt further onshore.

Because there are some naturally occurring divisions, some of the papers in the book address specific subregions while others discuss the whole region. The paper by Blanton et al. presents the wind stress climatology for the SAB noting both the subregional and seasonal variations. The papers by Pietrafesa et al. and Lee et al. discuss circulation in subregions of the SAB; the former paper deals with the area north of Cape Fear that is significantly influenced by Cape Fear, Cape Lookout, and Cape Hatteras, while the latter paper treats the area south of Cape Fear where such capes are absent. The tidal regime in the SAB is presented in another paper by Pietrafesa et al. The general hydrography and impact of Gulf Stream intrusions on nutrient supply to the SAB is discussed in the paper by Atkinson. Later papers discuss the biology and chemistry of the SAB. The response of phytoplankton to Gulf Stream and shelf processes is presented in the paper by Yoder, and zooplankton are discussed in the paper by Paffenhöfer. Microbiological processes are presented in the paper by Pomeroy. The results of extensive microinfaunal observations appear in the paper by Tenore. Processes that affect the distribution of trace metals are discussed in the paper by Windom and Smith. Finally, in the last paper by Bush et al. we discuss the future research plans for the SAB.

History of SAB Oceanographic Research

The history of oceanographic exploration in the South Atlantic Bight is not as extensive as that of other continental shelves. Observations and publications do, however, date back at least 200 years to *Franklin* [1786], who published observations on the flow of the Gulf Stream to provide sailing directions between New England and England. These directions included a previously published chart showing the general flow of the Gulf Stream north from the Florida Straits.

Abbe [1895] noted that the cuspate shape of the Carolina capes may have been caused by "back eddies" of the Gulf Stream. He also stated that southerly coastal currents had been observed by state geologists in South Carolina. *Bumpus* [1955] reviewed many older papers, which are recommended reading.

Bumpus [1973, p. 137] summarized his knowledge of circulation in the SAB by stating that there are

> two conflicting systems at play here, a geostrophic current interrupted by invasions of the Florida Current. The geostrophic current tends to flow southerly and does so successfully in May, during the late summer, and early autumn, from Frying Pan Shoals [Cape Fear] southward. It is interrupted frequently by invasions of the Florida Current riding up over the shelf carrying surface water northward. On those occasions when the recovery rate [surface drifter] is poor in the South Atlantic Bight, one can generally assume the surface water has been forced out by a meander of the Florida Current, and the drift bottles have been carried north past Cape Hatteras and out of the immediate system. It would appear that meanders are more frequent and more successful in accomplishing this total exchange of water north of Frying Pan Shoals.

Modern studies of the oceanography of the SAB began with circulation studies by Bumpus between 1949 and the 1960's [*Bumpus and Wehe,* 1949; *Bumpus,* 1955, 1973]. Other work included observations of biological indicators of circulation patterns [*Bumpus and Pierce,* 1955] and the effect of physical variations on shelf biology [*Parr,* 1933; *Pearse and Williams,* 1951; *Pierce,* 1953; *Williams,* 1948, 1949].

While the above studies were in progress, the dynamics of Gulf Stream circulation were also being investigated.

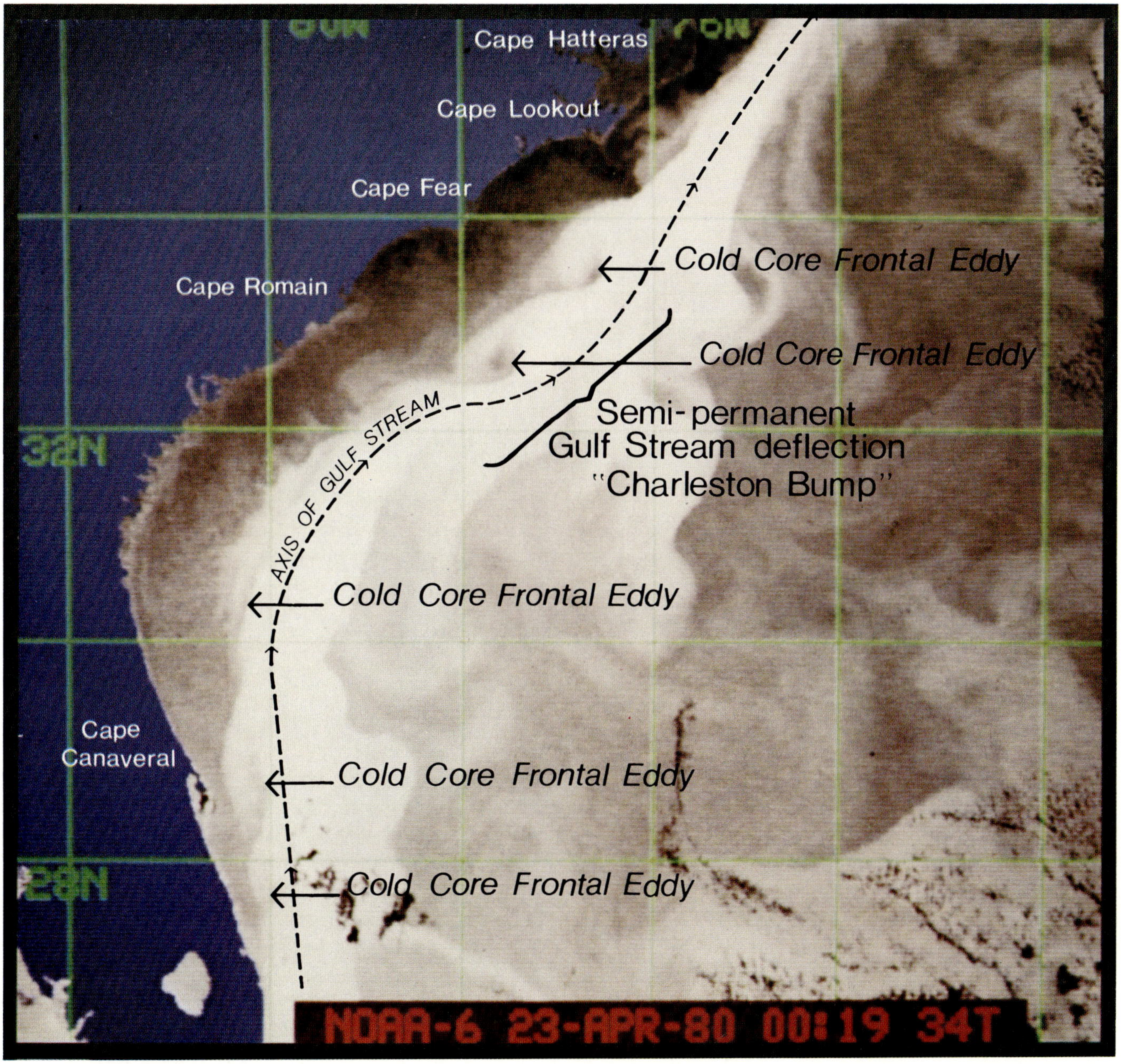

Plate 1. Infrared photograph of the SAB showing the Gulf Stream and adjacent shelf waters. Prominent dynamic features are noted. Image provided by O. Brown and R. Evans, University of Miami.

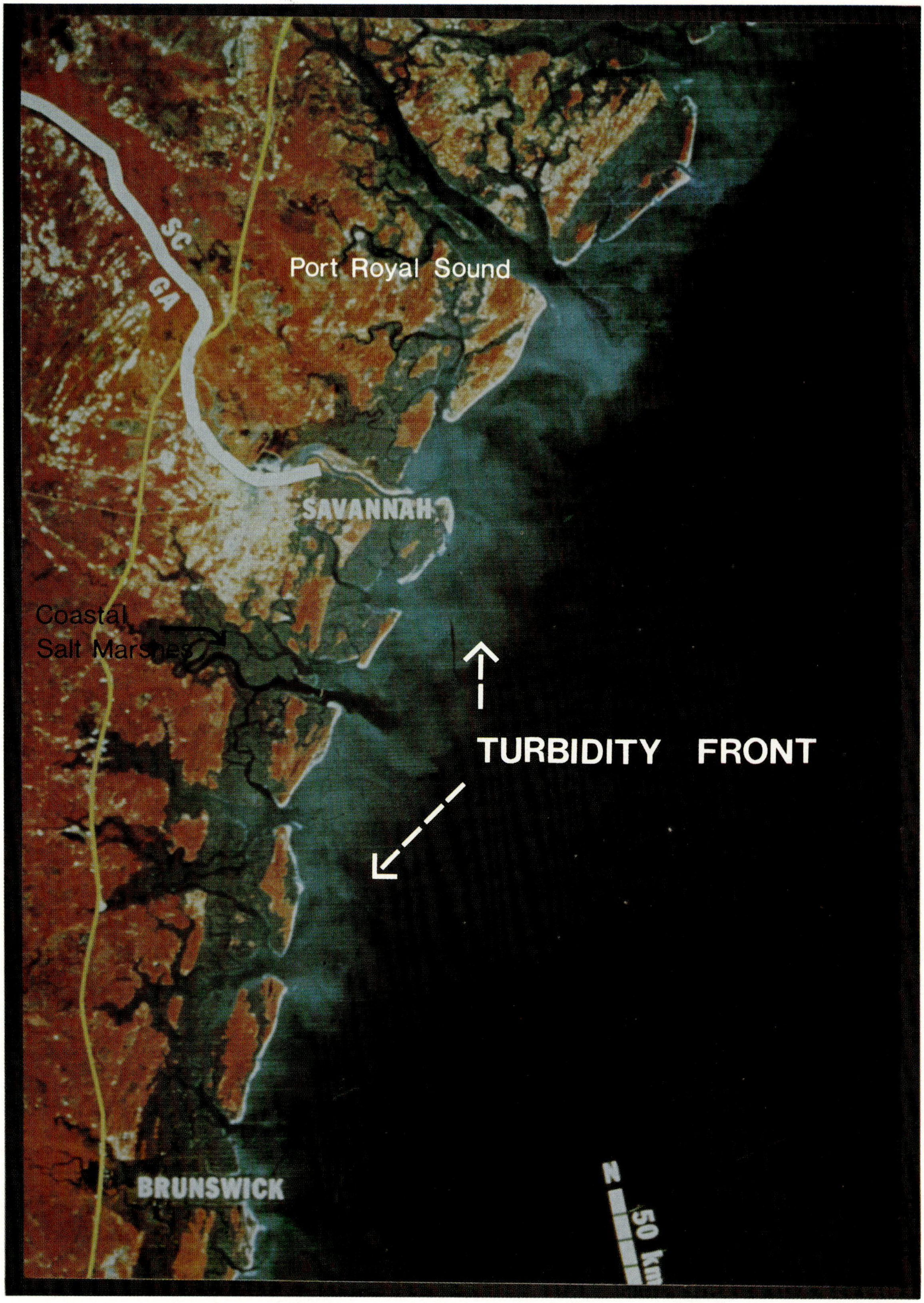

Plate 2. Infrared image of Georgia coast taken with thematic mapper aboard Landsat D. Image taken at 0943, November 9, 1982. Image provided by H. Kim, NASA Goddard Space Flight Center. The extensive salt marshes are represented by the light green areas behind the islands.

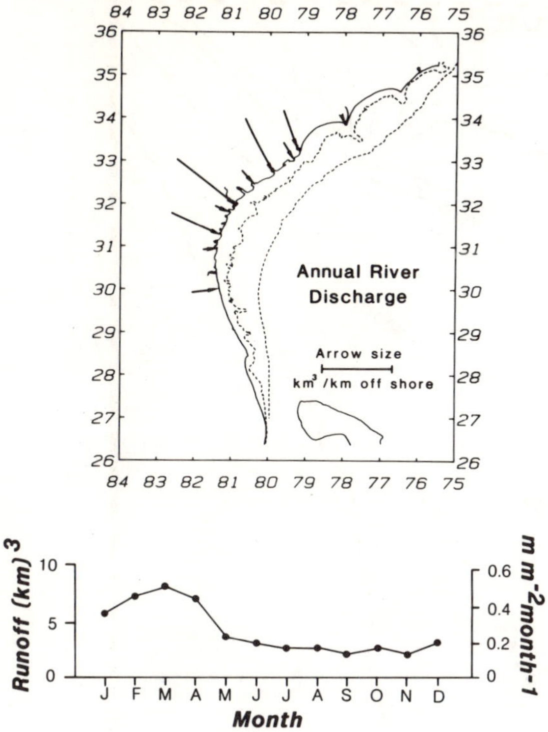

Fig. 2. Chart of SAB showing concentration of river flow and average monthly flow. Arrows on chart show annual average river flow as cubic kilometers of runoff per kilometer of shoreline. The kilometer of shoreline unit is calculated by summing one-half the distance to the adjacent gaged rivers. River flow is concentrated in the central portion of the SAB.

Important studies include those of *Webster* [1961] on Gulf Stream meanders off Onslow Bay, describing the meander process, and that of *Ford et al.* [1952] on the occurrence of a low-salinity band in the Gulf Stream front. In the 1960's, North Carolina shelf waters were studied for several years during all seasons, and the dominating influence of the Gulf Stream was identified [*Stefansson and Atkinson,* 1971; *Blanton,* 1971]. These earlier research efforts in SAB shelf waters and the adjacent Gulf Stream provided the impetus to study the influence of the Gulf Stream on shelf oceanography.

By the late 1960's, research in the SAB slowed. Conceptual models were quite advanced, but modern technology, such as satellite remote sensing and computerized shipboard data acquisition systems, was lacking. Into the early 1970's, researchers used what we now call older technology (i.e., bottles on wire), less-than-synoptic sampling, no satellite remote sensing, and few or no moored devices.

Our basic lack of knowledge was in areas related to the effect of the Gulf Stream on shelf waters and to the extent and nature of the coupling between coastal marshes and rivers with coastal and middle shelf waters. In 1973 and 1974, seasonal cruises between Jacksonville and Charleston provided new insights into the shelf mixing processes [*Atkinson et al.,* 1978*a*], the seasonal and latitudinal extent of runoff and intrusions of Gulf Stream water onto the shelf [*Atkinson,* 1977], and phytoplankton production [*Haines and Dunstan,* 1975]. In the mid-1970's the physical dynamics of the shelf were still not well known conceptually or quantitatively, and the effect of circulation on chemistry and geology was not fully understood. We first attempted to study the short-term effect of a Gulf Stream intrusion into shelf waters in 1975. It quickly became apparent that the effect of the Gulf Stream on biological processes on the shelf proper must be a significant, if not a dominant, controlling factor [*Atkinson et al.,* 1978*b*]. The plan for this study was derived from analyses of data from 1973–1974 cruises and the observations of others [*Green,* 1944; *Taylor and Stewart,* 1959; *Blanton,* 1971; *Bumpus,* 1973]. In the late 1970's, research increased with primary interest focused on Gulf Stream/shelf water interaction. By that time, modern technology was available to aid in observing and interpreting various controlling processes.

During the planning of the research discussed in subsequent papers of this book, a principal problem was to adjust sampling scales to the phenomena being studied. Because the time scale over which processes occur in the SAB is rather short, especially at the shelf break where Gulf Stream meanders pass every 5 to 10 days, it was important to repeatedly sample a large area in a short period of time. This requirement was almost impossible to fulfill because of slow ship speeds. An alternative was to study subregions in which a particular process was occurring. This required remote sensing technology to identify the appropriate study areas. By receiving surface temperature maps on board ship, we were able to direct sampling operations to specific locations where a process, typically the passage of a frontal eddy, was occurring. Much of our research was in the spring, usually in April when surface thermal contrast is high. Even with this assistance from remote imagery, large-scale studies required two ships, one mapping the shelf break area and another the remaining shelf. The ship sampling the larger area of the shelf was equipped with conventional CTD's and equipment for surface mapping of temperature, salinity, chlorophyll, nutrients, etc., requiring a minimum of station time. The other ship accommodated scientists whose studies required more station time. The advantage of this dual approach is that observations that require extended station time are also those that cannot create a large number of samples because of lengthy sample processing.

In many sampling efforts we attempted to repeatedly sample a few locations. With two ships we coordinated our movements such that when the mapping ship was away from a required sampling location, the other ship was directed there.

Two major field experiments demonstrate these strategies. Georgia Bight Experiments (GABEX) I and II took place in the spring of 1980 and the summer of 1981, re-

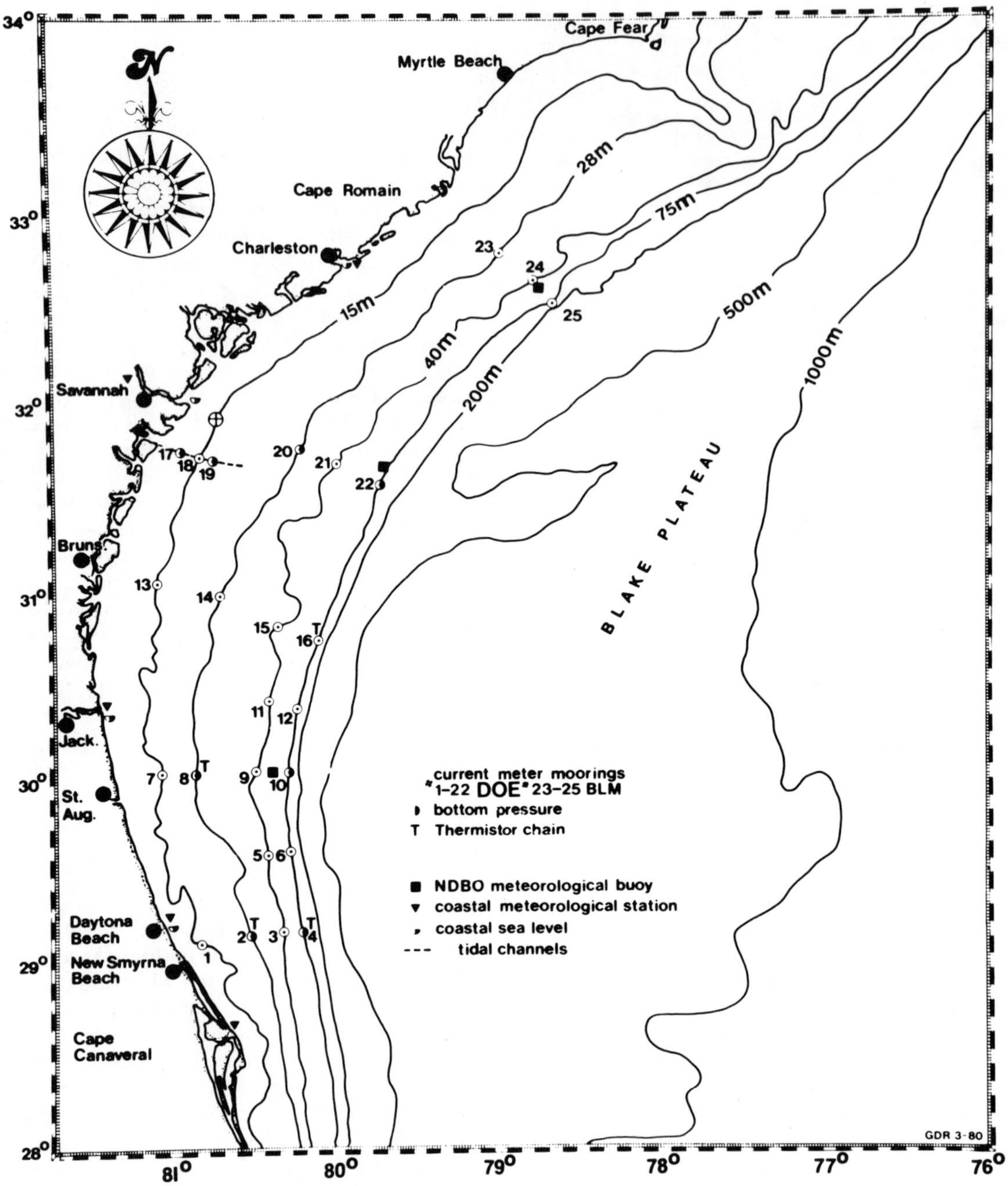

Fig. 3. Current meter locations during GABEX I and II. Moorings over the inner and middle shelf were deployed by L. Pietrafesa (North Carolina State University). Outer shelf moorings were deployed by T. Lee (University of Miami). Instruments on the Savannah Navigational Light Tower were maintained by J. Blanton (Skidaway Institute of Oceanography).

spectively. GABEX I was designed to study Gulf Stream shelf water interaction when shelf waters were vertically well mixed. GABEX II was designed to study the same waters during stratified summer conditions. These two field programs included shipboard observations and moored instruments and were designed with attention to both the spatial and temporal scales known to occur in the area and the various operational constraints that occur in oceanographic operation. The current meters were typically moored at spatial distances of 100 km alongshelf and 20

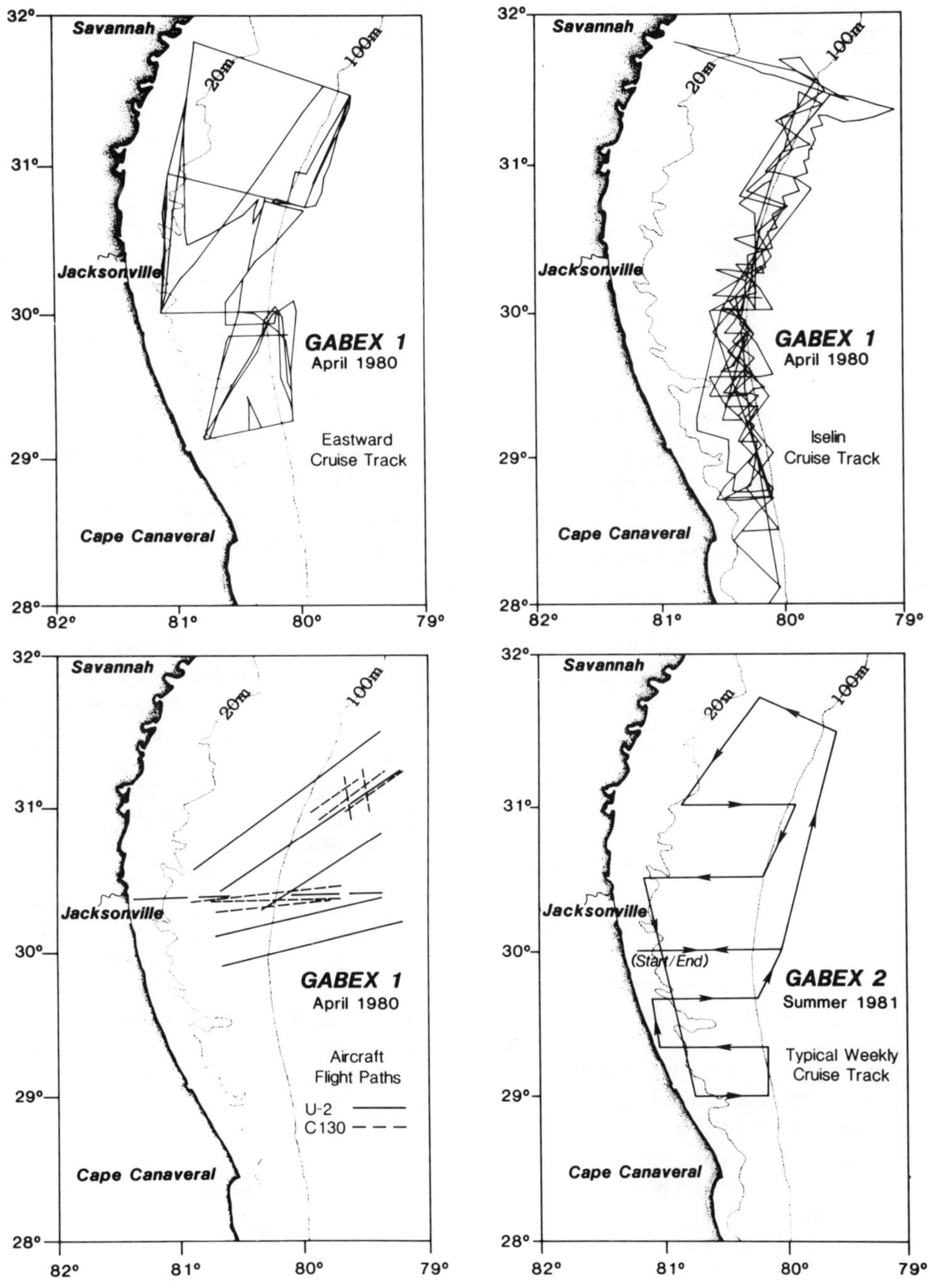

Fig. 4. Representative cruise tracks of the two ships and aircraft flight paths during GABEX I and II.

km cross shelf (Figure 3). During GABEX I, alongshelf mooring spacing varied from approximately 50 km in the outer shelf area to 100 km over the inner shelf. Vertical spacing varied from only two instruments, one near the surface layer and another in the bottom layer, to instruments at each 10-m depth interval. Moorings were in place several months before and after the cruises. The cruise plans for both studies were similar, using two ships to cover the area between Cape Canaveral and Savannah. One ship, usually called the mapping or fast ship, made synoptic hydrographic surveys, while the second ship, usually called the slow ship, made more time-consuming measurements in areas determined to be critical to biological processes. During GABEX I the fast ship (R/V *Columbus Iselin*) made

TABLE 1. General Characteristics of the South Atlantic Bight Continental Shelf Waters

	Value or Characteristic
Shelf width	50–200 km
Depth of shelf break	50 m
Surface area (coast to 60-m isobath)	90,600 km^2
Volume	
Coast to 20-m isobath	326 km^3
20-m to 40-m isobath	1312 km^3
40-m to 60-m isobath	722 km^3
Annual river runoff	84 km^3
Dynamic factors	
Florida/Georgia shelf*	
Inner shelf	
Winter	5 ± 2 toward 10°True
Summer	6 ± 3 toward 70°True
Middle shelf	
Winter	1–4 ± 10 toward north
Summer	1–4 ± 10–20 toward north
Outer shelf	
Winter	50 ± 30 toward north
Summer	30–40 ± 50 toward north
Winds (off Savannah, $m\ s^{-1}$)	
Winter	4 ± 4 toward NE
Summer	3 ± 2 toward NNE
Inertial radius for weakly stratified inner shelf waters	4–20 km
Tidal range	1–3 m (maximum off Georgia)

*Middepth currents; mean and standard deviation are in centimeters per second.

detailed observations over the outer shelf as Gulf Stream eddies propagated northward (Figure 4, upper left). The track of this ship was determined from ship observations, satellite-derived sea surface temperature observations, and airborne radiation thermometer observations obtained by the Coast Guard Oceanographic Unit. Temperature observations were received on board ship via satellite communications. The slow ship (R/V *Eastward*) mapped the inner and middle shelf, covering a predetermined pattern based on current meter locations, known spatial scales and synopticity consideration (Figure 4, upper right). The ships were at sea for 20 days, assuring observation of several storm and Gulf Stream events.

During GABEX II the sampling strategy was slightly different. Gulf Stream events are not easily observable during the summer because of the uniformly warm surface layer and also because the effect of Gulf Stream frontal events are felt across the whole shelf. The GABEX II ships systematically sampled the inner, middle, and outer shelf waters. The fast mapping ship (R/V *Cape Henlopen*) covered the area between Cape Canaveral and Savannah once per week (Figure 4, lower left). The slow ship (R/V *Blue Fin,* then R/V *Cape Florida*) concentrated on biological sampling in areas of highest interest (Figure 4, lower right). Both ships concentrated coverage in the area off St. Augustine, creating a time series of data at many stations across the shelf. Observations were extended over a 6-week period to ensure the observation of at least one major event. The papers in this book refer to these experiments as well as others, such as a series of coastal frontal studies (Front Flux) and various smaller-scale cruises.

Acknowledgments. The results described here were primarily derived from research efforts funded by the Department of Energy to the individual authors. The Minerals Management Service (formerly Bureau of Land Management) provided considerable funding for specific aspects of these studies. The National Science Foundation also funded some of the individual research projects for shorter time periods. We thank Helen McCammon, George Saunders, William Forster, and Charles Osterberg at the Department of Energy, Edward Wood, formerly of Bureau of Land Management, William Lang of the Minerals Management Service, and Evans Waddell of Science Applications, Inc., for their combined support and encouragement.

References

Abbe, C., Jr., Remarks on the cuspate capes of the Carolina coast, *Proc. Boston Soc. Nat. Hist., 26,* 489–497, 1895.

Atkinson, L. P., Modes of Gulf Stream intrusion into South Atlantic Bight shelf waters, *Geophys. Res. Lett., 4,* 583–586, 1977.

Atkinson, L. P., J. O. Blanton, and E. B. Haines, Shelf flushing rates based on the distribution of salinity and fresh water in the Georgia Bight, *Estuarine Coastal Mar. Sci., 7,* 464–472, 1978*a*.

Atkinson, L. P., G. A. Paffenhofer, and W. M. Dunstan, The chemical and biological effect of a Gulf Stream intrusion off St. Augustine, Florida, *Bull. Mar. Sci., 28,* 667–679, 1978*b*.

Bane, J. M., Jr., and D. A. Brooks, Gulf Stream meanders along the continental margin from the Florida Straits to Cape Hatteras, *Geophys. Res. Lett., 6,* 280–282, 1979.

Blanton, J. O., Exchange of Gulf Stream water with North Carolina shelf water in Onslow Bay during stratified conditions, *Deep Sea Res., 18,* 167–178, 1971.

Bumpus, D. F., The circulation over the continental shelf south of Cape Hatteras, *Eos Trans. AGU, 36,* 601–611, 1955.

Bumpus, D. F., A description of the circulation on the continental shelf of the east coast of the United States, *Prog. Oceanogr., 6,* 111–158, 1973.

Bumpus, D. F., and E. L. Pierce, The hydrography and the distribution of chaetognaths over the continental shelf off North Carolina, *Deep Sea Res., 3,* 92–109, 1955.

Bumpus, D. F., and T. J. Wehe, Hydrography of the western Atlantic: Coastal water circulation off the east coast of the United States between Cape Hatteras and Florida, *Rep. 49-6,* 11 pp., Woods Hole Oceanogr. Inst., Woods Hole, Mass., 1949.

Emery, K. O., and E. Uchupi, *Western North Atlantic Ocean: Topography, Rocks, Structure, Water Life and Sediments,* 532 pp., American Association of Petroleum Geologists, Tulsa, Okla., 1972.

Ford, W. L., J. R. Longard, and R. E. Banks, On the nature, occurrence and origin of cold low salinity water along the edge of the Gulf Stream, *J. Mar. Res., 11,* 281–293, 1952.

Franklin, B., A letter to Mr. Alphonsus LeRoy, member of several academies, at Paris, containing sundry maritime observations, *Trans. Am. Philos. Soc., 2,* 294–329, 1786.

Green, C. K., Summer upwelling, northeast coast of Florida, *Science, 100,* 546–547, 1944.

Haines, E. B., and W. M. Dunstan, The distribution and relation of particulate organic material and primary productivity in the Georgia Bight, 1973–1974, *Estuarine Coastal Mar. Sci., 3,* 431–441, 1975.

Legeckis, R., Satellite observations of the influence of bottom topography on the seaward deflection of the Gulf Stream off Charleston, South Carolina, *J. Phys. Oceanogr., 9,* 483–497, 1979.

Parr, A. E., A geographical-ecological analysis of the seasonal changes in temperature conditions in shallow water along the Atlantic coast of the United States, *Bull. Bingham Oceanogr. Collect. Yale Univ., 4,* 1–90, 1933.

Pashinski, D. J., and G. A. Maul, Use of ocean temperature while coasting between the Straits of Florida and Cape Hatteras, *Mariners Weather Log, 17,* 1–3 Jan. 1973.

Pearse, A. S., and L. G. Williams, The biota of the reefs off the Carolinas, *J. Elisha Mitchell Sci. Soc., 67,* 133–161, 1951.

Pierce, E. L., The chaetognaths over the continental shelf of North Carolina with attention to their relation to the hydrography of the area, *J. Mar. Res., 12,* 75–92, 1953.

Stefansson, U., and L. P. Atkinson, Nutrient-density relationships in the western North Atlantic between Cape Lookout and Bermuda, *Limnol. Oceanogr., 16,* 51–59, 1971.

Stefansson, U., L. P. Atkinson, and D. F. Bumpus, Hydrographic properties and circulation of the North Carolina shelf and slope waters, *Deep Sea Res., 18,* 383–420, 1971.

Taylor, C. B., and H. B. Stewart, Jr., Summer upwelling along the east coast of Florida, *J. Geophys. Res., 64,* 33–40, 1959.

Webster, F., A description of Gulf Stream meanders off Onslow Bay, *Deep Sea Res., 8,* 130–143, 1961.

Williams, L. G., Seasonal alternation of marine floras at Cape Lookout, North Carolina, *Am. J. Bot., 35,* 682–695, 1948.

Williams, L. G., Marine algal ecology at Cape Lookout, North Carolina, *Bull. Furman Univ., 31,* 1–21, 1949.

Wind Stress Climatology in the South Atlantic Bight

J. O. Blanton,[1] F. B. Schwing,[1] A. H. Weber,[2] L. J. Pietrafesa,[3] and D. W. Hayes[2]

Fluctuations in the position of synoptic scale pressure systems largely govern mean wind patterns in the South Atlantic Bight (SAB). These winds provide the primary forcing to circulation in the broad and shallow regions of the shelf removed from direct Gulf Stream influence. Winds over the SAB originate either from the Bermuda-Azores High in the North Atlantic, or the smaller Ohio Valley High, which dominates between August and January. Monthly mean wind fields derived from ship observations depict five seasonal wind regimes in the SAB. While these regions serve as useful indicators of month-to-month change of mean vectors, a great deal of year-to-year variability exists at a single point. Coastal wind regimes exist on three basic time scales: seasonal, synoptic weather system, and diurnal. Further offshore, the diurnal scale is insignificant. Important spatial differences occur between land and ocean observations over a distance of only 50 km. Energy data suggest a narrow lateral boundary layer separates land and sea wind regimes. Caution is needed when using over-land wind data to describe oceanic winds.

Introduction

The surface of the coastal ocean responds relatively quickly to changes in wind and air temperature due to the relatively broad width and shallow depth of the continental shelf of the South Atlantic Bight (SAB) (Figure 1). For example, changes in water temperature on the inner shelf closely follow changes in air temperature on both the seasonal time scales (>1 month) and synoptic scales (3–20 days) [*Atkinson et al.*, 1983; *Wittman,* 1982]. Ocean currents on the inner and middle shelf respond in less than 1 day to changes in wind stress [*Schwing et al.*, 1983; *Lee and Brooks*, 1979]. Rainfall accompanying synoptic scale weather patterns influences the salinity and density regime of the inner shelf which, in turn, causes vertical changes in horizontal currents. Although maximum rainfall occurs in coastal regions in summer, maximum freshwater discharge to the ocean usually occurs in March or April because most of the drainage areas of the southeastern United States are in the mountain and piedmont areas which receive most of their rainfall in early spring. The influence of freshwater discharge on the monthly salinity regime of the inner shelf has been discussed by *Atkinson et al.* [1978] and *Blanton and Atkinson* [1983].

In this paper, we concentrate on the climatology of wind stress and the associated synoptic scale systems which influence it. We have drawn largely from existing studies of the wind regime by *Weber and Blanton* [1980], *Wendland and Bryson* [1981], and *Weisberg and Pietrafesa* [1983]. Important earlier studies are those of *Wexler and Namias* [1938], *Rossby* [1941], *Jacobsen* [1974], *Bryson and Hare* [1974], *Ruzecki* [1974], and *Saunders* [1977]. The fluctuations in the Icelandic Low, the Bermuda-Azores High, and the Ohio Valley High largely govern the mean wind patterns in the SAB. The fluctuations in the mean pattern provide the dominant forcing to the shelf circulation in regions removed from direct Gulf Stream influence. We also describe variability in terms of time and space scales of wind measured at discrete points. In all cases, we adhere to the oceanographic convention when discussing wind direction, i.e., wind vectors point downwind.

Monthly Mean Wind Fields

Airstream Sources

Bryson and Hare [1974] and *Wendland and Bryson* [1981] have analyzed streamline patterns drawn from monthly wind vectors. The differences in streamlines from month to month are due mostly to relocation of principal centers of action such as the Azores-Bermuda High (hereinafter called Azores High) and the Icelandic Low. Their work clearly shows that winds blowing over the SAB originate either from the Azores High in the North Atlantic or from a smaller-scale anticyclone centered over the Ohio Valley (hereinafter called Ohio Valley High), cover the southern portion of the United States east of 110°W, and result in mean eastward wind stress across the SAB. A confluence of streamlines stretches southwest to northeast across the southern portion of SAB and separates the modified Pacific or Arctic air mass originating out of the Ohio Valley High

[1] Skidaway Institute of Oceanography, Savannah, Georgia.
[2] Savannah River Laboratory, Aiken, South Carolina.
[3] Department of Earth and Atmospheric Sciences, North Carolina State University, Raleigh.

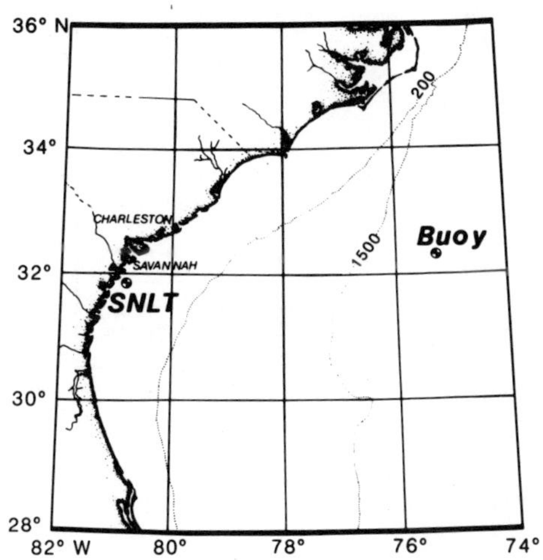

Fig. 1. Location map showing principal observations points used in this study.

from tropical maritime air originating from the Azores High (see Figure 1 of *Wendland and Bryson* [1981] for details).

During spring (March–May) the prominent influence of the Ohio Valley High is replaced by the Azores High. Northward flow of warm humid air originates as westward flow on the southern flank of the Azores High. This flow turns northward over the western North Atlantic and the Gulf of Mexico. The confluence zone between this air mass and the dry air from the Ohio Valley High disappears from the region during this time (see Figure 2 of *Wendland and Bryson* [1981]).

The northward flow of air strengthens in summer (June–August) as the Azores High strengthens and shifts westward. The airstreams which pass into the Gulf of Mexico turn northward and flow northeastward over land into the SAB and join the airstreams flowing northward over the ocean. This pattern (see Figure 3 of *Wendland and Bryson* [1981]) produces a confluence zone along which frequent rain occurs over land [*Bryson and Hare*, 1974].

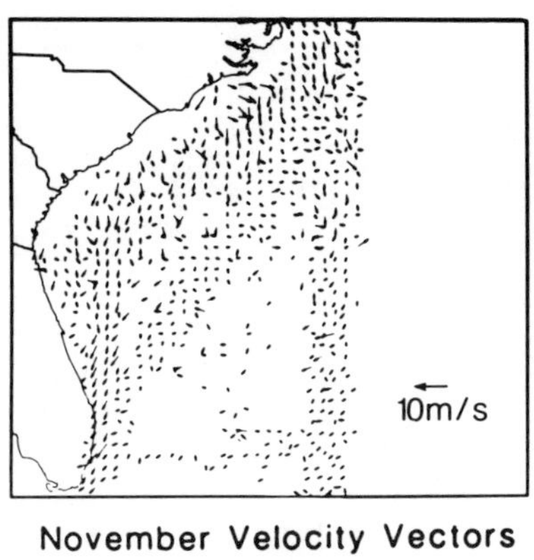

Fig. 2. Mean monthly wind vectors for November (1945–1963).

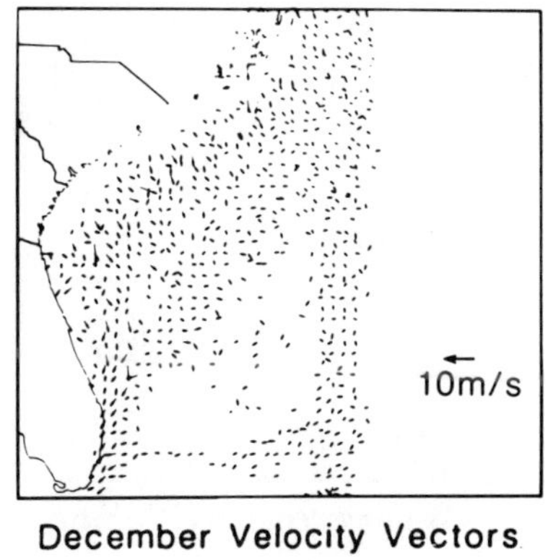

Fig. 3. Mean monthly wind vectors for December (1945–1963).

During autumn (September–November), the penetration of tropical air in the SAB rapidly collapses and is replaced by air originating from the Ohio Valley High. The airstreams flow out of the eastern flank of the Ohio Valley High and produce strong mean southwestward stress over the SAB. The Azores High shifts eastward and warm moist air flows northward only over the Blake Plateau and offshore (see Figure 4 of *Wendland and Bryson* [1981]).

Mean From Ship Observations

Marine weather observations from ships have been analyzed to produce monthly mean wind fields for the SAB [*Weber and Blanton*, 1980]. We have updated the analyses to include ship observations from the Blake Plateau. The data source and editing procedures are the same used by *Weber and Blanton* [1980]. We report here the analyses for the period 1945–1963.

Monthly mean wind vectors depict five seasonal wind regimes for the SAB [*Weber and Blanton*, 1980]. During winter (November–February) winds exert southeastward stress over the northern portion of the SAB (Figures 2–5). The wind stress gradually shifts southward toward more southerly latitudes. Winds are stronger in the northern

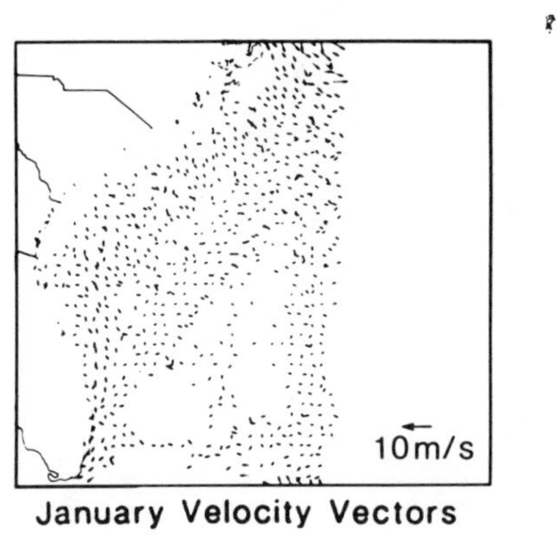

Fig. 4. Mean monthly wind vectors for January (1945–1963).

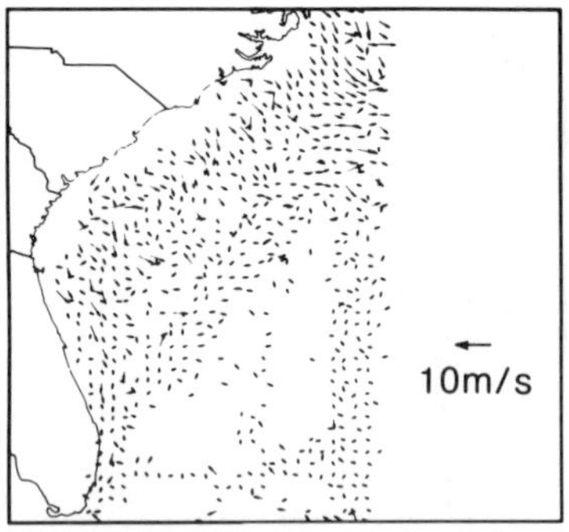

February Velocity Vectors

Fig. 5. Mean monthly wind vectors for February (1945–1963).

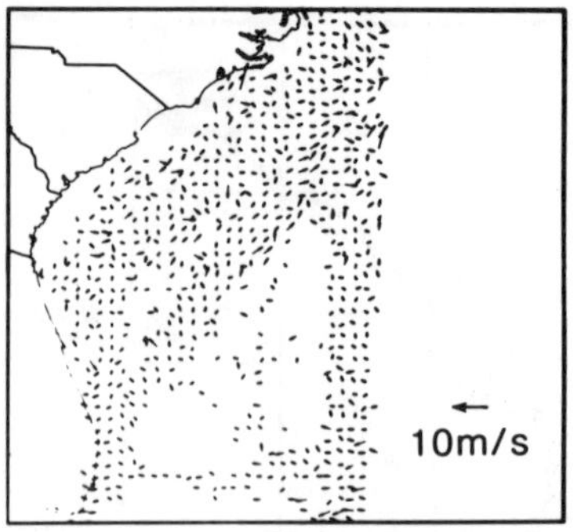

April Velocity Vectors

Fig. 7. Mean monthly wind vectors for April (1945–1963).

half of the SAB. The monthly analyses of *Bryson and Hare* [1974] confirm this general picture, except that the spatial resolution is much coarser (3° latitude versus 0.5° latitude in the work of *Weber and Blanton* [1980]). Both studies indicate a discontinuity in the wind field over the Blake Plateau, represented in Figures 2–5 as confused orientation of vectors which tend to become more organized in the northern half of the Blake Plateau as the winter season progresses. Wind strengths are generally weaker beyond the shelf break of the SAB.

The spring transition (March–May, Figures 6–8) shows a gradual shift to eastward and northeastward stress in the central portion of the SAB nearshore [*Weber and Blanton*, 1980]. Westward stress occurs off southern Florida in agreement with *Bryson and Hare* [1974]. Flow is more organized over the Blake Plateau as the season progresses, and westward stress in the south gradually shifts to northward and northeastward stress, also in agreement with *Bryson and Hare* [1974].

During a portion of the summer (June–July, Figures 9–10), wind is westward and southwestward along the southern half of Florida. Stress is more northward and northeastward in the northern half of the Bight and over the Blake Plateau. Stress appears most organized in July over both the SAB and Blake Plateau where winds are strongest. The organized field of northward stress continues in August over the Blake Plateau but breaks down in the SAB (Figure 11). August was considered a transition month for the SAB by *Weber and Blanton* [1980]. The August analysis by *Bryson and Hare* [1974] does not reflect this disorganization on the SAB. However, the Ohio Valley High is formed during this month and generates airstreams opposing the ones coming from the Azores High. It seems likely that the weak and erratic mean winds over the SAB in August result from the averaging of these opposing airstream regimes.

The autumn regime (September–October, Figures 12–13) has been designated "mariners' fall" by *Weber and Blanton* [1980]. This regime exerts strong southwestward wind stress in the SAB but does not extend to the Blake Plateau, where the mean stress is smaller and mostly westward. This pattern appears in the work of *Wendland and Bryson* [1981] as southwestward flow over the SAB. This flow originates from the Ohio Valley High and collides with westward to northwestward flow over the Blake Plateau which originates from the Azores High. The demarcation between the opposing airstreams is depicted clearly in Figures 12 and 13.

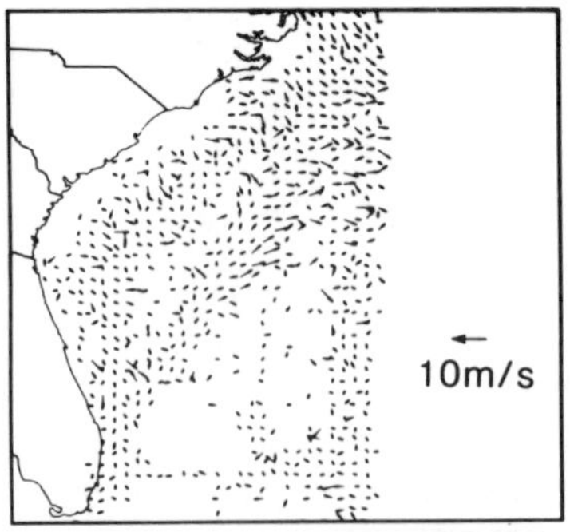

March Velocity Vectors

Fig. 6. Mean monthly wind vectors for March (1945–1963).

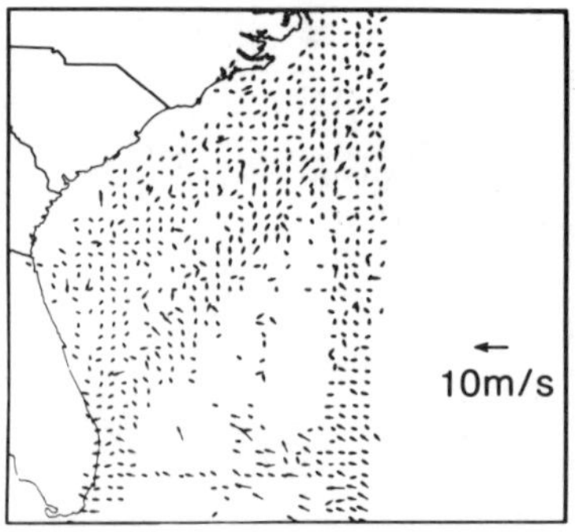

May Velocity Vectors

Fig. 8. Mean monthly wind vectors for May (1945–1963).

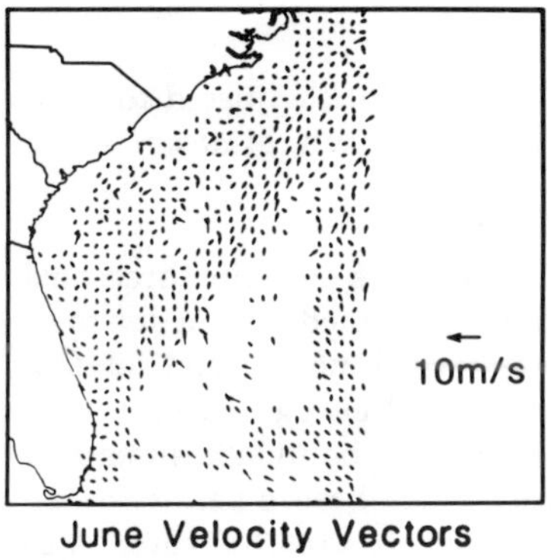

Fig. 9. Mean monthly wind vectors for June (1945–1963).

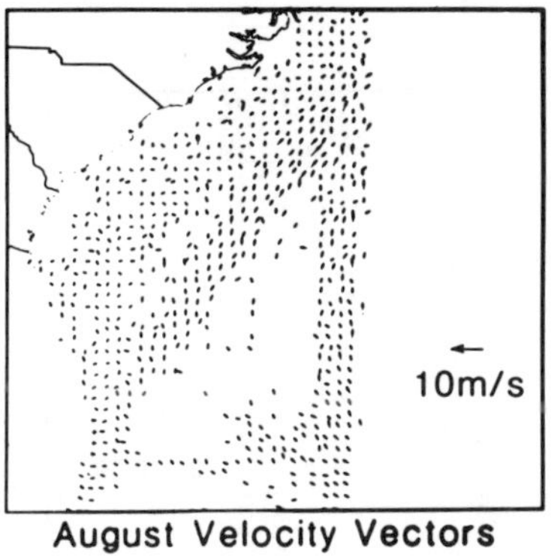

Fig. 11. Mean monthly wind vectors for August (1945–1963).

Interannual Variability

In the 19-year climatology depicted in Figures 2–13, the standard deviations of wind speeds usually exceed the means themselves, which makes the definition of a regime of vectors somewhat subjective and more a matter of convenience. While the regimes serve as useful reference points in the month-to-month change of the mean vectors, it can be difficult to identify a given monthly mean for a given year with a "regime" based on a several-year mean. We chose an 8-year set of wind data from Savannah, Georgia, to illustrate this difficulty.

We have defined monthly mean stress vectors for the years 1957–1964 by converting the daily vector means of wind speed from climatological summaries to stress using the quadratic speed law and a drag coefficient of 2×10^{-3} (Figure 14). We have included from *Weber and Blanton* [1980] a dotted vector for wind direction only, taken from a cell immediately offshore of the South Carolina–Georgia border, a cell that contained an average of 80 observations per month. Except in August, the 19-year mean wind direction and the 8-year vector mean stress direction agree within 45°.

In no single month do all eight vectors fall within the same 90° quadrant. During mariners' fall (September–October), vectors are scattered more or less uniformly around the mean. In transition months such as August, some years reflected a summertime regime while others indicated (perhaps) the onset of mariners' fall. February contained the greatest interannual variation in magnitude. While March is depicted as a transition month in the mean wind field, the eight means for March have remarkable stability.

While the mean wind fields from ship observations were computed from a 19-year data set of 1945–1963 [*Weber and Blanton*, 1980], the fields derived from a subsequent 11-year data set (1963–1973) were essentially identical and the regimes so defined remained unchanged. However, the 8-year means (1957–1964) from Savannah (Figure 14) illustrate the great year-to-year variability exhibited by data at a single point. This problem results, in part, from mixing interannual (temporal) variability that may reflect global weather patterns with spatial variability surrounding the point.

Time and Space Scales in Wind Fields

Oceanographers frequently resort to coastal wind velocity data to determine the wind stress over the ocean. Wind velocity over a land or sea surface is a complicated function

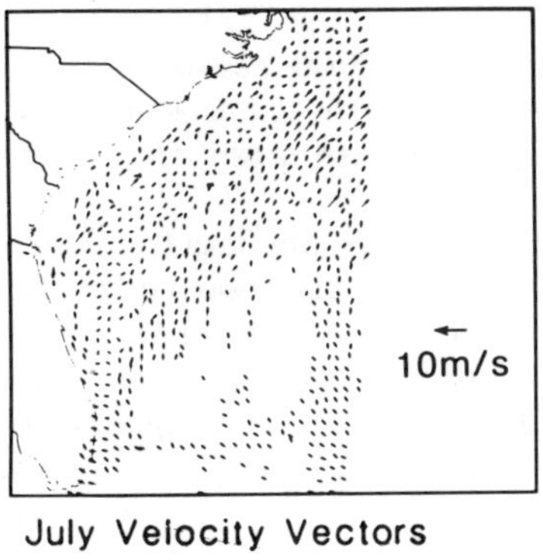

Fig. 10. Mean monthly wind vectors for July (1945–1963).

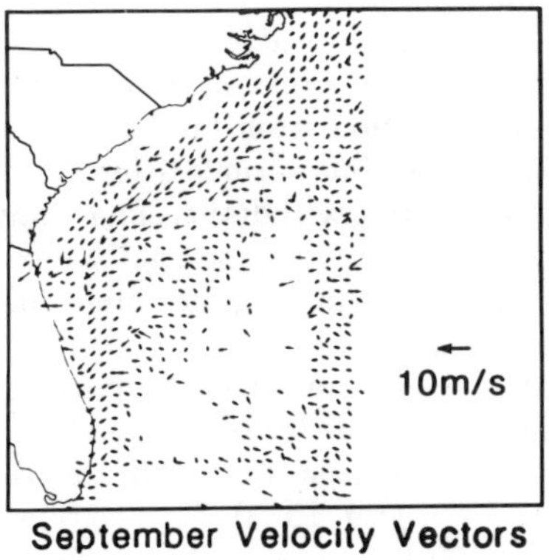

Fig. 12. Mean monthly wind vectors for September (1945–1963).

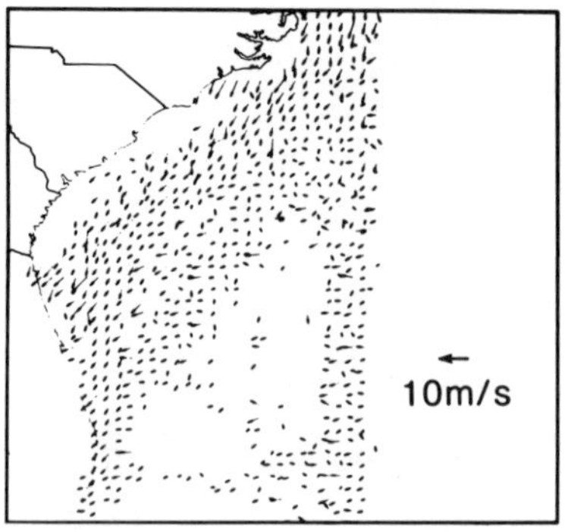

Fig. 13. Mean monthly wind vectors for October (1945–1963).

of stress at the surface, the height above the surface, and the thermal stability of the air [*Kraus,* 1972]. To accurately estimate over-ocean winds from over-land wind velocity data requires a knowledge of the velocity shear and the thermal stability at both locations, a requirement that is seldom met in practice. Wind speeds are usually greater over ocean than over land, and there is often a shift in direction which can lead to serious errors when the vector is resolved into components. Thus it is difficult to adjust land data to accurately reflect sea surface stress.

There are several influences which cause winds over the ocean to vary from those over land (for example, see *Richards et al.* [1966]). They vary due to differences in frictional effects between land and ocean as well as due to differences in atmospheric stability created by air-sea temperature differences. The length of over-ocean fetch also influences the variation [*SethuRaman and Raynor*, 1980], but its effect is apparently limited to scales of about 50 km [*Richards et al.*, 1966]. Most studies are concerned with differences in wind speed over land and over sea, but there are also important differences in direction [*Hamblin and Elder*, 1973; *Schwing and Blanton*, 1984] that can be attributed to differences in boundary friction and stability. The problem of directional differences will be discussed later.

When the adjustment is attempted over scales of 1000 km or greater, synoptic scale variation in the wind regime may make extrapolation from land to sea winds almost impossible. We know of no serious attempts to extrapolate over synoptic scales but will discuss one study which compares winds over a 500-km distance, a distance large enough that major changes in wind regimes might pose a problem under some conditions.

In this section, we summarize a comparison of wind data separated over a 500-km distance between Charleston, South Carolina, and an ocean station over the Blake Plateau [*Weisberg and Pietrafesa*, 1983]. We also compare wind velocity data over 50 km between Savannah, Georgia, and an ocean tower. In both comparisons, there are important differences in the wind regimes over land and ocean. In neither case are the data adjusted to compensate for different shear and stability regimes.

Large-Scale Differences

Coastal wind regimes contain three basic times scales: a seasonal modulation, modulation due to synoptic weather systems, and a diurnal modulation related to sea breezes. A corresponding wind energy spectra over the ocean and far removed from land boundaries has two of these three scales, the diurnal sea breeze being absent.

Seasonal modulations are illustrated by comparing mean wind vectors and monthly changes of variance about the mean (Figure 15). Over a 47-year period, the mean winds at both stations (Figure 1) blow northward from March until August or September. Speeds at the ocean buoy are higher by a factor in excess of 2 compared with those at the coast. During the remainder of the year, the means are dissimilar. For example, between September and November the Charleston data are similar to the monthly means at Savannah and display the expected year-to-year variability. The mean vectors at the buoy, however, do not agree with the coastal stations. This also suggests that the fall weather regime over the shelf of the SAB may not extend seaward over the Blake Plateau, in agreement with monthly means (Figures 12–13).

The monthly mean wind fields for August, September, and October (Figures 11–13) show a consistent pattern. In August, the summer wind regime breaks down over the shelf, while flow is well organized and northward over the Blake Plateau. By September and October mariners' fall is clearly established over the shelf, while the wind field over the Blake Plateau is less organized and remains so throughout autumn and winter. This data set supports the findings of *Weisberg and Pietrafesa* [1983]. We suspect that the convective regime over the shelf region described below is responsible for setting up the two different regimes in autumn.

There is a clear seasonal signal in variance over the Blake Plateau (Figure 15). Maximum variance occurs between November and February in both components and is roughly 4 to 6 times that at the coastal station. During summer, variance over the Blake Plateau is a minimum and approaches that of the coastal station. The monthly variances at the coastal stations seem to be more uniform throughout the year with a tendency for low variance in summer.

Rotary spectral analyses of data from the Charleston and the offshore buoy concisely describe the characteristics of the variance in the synoptic scale [*Weisberg and Pietrafesa*, 1983]. At Charleston significantly more energy is present in winter in the anticyclonic component than in the cyclonic one. Maximum anticyclonic energy occurs from December to March; minimum energy occurs from June to September and the velocity variations are more rectilinear. The ellipses at the coast are aligned alongshore in summer and winter.

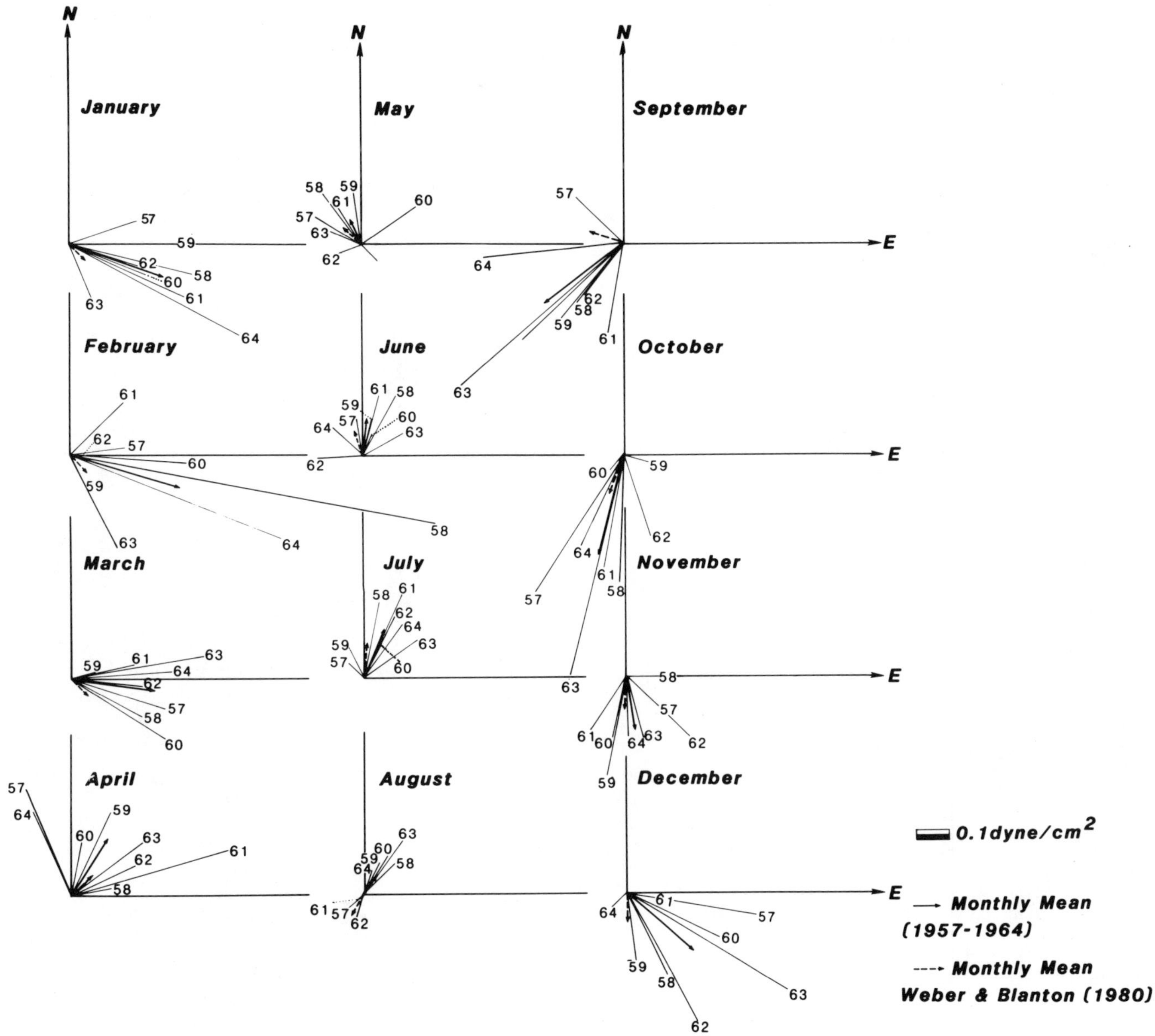

Fig. 14. Mean monthly wind stress vectors and eight individual monthly vectors from the Savannah Municipal Airport for the period 1957–1964. Means were computed from vector averages for each day. The dotted vector shows only the direction of a monthly mean vector from a offshore location taken from *Weber and Blanton*. [1980].

Five hundred kilometers offshore in winter *Weisberg and Pietrafesa* [1983] found that the larger variance at the synoptic scale is principally characterized by coherent anticyclonic elliptical motions similar to those at the coast (Table 1). During summer, the motion is incoherent and the alongshore alignment of the ellipse is probably not significant. However, winter ellipses are aligned in the onshore-offshore directions, a fact which the authors found puzzling.

Only the coastal station contained significant diurnal energy in spring and summer [*Weisberg and Pietrafesa*, 1983]. The motion was aligned cross shore and was essentially rectilinear. The diurnal energy maxima corresponded to minimum variance in summer (Figure 15*b*). Thus the authors cautioned that while total variance at the coastal station does not appear to modulate seasonally, the independent processes (diurnal and synoptic scale) comprising the total do.

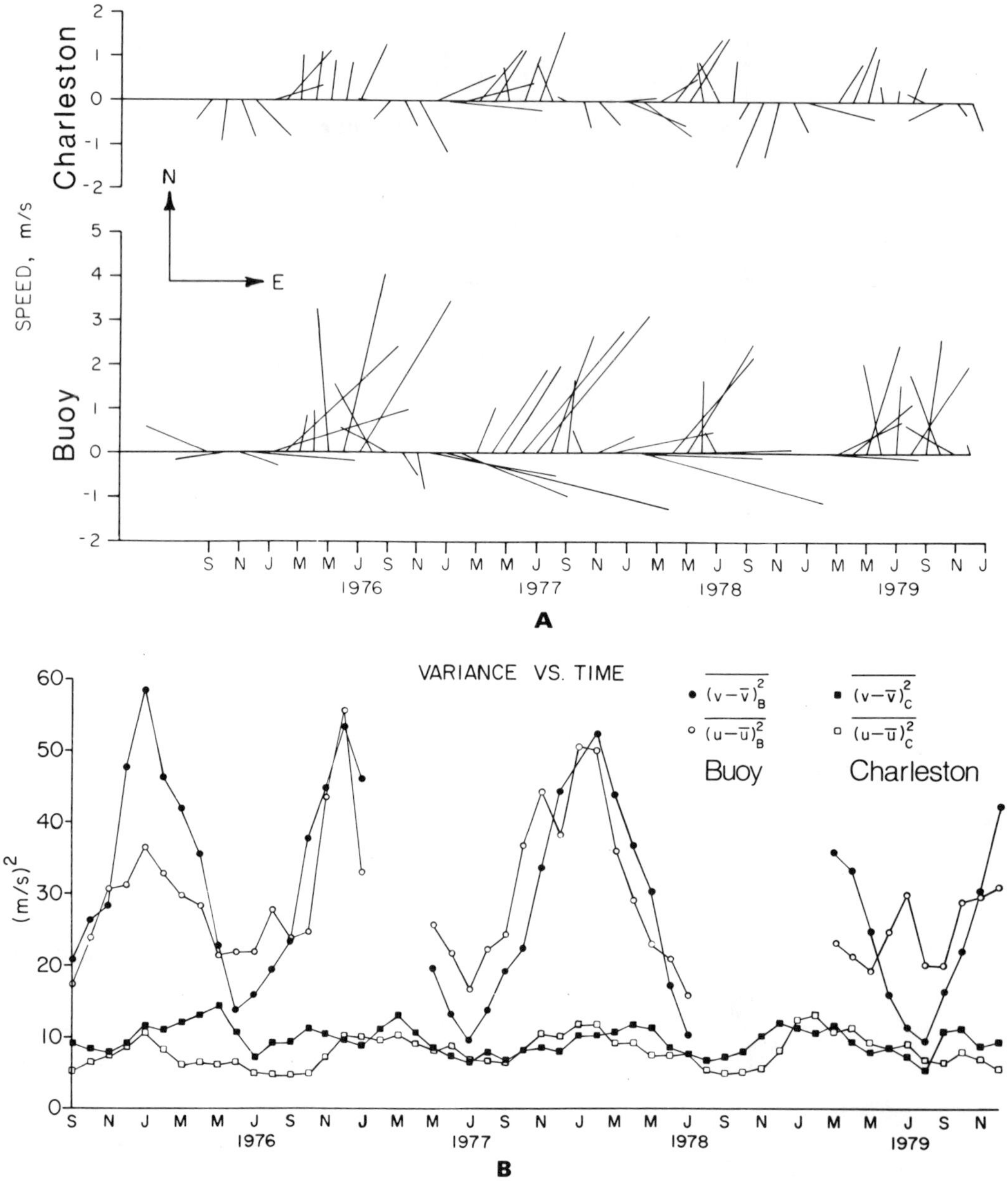

Fig. 15. A comparison of wind velocity monthly means and variances between Charleston, South Carolina, and an offshore buoy 500 km away: (*a*) monthly means and (*b*) variances. Figures 15*a* and 15*b* appear as Figures 7 and 8, respectively, of *Weisberg and Pietrafesa* [1983].

Small-Scale Differences

We show in this section that important differences may occur between land and sea stations over a distance of only 50 km. We have compared spectra of wind data for a winter and summer period over land and over water near Savannah, Georgia (Figure 1). Spectra for the two locations (Figure 16, Table 2) show two principal frequency bands: a synoptic band of 0–0.25 cycles per day and a diurnal band. At synoptic frequencies, the kinetic energy over the ocean was approximately 4 times that over land (in agreement with *Weisberg and Pietrafesa* [1983]). Moreover, simultaneous wind data at a point over land but only 20 km from the ocean platform (not shown) showed that the kinetic energy was still only 25% of that found over the ocean. Therefore the transition from low kinetic energy over land to high kinetic energy over the ocean occurred abruptly within a distance of 0–20 km from the coast, suggesting a narrow lateral boundary layer.

TABLE 1. Qualitative Comparison of Synoptic Scale (2–30 Days) and Diurnal Variance in Wind Velocity Between Charleston, South Carolina, and an Offshore Buoy 500 km Eastward

	Coastal	Open Ocean
Shape of velocity signal		
Winter	anticyclonic; elliptical	anticyclonic; elliptical
Summer	rectilinear	incoherent
Orientation of principal axis		
Winter	alongshore	onshore-offshore
Summer	alongshore	alongshore (but incoherent)
Diurnal signal		
Winter	none	none
Summer	rectilinear; cross shelf	none

See details in the work of *Weisberg and Pietrafesa* [1983].

The winter data (Table 2) show that seabreeze made up 4–5% of the total variance at both locations. Spectral characteristics of the seabreeze oscillations were statistically coherent only at the ocean station. Here the motion was elliptical with an orientation nearly perpendicular to the shoreline. Oscillations in the synoptic band were statistically incoherent at both locations. The relatively short data set in winter exhibited very little statistically steady behavior.

The seabreeze for the summer data set (Table 2) made up 11% of the total variance over land and 18% of the total over ocean. The total variance dropped by a factor of 2 between winter and summer. The seabreeze oscillations were statistically coherent at both locations. Over land, they were rectilinear (rotary coefficient, 0.17) while they were more elliptical (rotary coefficient, 0.55) over the ocean.

Oscillations at synoptic frequencies were also coherent at both locations in summer. The energy was again about 4 times higher over the ocean but the characteristics of the oscillations were similar (compare ANG and ROTC values at both locations).

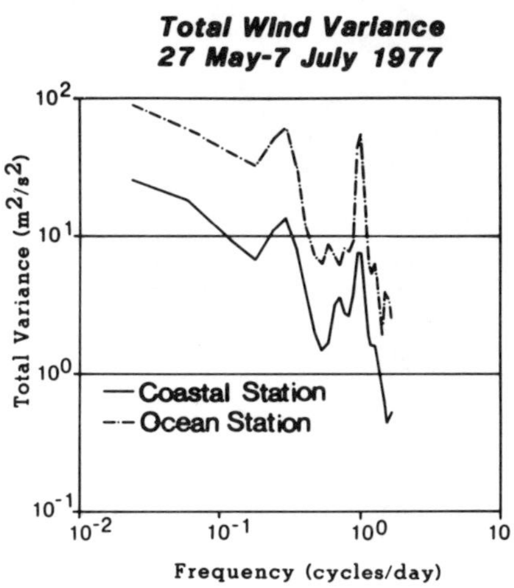

Fig. 16. Total energy density versus frequency for a land (SAV) and ocean (SNL) station separated by 50 km. Hourly data were smoothed with a 10-hour low-pass filter.

The difference between coastal and ocean winds that makes the strongest impression is the fact that mean vector magnitudes over ocean are about double those over land (Table 3). However, differences in direction are more difficult to understand, and the accuracy of shore-parallel and cross-shore components derived from direction data on land may be seriously degraded when they are assumed to represent ocean wind stress. To illustrate this difficulty, we will elaborate on the data set used to derive Figure 16.

The mean difference in direction for the coastal and oceanic data in Figure 17 is 0.5° ± 37° (Table 4). While the mean is not significantly different from zero, the high standard deviation suggests differences that can be large from time to time. A 3 day sequence illustrates this point (Figure 17). The overall trend suggests that winds blowing landward on March 12 were shifting to winds blowing seaward on March 14. Differences on March 12 never exceeded 20°. The next day, the difference reached 41°; on the last day, the difference was 112°. From a statistical viewpoint, enough data may suggest that the average difference will approach zero. But for relatively short episodes (a 3- or 4-day experiment) differences are too large to ignore.

We approached the problem another way be averaging the 1900-, 0100-, 0700-, and 1300-hour vectors for the entire data set (Table 4). The average difference during darkness (1900 and 0100 hours) was 15° ± 35° and 12° ± 29°, respectively. The positive differences mean that the coastal wind vector is usually 12°–15° counterclockwise of the ocean wind. The opposite occurs during daylight (1200 and 1800 hours). The coastal wind vector is usually 7°–18° clockwise of the ocean winds (Table 4).

The fact that the direction differences have a clear diurnal signal (Table 4) suggests that afternoon convection and nighttime subsidence over land may contribute to the difference. Moreover, since the energy in the diurnal band is greater in summer, one may have to be particularly cautious when using over-land wind in summer. Frictional effects due to the presence of land diminish the wind speed by a factor of 2. These processes are causal in effect and

TABLE 2. Comparison of Spectra Between Savannah, Georgia (Land), and an Oceanic Platform (Sea) 50 km Eastward for Winter (February 18 to March 20, 1977) and Summer (May 26 to July 9, 1977)

	Winter		Summer	
Frequency Band	Land (SAV)	Sea (SNL)	Land (SAV)	Sea (SNL)
0–0.25 cycle/day				
Variance, $m^2 s^{-2}$	11.85	44.82	5.07	20.13
Percent total variance	68	68	60	58
ANG, deg	96	93	78	73
ROTC	0.31	0.46	0.19	0.20
CMIN	0.14	0.28	0.09	0.04
CMAX	0.36	0.42	0.64	0.26
1 cycle/day				
Variance, $m^2 s^{-2}$	0.90	2.80	0.91	6.10
Percent total variance	5	4	11	18
ANG, deg	135	160	129	164
ROTC	~0	0.75	0.17	0.55
CMIN	~0	0.59	0.06	0.63
CMAX	0.21	0.61	0.52	0.82

ROTC is rotary coefficient (positive clockwise), ANG is angle of principal axis (degrees clockwise from north), CMIN is minimum coherence squared, and CMAX is maximum coherence squared. Data used were four values per day, smoothed with a 10-hour low-pass filter. Rotary values were calculated from *Fofonoff* [1969].

undoubtedly create differences between the wind regime measured over land and that measured over the adjacent ocean.

These differences may become especially important when actual wind data are input to a numerical model for shallow water circulation. If one were to create a numerical model of shallow water circulation and let the model run for a month, the momentum change imparted to a water column would simply be the accumulated impulse/unit area over the month. We have calculated the function of impulse/unit area based on simultaneous hourly wind stress values derived from the coastal station and the oceanic platform. Three impulse functions were computed: (1) unadjusted coastal impulse, (2) adjusted coastal impulse, and (3) oceanic impulse. The second function was computed by doubling the wind speed for the coastal data. These impulse curves (Figure 18) reveal large differences in wind stress regimes between unadjusted coastal data and oceanic data. By doubling the magnitude of coastal wind speed, the discrepancies are greatly diminished. Without this adjustment, the shore-parallel data (Figure 18*a*) indicate that the momentum change imparted to the water column after 1 month was probably 4 times that indicated by using adjacent coastal data. Moreover, some wind events such as the negative alongshore stress from March 4–6 are not clearly evident in coastal wind data.

The differences between adjusted coastal and oceanic data do not appear significant in the shore-parallel component.

TABLE 3. Statistical Comparison of Wind Speeds (Mean, Standard Deviation in Meters per Second) of Three Nearby Land Stations and the Oceanic Platform (SNL)

				Vector Mean	
Location	East	North	Scalar Mean	Speed	Direction
Winter					
SAV	0.9 ± 3.4	0.7 ± 2.4	4.1 ± 2.0	1.2	53
SVN	0.6 ± 2.7	0.7 ± 2.3	3.1 ± 2.3	0.9	37
SKD	0.3 ± 3.9	1.5 ± 3.2	5.4 ± 2.2	1.5	11
SNL	0.8 ± 6.0	1.8 ± 5.5	8.0 ± 3.1	2.0	26
Summer					
SAV	1.1 ± 2.5	0.5 ± 1.4	2.8 ± 1.4	1.1	66
SVN	1.0 ± 2.2	0.6 ± 1.5	2.4 ± 1.6	1.1	55
SNL	2.1 ± 4.6	2.3 ± 3.7	6.0 ± 2.8	3.1	42

Directions are measured in degrees clockwise from north. Summer data are not available for SKD. Data periods and filter applied are the same as stated for Table 2. SAV is Savannah Municipal Airport, SVN is Hunter Army Airfield, SKD is Skidaway Institute of Oceanography, and SNL is Savannah Navigational Light Tower. See Figure 1 for locations.

TABLE 4. Summary of Direction Differences Δα Between Savannah Airport (SAV) and Oceanic Platform 10 km Offshore (SNL), February 19 to March 21, 1977

	Averages by Hour				All Hours,
	1900 Hours	0100 Hours	0700 Hours	1300 Hours	Total
No. of observations	29	29	29	28	115
No. with >0	24	23	13	9	69
Average	15	12	−7	−18	0.5
Standard deviation	35	29	25	47	37

Times are local; Δα = direction (SAV − direction (SNL).

However, one would come to a different conclusion regarding the shore-normal component (Figure 18*b*) even though an adjustment to speed had been made. Over the month, net wind stress was onshore according to oceanic winds, but coastal winds indicated offshore stress. This was due to cumulative errors in direction.

Data illustrating differences on a 50-km scale were obtained during a transition period between winter and summer wind regimes. This may not be a fair test. *Weisberg and Pietrafesa* [1983] found organized and coherent motion over a 500-km scale in winter. The coherence was only marginal in summer. We suspect that incoherencies over 50 km would be evident in summer wind data, a period during which strong land heating sets up a large lateral change in relative humidity and vertical convection over land, i.e., the seabreeze regime.

Summary

The wind regimes in the SAB respond to the buildup and decay of the Ohio Valley High alternating with the buildup and shifting of the Azores High. While August winds would normally be a part of the summer regime, in fact an abrupt transition can be expected to occur signifying a buildup of the Ohio Valley High and an eastward migration of the Azores High. This sets up the fall regime which is the most dynamic period of the year and helps to drive coastal currents southward during September, October, and November [*Weber and Blanton*, 1980; *Blanton*, 1981; *Atkinson et al., 1983*]. The regime is persistent and usually exerts more impulse in September and October than any other 2-month period. This is due not only to higher wind speeds but to more efficient drag of cool air over warm continental shelf water.

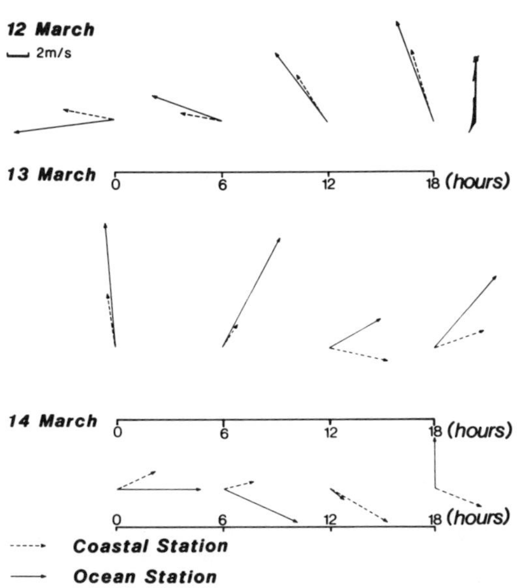

Fig. 17. A 3-day set of 6-hourly vectors from hourly data smoothed with a 10-hour low-pass filter. Solid line vector, ocean tower; dashed line vector, airport location. Times are UT.

The expanded analysis of ship observations (Figures 11 and 12) shows that a distinct fall regime of southward winds is established over the shelf, while the wind field over the Blake Plateau is less organized and remains so throughout autumn and winter. This suggests that the fall regime is associated with processes occurring over a shelf width scale, not a synoptic scale. We speculate that vigorous convection over the shelf in autumn results from warming of relatively cool dry air originating from the Ohio Valley High as it is advected over warm continental shelf waters. Vigorous convection over the shelf would lower sea level pressure and create a pressure gradient force from land to sea. The geostrophic balance of this pressure gradient force with the Coriolis force would be accompanied by southward flow along the coast.

The December–March mean wind fields (Figures 3–6) indicate a southeastward to southward flow over the SAB [*Weber and Blanton*, 1980]. It is important to realize that these means are made up of fluctuations due to the passages of major weather systems and the semipermanent Azores and Ohio Valley highs. They produce no circulation of themselves but are simply the climatological results of those fluctuations. A "typical" cycle of weather during winter begins with a low-pressure trough in central Canada. The resulting pressure field results in northeastward winds along the SAB. As the trough in Canada migrates eastward, the pressure gradient rotates by about 90° [*Rossby*, 1941] and the wind vector in the SAB turns clockwise to southeastward flow. As long as southeastward winds are slightly stronger and/or more persistent than the intervening northeastward winds, a net southeastward vector is produced. Obviously, interannual variability contributes

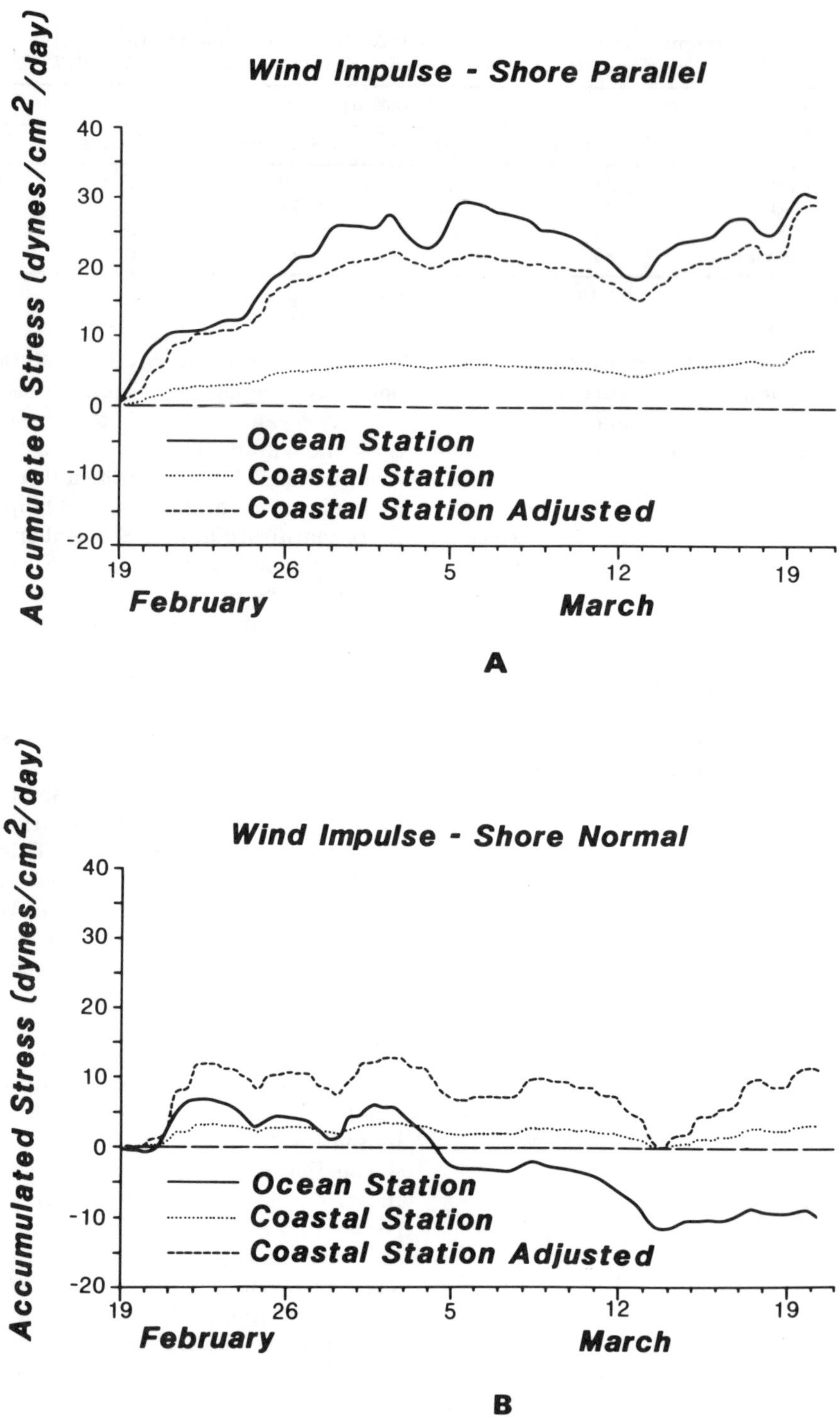

Fig. 18. Wind impulse/unit area computed by summing hourly wind stress values, February 19 to March 21, 1977. Drag coefficient $C_D = 2 \times 10^{-3}$. (*a*) Shore-normal component; (*b*) shore-parallel component. See text for details on the three functions plotted.

to large scatter in monthly winds averaged over many years. In years when the zonal circulation over Canada is more persistent, mean winds over the SAB would be northeastward, which might explain the persistent northeastward wind observed in Charleston, South Carolina, and over the Blake Plateau in the winters of 1976 through 1979 [*Weisberg and Pietrafesa*, 1983].

The wind regime has a clear diurnal signal during spring and summer that is present at coastal and near-coastal stations 20–30 km offshore but not present over the Blake Plateau far removed from land. We have little knowledge on the offshore extent of this important regime and the manner in which it dissipates with distance offshore. While there have been many studies of the dynamics of seabreezes at the coast and over land, only a few, such as *Johnson and O'Brien* [1973] have synthesized simultaneous observations over ocean and land. There is a definite need to better define the changes of the coastal atmospheric boundary layer in the offshore direction [*Weisberg and Pietrafesa*, 1983].

We recommend caution when using over-land wind data as a substitute for oceanic wind data. During some seasons, it was possible for *Weisberg and Pietrafesa* [1983] to transform coastal winds to wind data more representative of open ocean conditions. The technique was unsuccessful in summer. Again this is due to our lack of knowledge of the transition of the atmospheric boundary layer from a coastal regime to an open ocean regime.

Some adjustment to over-land wind data is mandatory when it is to be used in models, as Figure 18 suggests. Use of land winds in simple hydrodynamic models of coastal currents yields unrealistic bottom friction parameters [*Schwing and Blanton*, 1984]. While increasing the speeds of land wind vectors improves the model results, there are directional differences which produce cumulative errors in wind impulse delivered to coastal waters.

Acknowledgments. We wish to thank Jerome Namias (Scripps Institute of Oceanography) and Reid Bryson (University of Wisconsin) for guiding us to studies of airstream sources. Lester Lamhut performed programming assistance, Suzanne McIntosh and Anna Boyette drafted the illustrations, and Cindy Miller typed the manuscript. Work was supported by Department of Energy contracts DE-AS09-80-EV-10331 (J.O.B.), DE-AC09-76SR-00001 (A.H.W. and D.W.H.), DE-AS09-76-EY-00902 (L.J.P.), and DE-AS09-7600901 (D.W.M.).

References

Atkinson, L. P., J. O. Blanton, and E. B. Haines, Shelf flushing rates based on the distribution of salinity and freshwater in the Georgia Bight, *Estuarine Coastal Mar. Sci., 7,* 464–472, 1978.

Atkinson, L. P., T. N. Lee, J. O. Blanton, and W. S. Chandler, Climatology of the southeastern United States continental shelf waters, *J. Geophys. Res., 88,* 4705–4718, 1983.

Blanton, J. O., Ocean currents along a nearshore frontal zone on the continental shelf of the southeastern United States, *J. Phys. Oceanogr. 11,* 1627–1637, 1981.

Blanton, J. O., and L. P. Atkinson, Transport and fate of river discharge on the continental shelf of the southeastern United States, *J. Geophys. Res., 88,* 4730–4738, 1983.

Bryson, R. A., and F. K. Hare, The climates of North America, in *World Survey of Climatology,* vol. 11, pp. 1–47, Elsevier, New York, 1974.

Fofonoff, N. P., Spectral characteristics of internal waves in the ocean, *Deep Sea Res., 16 suppl.,* 59–71, 1969.

Hamblin, P. F., and F. C. Elder, A preliminary investigation of the wind stress field over Lake Ontario, *Proc. Conf. Great Lakes Res., 16,* 723–734, 1973.

Jacobsen, J. P., A socio-economic environmental baseline summary for the South Atlantic region between Cape Hatteras, North Carolina and Cape Canaveral, Florida, in *Physical Oceanography,* vol. 1, 247 pp., Virginia Institute of Marine Science, Gloucester Point, 1974.

Johnson, A., Jr., and J. J. O'Brien, A study of an Oregon seabreeze event, *J. Appl. Meteorol., 12,* 1267–1283, 1973.

Kraus, E. B., *Atmosphere-Ocean Interaction,* pp. 139–141, Clarendon, Oxford, 1972.

Lee, T. N., and D. A. Brooks, Initial observations of current, temperature and coastal sea level response to atmospheric and Gulf Stream forcing on the Georgia shelf, *Geophys. Res. Lett., 6,* 321–324, 1979.

Richards, T. L., H. Dragert, and D. R. McIntyre, Influence of atmospheric stability and over-water fetch on winds over the lower Great Lakes, *Mon. Weather Rev., 94,* 448–453, 1966.

Rossby, C. G., The scientific basis of modern meteorology, in *Climate and Man, Yearbook of Agriculture,* pp. 599–654, U.S. Department of Agriculture, Washington, D. C., 1941.

Ruzecki, E. P., A socio-economic environmental baseline summary for the South Atlantic region between Cape Hatteras, North Carolina and Cape Canaveral, Florida, in *Climatology,* vol. 2, 211 pp., Virginia Institute of Marine Science, Gloucester Point, 1974.

Saunders, P. M., Wind stress on the ocean over the eastern continental shelf of North America, *J. Phys. Oceanogr., 7,* 555–566, 1977.

Schwing, F. B., and J. O. Blanton, The use of land- and sea-based wind data in a simple circulation model, *J. Phys. Oceanogr., 14,* 193–197, 1984.

Schwing, F. B., B. Kjerfve, and J. E. Sneed, Nearshore coastal currents on the South Carolina continental shelf, *J. Geophys. Res., 88,* 4719–4729, 1983.

SethuRaman, S., and G. S. Raynor, Comparison of mean wind speeds and turbulence at a coastal site and an offshore location, *J. Appl. Meteorol., 19,* 15–21, 1980.

Weber, A. H., and J. O. Blanton, Monthly mean wind fields for the South Atlantic Bight, *J. Phys. Oceanogr., 10,* 1256–1263, 1980.

Weisberg, R. H., and L. J. Pietrafesa, Kinematics and correlation of the surface wind field in the South Atlantic Bight, *J. Geophys. Res., 88,* 4593–4610, 1983.

Wendland, W. M., and R. A. Bryson, Northern hemisphere airstream regions, *Mon. Weather Rev., 109,* 255–270, 1981.

Wexler, H., and J. Namias, Mean monthly isentropic charts and their relation to departures of summer rainfall, *EOS Trans. AGU, 19,* 164–170, 1938.

Wittman, P., Winter-spring transition along the Carolina Capes, M.S. thesis, 105 pp., N. C. State Univ., Raleigh, 1982.

Physical Oceanographic Processes in the Carolina Capes

L. J. Pietrafesa, G. S. Janowitz, and P. A. Wittman

Department of Marine, Earth, and Atmospheric Sciences, North Carolina State University, Raleigh

Oceanographic processes in the Carolina Capes region of the South Atlantic Bight are influenced by the cuspate shore line and prominent shoals that extend to the shelf break. Meteorological, hydrographic, and moored current data indicate that synoptic scale wind and Gulf Stream events are the primary driving forces in the region. Topographically enhanced upwelling may be important at the shelf break.

Introduction

The South Atlantic Bight (SAB) is geomorphically divided into a southern region, extending from Cape Canaveral, Florida, to Cape Romain, South Carolina, and characterized by a smoothly curved topography, and a northern region, extending from Cape Romain to Cape Hatteras, North Carolina, and characterized by complex topography and prominent shoals that project from the coast out to the shelf break. This northern region of the SAB is called the Carolina Capes region and includes Long, Onslow, and Raleigh bays and Frying Pan, Cape Lookout, and Diamond shoals. While the origin of these shoals is not known, they play an integral role in modifying physical processes, seemingly in a manner that helps maintain their existence [*Pietrafesa*, 1983]. The physical oceanography of the Carolina Capes region is the subject of this paper.

The Carolina Capes region is bounded to the west by low-lying barrier islands in Raleigh Bay and upper Onslow Bay and by a marshy coastal plain in Long Bay (Figure 1). In Raleigh and Onslow bays, little freshwater is provided through the inlets that connect bar-built sounds to the coastal ocean. South of Frying Pan Shoals the Cape Fear, Pee Dee, North Santee, South Santee, and Cooper rivers provide periodic intrusions of freshwater into Long Bay.

The Gulf Stream front is the seaward boundary of the Carolina Capes region. The northeasterly flowing Gulf Stream directly affects physical processes year round through short-term frontal events. The region between 32° and 33.5°N latitude is a recirculating zone known as the Charleston Trough. Since the front abuts the outer continental margin of the SAB, there is no slope water south of Cape Hatteras, and the shelf break is directly coupled with deep North Atlantic Central Water. The principal features of the Gulf Stream western boundary are clearly revealed in satellite infrared imagery and are schematically illustrated in Figure 2. Disturbances form along the western edge of the Gulf Stream, the cyclonically sheared front, and are advected to the north along the shelf break and slope region as both stable and unstable long waves. Stable waves are manifested as 100- to 300-km-long meanders of the front, with cross-shelf amplitudes of 10–100 km in the Carolina Capes. Occasionally, these waves grow and break backward onto the shelf as the onshore portion of the wave crests fold back, forming filaments. At 32°N a topographic feature known as the Charleston Bump or Rise (actually Hoyt's Hill) causes the Gulf Stream to deflect eastward [*Pietrafesa et al.*, 1978; *Brooks and Bane*, 1978], forming a quasi-permanent 10- to 150-km meander. The Gulf Stream returns to the shelf break at about 33.5°N. This deflection process is coupled to the generation of topographic Rossby waves [*Rooney et al.*, 1978; *Chao and Janowitz*, 1979] which try to propagate south but are advected to the north.

Physical processes in the SAB exhibit synoptic (2 days to 2 weeks) and seasonal (greater than 2 weeks) variability. Synoptic variability is caused principally by wind forcing and Gulf Stream frontal effects, while seasonal variability is caused principally by large-scale atmospheric events. This paper will separately examine the seasonal and synoptic variability of the Carolina Capes region.

Seasonal Variability

Wind seasons in the Carolina Capes are defined as winter (from November into February), spring (from mid-February to mid-May), summer (from late May through July), and fall (from August through October). There are, in general, preferred wind directions, magnitudes, and levels of variability during these four seasons [*Weisberg and Pietrafesa*, 1983]. Shelf break wind stresses are typically 2 to 7 times larger than coastal wind stresses. Consequently, horizontal spatial gradients of the wind stress vector, especially the offshore gradient of the alongshore component, are not negligible and may contribute to the mean circulation much as the wind stress vector does directly.

Weisberg and Pietrafesa [1983] showed that the pressure, temperature, and wind velocity at Charleston, South Carolina, are coherent with those at Wilmington and at Cape Hatteras, North Carolina, with little or no phase difference and are characterized principally by short-term fluctuations caused by the passage of high and low-pressure sys-

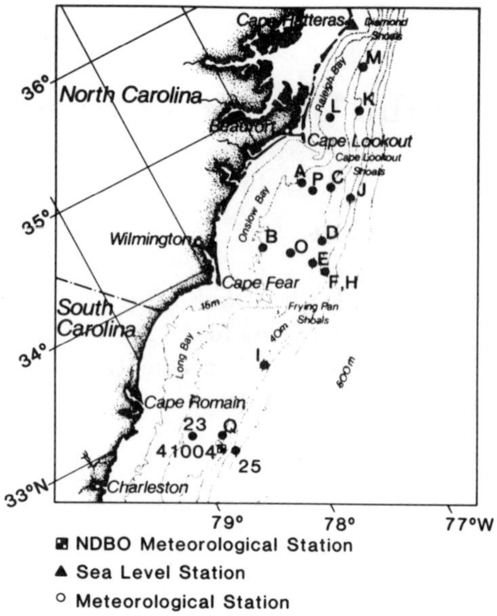

Fig. 1. The Carolina Capes study region showing locations of current meter moorings, coastal meteorological stations, and NDBO buoy 41004. Current meter moorings were supported by the U.S. Department of Energy, the U.S. Department of the Interior, and National Oceanic and Atmospheric Administration/University of North Carolina Sea Grant.

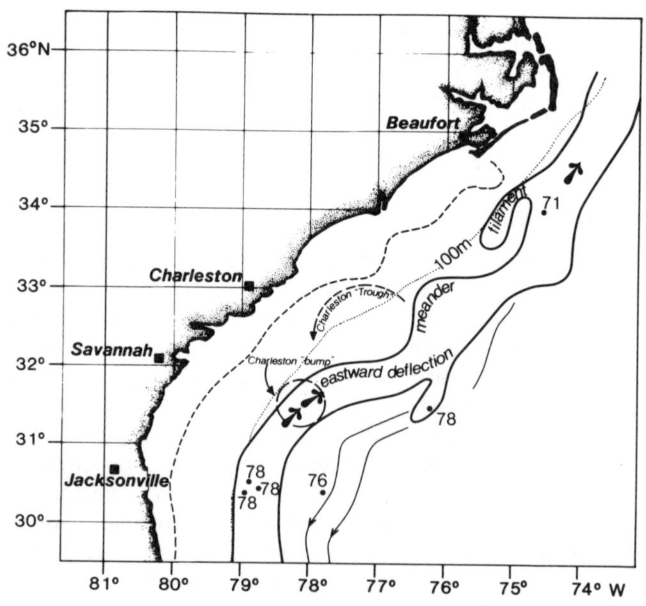

Fig. 2. Schematic representation of Gulf Stream frontal meanders and filaments as revealed by satellite infrared imagery. The eastward deflection of the Gulf Stream at the Charleston Bump and the location of the Charleston Trough are also shown. The dotted line indicates the 100-m isobath.

tems. Amplitudes of the pressure variations decrease from winter to summer as a consequence of the northward migration of the polar front and the strengthening of the Azores-Bermuda high-pressure system. This decrease in atmospheric variability is also reflected in the wind and air temperature records with 10° to 20°C fluctuations present in winter and only 5° to 10°C fluctuations in summer.

Seasonal variations in the thermohaline structure of Carolina Capes shelf water were first described by *Bumpus* [1955] and *Bumpus and Pierce* [1956] using ocean station data collected during May and June of 1953 and January of 1954. Major hydrographic surveys encompassing the entire SAB were conducted between 1953 and 1955 [*Anderson et al.*, 1956*a*, *b*, 1957*a*, *b*, 1958*a*, *b*, 1959*a*, *b*, *c*]. From June 1965 to August 1967, hydrographic cruises were conducted in Raleigh and Onslow bays and described by *Stefansson et al.* [1971]. All of these data were summarized by *Jacobsen* [1974]. *Kirshen* [1979] updated and expanded this work by including all the data on the Naval Oceanographic Data Center (NODC) files for the SAB. These data were spatially and seasonally averaged.

Seasonally averaged temperature, salinity, and density differences were computed in the cross-shelf, alongshelf, and vertical directions from *Kirshen* [1979] and are presented in Table 1. The data are from cross-shelf sections A, B, C, and D in Raleigh, Onslow, and Long bays (Figure 3). The horizontal distance between the regions used in computing the differences was approximately 100 km in both the cross-shelf and alongshelf directions. The vertical differences were taken between the surface and 50 m.

Surface isotherms tend to parallel the coastline throughout the year in the Carolina Capes. The alongshelf thermal gradients are generally an order of magnitude smaller than their cross-shelf counterparts, with temperature increasing toward the south. During winter, shelf waters are much cooler than Gulf Stream waters, producing a cross-shelf temperature gradient of approximately 8°C. Throughout the winter the shelf waters are isothermal due to strong wind mixing. In the spring the cross-shelf temperature gradient weakens and a vertical temperature gradient begins to develop. By summer a strong thermocline develops, and the temperature difference between the inner shelf and the Gulf Stream front is approximately 1°C. The annual cycle of surface temperature across the shelf [*Stefansson et al.*, 1971] has the greatest range nearshore and the smallest range at the Gulf Stream front. As autumn approaches, winds increase and air temperature drops, causing surface cooling and strong vertical mixing. Thus maximum vertical differences occur in summer and maximum cross-shelf difference in winter. Bottom temperatures near the coast are affected by air temperatures and wind mixing, while bottom temperatures near the shelf break are affected by intrusions of Gulf Stream waters [*Blanton*, 1971].

Carolina Capes shelf water is relatively more saline than that of the Middle Atlantic Bight to the north and the Georgia Bight to the south (Figure 4) due to the proximity

TABLE 1. Cross-Shelf, Alongshelf, and Vertical Gradients of Temperature, Salinity, and Density Determined From *Kirshen* [1979]

	Cross Shelf				Alongshelf		Vertical		
	A4-A2 (0 m)*	B4-B1 (0 m)	C4-C1 (0 m)	D4-D1 (0 m)	C4-D4 (0 m)	C4-D4 (50 m)	B3 (0–50 m)	C3 (0–50 m)	C4 (0–50 m)
				Temperature Differences, °C					
Winter	13.71	6.54	7.96	6.08	−0.78	−1.82	0.27	0.61	1.42
Spring	1.35	1.76	4.49	5.11	−0.13	−0.22	0.93	2.42	1.21
Summer	1.13	1.82	1.08	0.69	−0.03	−0.76	1.36	5.06	4.06
Fall	4.22	7.43	8.41	5.82	0.23	−0.88	0.06	−0.59	0.58
				Salinity Differences, ppt					
Winter	4.43	1.08	0.66	0.65	−0.03	−0.08	−0.91		−0.06
Spring	4.65	0.31	1.35	2.09	0.11	−0.10	−0.24		−0.10
Summer	5.06	0.08	1.90	0.93	0.27	−0.08		−0.49	−0.08
Fall	4.22	0.73	0.64	0.03	0.06	0.06	−0.61	−0.01	0.06
				Sigma-t Differences					
Winter	0.99	−0.20	−1.49	−0.95	0.19	0.46	−0.06	−0.13	0.19
Spring	2.15	0.06	−0.15	−0.12	−0.23	−0.09	−0.85	−0.76	−0.47
Summer	2.35	−0.36	−0.15	0.48	−0.04	0.66		−0.66	−1.47
Fall	3.21	−0.14	−1.56	−1.49	−0.03	−0.03	0.24	−0.45	

Data regions A, B, C, and D are shown in Figure 3. Cross-shelf and alongshelf length scales are 100 km; vertical scale is 50 m.
*Depth.

of the Gulf Stream and relatively low river runoff along the Carolina Capes [*Bumpus*, 1955; *Stefansson et al.*, 1971]. The salinity distribution along the Carolina shelf is highly variable, ranging from 30 to 36‰ [*Stefansson et al.*, 1971]. In general, surface isohalines parallel the coast, increasing seaward to the Gulf Stream front [*Jacobsen*, 1974]. Minimum salinities occur in spring when river runoff is at a maximum, evaporation is at a minimum, and wind-driven intrusions of alongshore, southward flowing freshwater are frequent.

Fresh Virginia coastal water can move around Cape Hatteras after periods of prolonged northeasterly winds and can lower salinities as far south as Cape Fear [*Bumpus*, 1955; *Stefansson et al.*, 1971]. These events occur most frequently between September and March [*Jacobsen*, 1974]. Seasonal variations in the surface salinity fields are not as well defined as those in the temperature fields. Cross-shelf salinity gradients are greatest off Charleston during the spring, in Onslow Bay during the summer, and in Raleigh Bay during the winter. Cross-shelf gradients at Cape Hatteras do not vary seasonally. Alongshelf salinity gradients are strongest in the Cape Hatteras region at the boundary between Carolina Capes shelf water and Virginia coastal water. The vertical salinity field follows a seasonal trend similar to that of the vertical temperature field. Maximum vertical gradients and minimum horizontal gradients are found during the summer, while the reverse is true during the winter (Table 1).

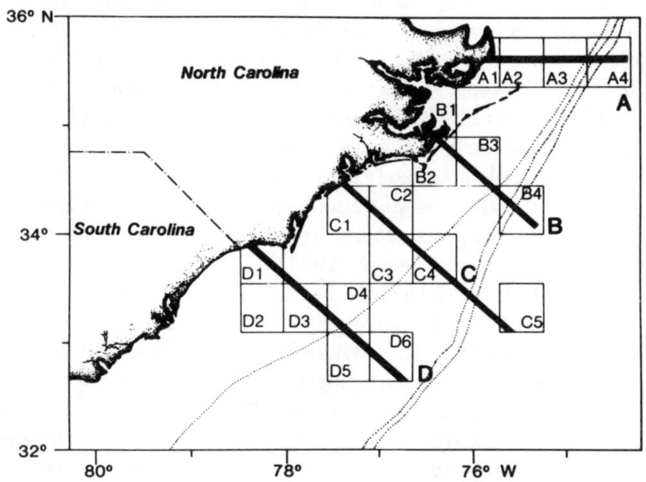

Fig. 3. Cross-section designations showing data regions [from *Kirshen*, 1979].

Densities of Carolina Capes shelf water are relatively high in comparison with Virginia coastal water and Georgia shelf water because of higher salinities (Figure 4). Surface density is high in Onslow Bay throughout the year, decreasing to the north in Raleigh Bay and to the south in Long Bay. In the spring and summer the cross-shelf density gradient in Onslow Bay weakens and virtually disappears. The weak cross-shelf density gradient during the summer results from opposing cross-shelf temperature and salinity fields. During winter a stronger cross-shelf density gradient exists, with density decreasing in the offshore direction south of Cape Hatteras. The cross-shelf density gradient during the winter is controlled by the temperature field, though the cross-shelf salinity field somewhat counteracts the effect of temperature on density [*Bumpus*, 1955]. Density fields are vertically homogeneous during winter and stratified during summer. The transition from winter conditions usually occurs during May, starting at Cape Romain and migrating north to Hatteras. This tran-

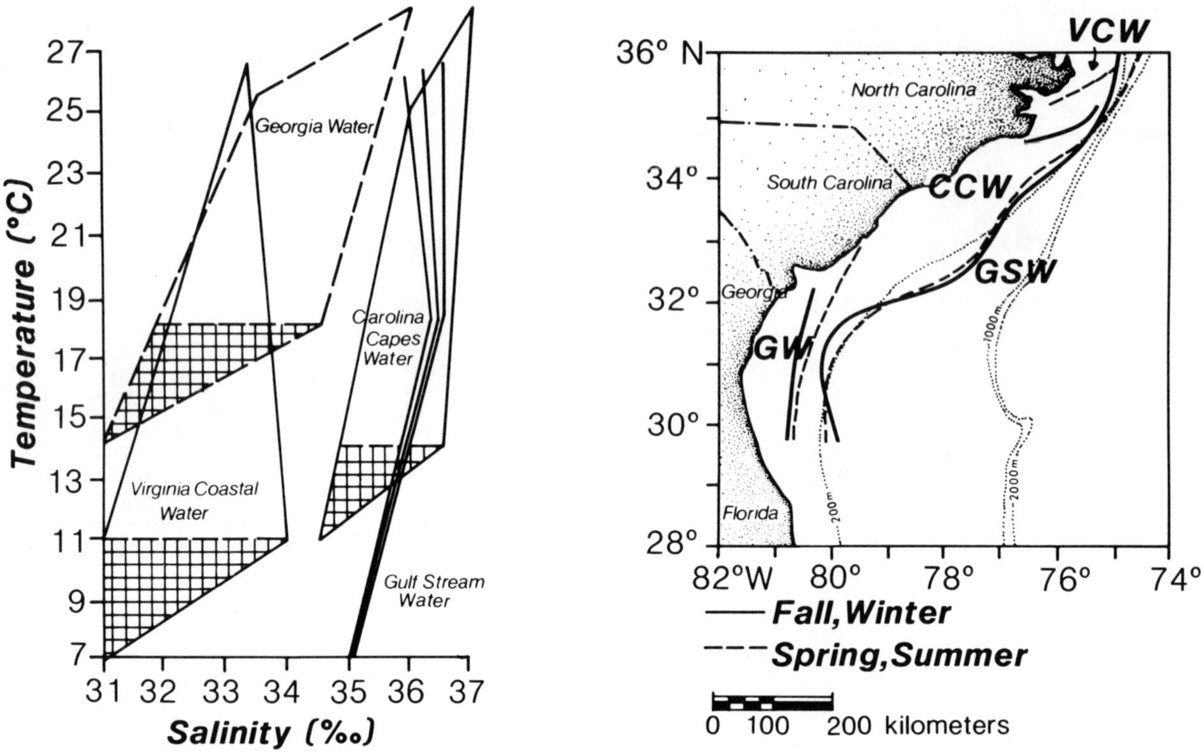

Fig. 4. Temperature-salinity (T/S) relationships in the Carolina Capes region. The left panel shows T/S relations for Georgia, Virginia coastal, Carolina Capes, and Gulf Stream water types. The right panel shows the geographic locations of the water types.

sition is controlled by surface heating resulting from a slight increase in solar radiation, a large increase in air temperature, and the loss of heat from air to water [*Wittman et al.,* 1985].

Heat fluxes were calculated for 1978 from sea and air temperature data measured at NDBC buoy 41004. Sensible heat flux was greatest during winter when sea and air temperature differences were greatest and wind speeds were strong. During April and May, sensible heat flux became positive, indicating an oceanic heat gain, and decreased by an order of magnitude. Latent heat flux was also greatest during winter when cold dry continental air masses moved over the warmer ocean surface. Latent heat flux was approximately twice as large during winter as during summer. Solar radiation also increased by a factor of 2 from winter to summer. During winter, heat loss by sensible heat, latent heat, and long-wave radiation was greater than heat gain by solar radiation. During spring, heat loss by latent heat and long-wave radiation was greater than heat gain by sensible heat and solar radiation. Maximum heating occurred in May.

During January and February, air and water temperatures are the lowest of the year in the Carolina Capes [*Pietrafesa*, 1981]. During late winter the heat content of North Atlantic Central Water down to 100 m is at its yearly minimum, and the water column height of the entire North Atlantic is depressed. Atmospheric cooling of the coastal ocean also occurs with the regular passage of cold fronts. As a consequence, sea level is at its yearly minimum, at 20–25 cm below mean annual datum. With the reduction of cold front activity in February, sea level rises across the SAB shelf. Mean sea level at Charleston shows a strong relationship to the direction and magnitude of coastal winds. In effect, northerly and easterly winds cause a rise in coastal sea level, while southerly and westerly winds cause coastal sea level to fall. The distance offshore to the Gulf Stream front does not appear to be directly related to sea level at the coast [*Pietrafesa et al.*, 1981].

Current meter data collected during 1977–1978 at moorings P and Q in Onslow Bay and off Cape Romain, respectively, show that in Onslow Bay surface and bottom currents display a mean northeastward annual and seasonal flow (Figure 1). In contrast, annual mean flow in Long Bay is toward the southwest at surface and middepth and toward the south at the bottom. The large ranges and standard deviations indicate large seasonal variability. Mean current directions for Q surface and middepth are toward the northeast during spring and summer and toward the southwest during fall and winter. The seasonal reversal of currents at Q and the persistent northeastward drift at P are consistent with drifter studies [*Bumpus*, 1973]. The mean southerly flow in Long Bay is opposite to that reported for middle to outer shelf waters both to the north [*Pietrafesa*, 1983] and south [*Lee and Atkinson*, 1983].

Throughout the Carolina Capes the alongshore velocity component is much greater than the cross-shelf component. Current velocities are greater than expected from Ekman theory for direct wind forcing. The surface instruments at Q (10 m) and P (14 m) are probably immediately below or within the transition zone between the top frictional layer and the geostrophic interior flow. Since the wind drift current decreases exponentially with depth, the influence of the wind is probably overshadowed by the interior flow near the transition zone. The surface Ekman layer depth ranges from 20 m in winter to 15 m in summer, while the bottom Ekman layer depth ranges from approximately 2.0 m in the winter to 1.5 m during the summer. Alongshore coastal currents can also be indirectly forced by wind-induced sea level fluctuations across the shelf. Calculated geostrophic currents compare well in both magnitude and direction with observations at Q surface and middepth, assuming a linear cross-shelf sea surface slope with zero value at 100 km offshore.

Mean monthly alongshore currents at P surface and bottom are toward the northeast except for reversals in February, July, and October, which is consistent with wind reversals during these months. As at Q, direct wind forcing of the mean monthly currents at 17 m is not likely. The bottom alongshore current at P is coherent with that at the surface, with a decrease in magnitude. The cross-shelf current is offshore during the winter and onshore during the summer, due to upwelling favorable winds during the summer and a compensating onshore flow in the bottom Ekman layer.

From late spring to early fall, vertical current shear is negligible, reflecting weak cross-shelf density gradients. The flow is barotropic in the upper water column. Shear during winter and early spring reflects the greater cross-shelf density gradient. The mean monthly current at Q bottom is to the south throughout most of the year, with a counterclockwise veering of geostrophic currents in the bottom Ekman layer. Summertime upwelling, due to southerly winds, can cause onshore flow and onshore veering of the current in the bottom Ekman layer [*Janowitz and Pietrafesa*, 1980]. The vertical shear of alongshore velocity changes sign with the change in direction of the alongshore component of the monthly wind vector. The alongshore vertical velocity gradient at midshelf in Onslow Bay compared with the cross-shelf density gradient during February and March [*Curtin*, 1979*a*, *b*] indicates that the observed vertical shear is too large to be accounted for by the density gradient. Since the bottom current meter lies in the bottom Ekman layer, friction may be responsible for the vertical velocity shear.

Gulf Stream frontal events affecting the outer and middle shelf can result in baroclinic currents driven by the density gradients associated with the Gulf Stream [*Pietrafesa*, 1981]. The contribution of Gulf Stream forcing to the mean monthly alongshore current is probably greater at P than at Q because of the proximity of the Gulf Stream off Onslow Bay.

Pietrafesa and Janowitz [1979*a*] found that the buoyancy flux over the shelf waters can force currents of the order of 20 cm/s. Evaporation, precipitation, river runoff, and atmospheric heating and cooling cause density variations in the cross-shelf direction which can drive an interior alongshore flow. Heating of the ocean by the atmosphere occurs only during the spring when water temperatures are cooler than those of the atmosphere. Because of the sloping bottom, cooling occurs more rapidly on the inner shelf than the outer shelf. Nearshore water will be denser than that offshore, producing a negative offshore density gradient which drives a geostrophic interior current toward the northeast. Since mean monthly currents at Q flow toward the south during the fall and winter, density-driven currents are likely overshadowed by wind forcing. At P the currents also seem to respond to wind and Gulf Stream forcing on a synoptic time scale.

Mean monthly bottom currents at Q are generally to the south. A comparison of the currents at Q bottom with coastal winds and sea level at Charleston and with buoy winds collected near Q suggests that bottom currents are only occasionally related to atmospheric winds and coastal sea level. This is not true of near-surface currents, which follow coastal sea level. *Pietrafesa and Janowitz* [1980] postulated that the shelf break region off Charleston is part of a large counterclockwise rotating gyre that has the inshore side of the Gulf Stream as its southern, eastern, and northern boundaries and is closed on its inshore side by a southerly flowing current. The persistent southward bottom currents at Q may be a manifestation of the inshore side of this gyre (Figure 2).

Pietrafesa et al. [1978] presented evidence of an eastward, offshore deflection of the Gulf Stream front at 32°N, 79°W over the site of the Charleston Bump, and *Rooney et al.* [1978] established a theoretical basis for the permanent deflection of the front at that site. The topographic feature causes an imbalance in vorticity, resulting in an offshore motion of the jet. As the Gulf Stream moves to deeper water, the jet moves back toward the shelf break affecting the generation of topographic Rossby waves. The deflection region is the trough of such a wave. The cyclonic circulation about this trough provides the mechanism for driving the southerly flow observed at Q bottom, which results from an offshore drop in sea level coupled with an onshore rise in isopycnals. The net pressure gradient, which is directed seaward, geostrophically drives a current to the south. *Singer et al.* [1983] observed a shoreward rise of isopycnals and a doming up of the isopycnals to the northwest of the Charleston Bump. The pressure gradients associated with these isopycnal structures are not sufficient to drive a southerly current and may in fact be a consequence of the current. The cross-shelf sea surface structure provides a pressure gradient force responsible for driving the southerly flow.

Synoptic Variability

Wind stress in the Carolina Capes is strongly coupled to inner and middle shelf dynamics, with both alongshore and cross-shelf currents responding to synoptic atmospheric forcing. This is also the case in the Georgia Bight [*Lee and Brooks*, 1979]. Sea level and wind stress data for Onslow and Long Bays suggest that alongshore winds mechanically force sea level to rise and fall during periods of greatest wind energy, typically 10, 5, 3.3, and 2.2 days. Comparisons of Onslow Bay alongshore winds with sea level and of Charleston alongshore winds with sea level are virtually identical.

Coastal winds and sea level are coherent with alongshore current fluctuations in the inner and middle shelf regions of Onslow Bay and Long Bay. However, such coherency is not found on the outer shelf. Coherency between alongshore winds and alongshore currents seaward of the middle shelf in Long Bay tends to be higher than that in Onslow Bay. The same is true for coherency between sea level and currents. The Long Bay shelf, while greatly influenced by Gulf Stream frontal events, is also influenced more by wind forcing, since the Gulf Stream is further offshore than elsewhere in the Carolina Capes. Alongshore wind stress τ^y is generally more energetic than cross-shelf stress τ^x over the 2- to 20-day period in the Carolina Capes [*Weisberg and Pietrafesa*, 1983]. Consequently, the coherency between τ^y and both shelf currents and coastal sea level is likely to be greater than that of τ^x with the same parameters.

Middle to inner shelf currents and sea level show a time dependent response to the total wind stress vector over variable topography [*Janowitz and Pietrafesa*, 1980; *Chao and Pietrafesa*, 1980]. Winds from the west to south cause a fall in coastal sea level, while winds from the north to east cause a rise. Thus an alongshore wind directed to the northeast drives a surface flow which is initially balanced by an interior onshore flow. An alongshore flow resulting from the Coriolis force associated with this onshore flow and a bottom frictional layer with onshore flow are set up. Since, in shallow waters, the initial flow is stronger, as is the vertical velocity, isopycnals initially rise more quickly. The alongshore jet has a negative cross-shelf gradient which, through the action of the bottom layer, suppresses the rise of isopycnals. Since the spinup process occurs first in shallower waters, the suppression of the rise creates the impression of an isopycnal bulge moving from shallow to deeper waters. This is the fundamental wind-induced upwelling process in the Carolina Capes [*Janowitz and Pietrafesa*, 1980]. With a negative alongshore wind the process is reversed, with southerly offshore bottom flow, and a drop in coastal sea level. The effect of an alongshore pressure gradient, set up by the wind such that sea level rises in the direction of the wind, is to drive currents counter to the direction of the wind and to support an interior cross-shelf flow directed to the left of the wind.

During late fall to early spring, both momentum and buoyancy flux exchanges at the surface of the coastal ocean can drive coastal currents. During the fall, when cold air is convected offshore, vertical mixing creates a more homogeneous density field. Atmospheric cold fronts tend to move offshore in the SAB, cooling surface and shallow waters. As a result, isopycnals tend to slope downward in the offshore direction [*Kirshen*, 1979], creating a negative cross-shelf density gradient. This causes a southerly flow throughout the water column with the flow increasing with increasing depth. In the spring, warmer air moves offshore and the coastal ocean begins to heat up. This warming process causes the isopycnals to slope downward and seaward across the shelf, generating a northerly flow in the water column. Predictions of wind-driven alongshore currents typically underestimate measured currents at a site off Charleston.

Topographic effects are expected to be important in the Carolina Capes where Frying Pan, Lookout, and Diamond shoals extend to the shelf break. Cold bottom waters found on the north sides of these shoals [*Blanton et al.*, 1981] may result from induced upwelling [*Blanton et al.*, 1981] the interaction of the Gulf Stream front and bottom topography [*Janowitz and Pietrafesa*, 1982]. Topography also can interact with an upwelling favorable wind to produce alongshore bands of upwelling (Figure 5). *Janowitz and Pietrafesa* [1980] show that upwelling is augmented where the

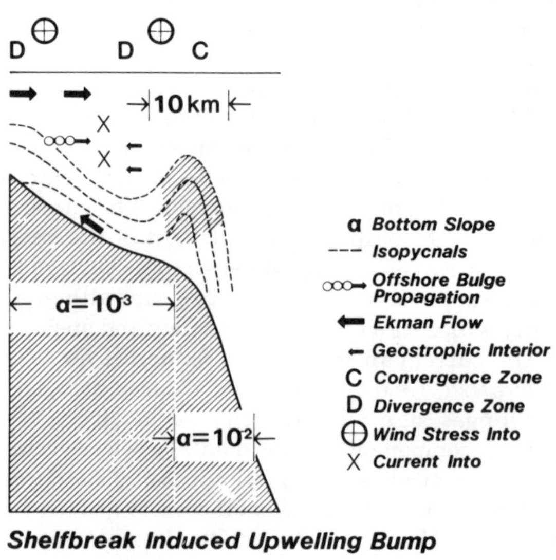

Fig. 5. Schematic diagram of shelf-break-induced upwelling caused by a sharp shelf break interacting with an upwelling favorable wind.

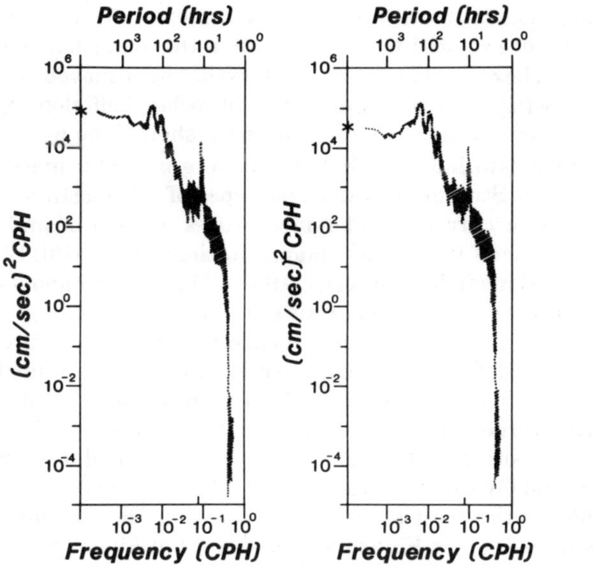

Fig. 6. Kinetic energy spectra for cross-shelf (left panel) and alongshelf (right panel) current components at 17 m below the surface on the 45-m isobath in Onslow Bay.

cross-shelf bottom slope changes abruptly. The Carolina Capes shelf break is a prime site for topographically enhanced upwelling, since there is a sharp break in the bottom slope at a depth of about 50 m. These upwelling bands are only a few kilometers wide in the offshore direction, with an alongshore extent dependent on the gradient of the bottom topography. *McClain et al.* [1984] present coastal zone color scanner (CZCS) imagery showing these shelf break zones, while *Thompson and O'Brien* [1973] and *Brink* [1982] report similar features on the Oregon coast.

Current kinetic energy density spectra for Onslow Bay indicate considerable energy in the 2-day to 2-week period associated with Gulf Stream events, with less energy present in the 2-week to 3-month period (Figure 6). Dominant events have time scales of less than 2 weeks. The low-frequency dropoff in alongshore current energy, contrasted with the slight rise in cross-shelf current energy, suggests a limitation in the response of middle to outer shelf alongshore currents to external seasonal time scale forcing. Inner shelf alongshore currents do not appear to be response limited as are outer shelf alongshore currents.

Current fluctuation kinetic energy distributions, shown in Figure 7, depict velocity fluctuations in terms of the percent of total energy at a particular site. Random forcing by the atmosphere and Gulf Stream dominates current activity at the middle to outer shelf. Both the total current fluctuations and the energy fractions are relatively constant from season to season and year to year. There is, in general, a decrease in current activity from surface to bottom. Near-surface currents decrease in the winter, while those near bottom increase. Finally, there is a decrease in low-frequency variance at middle to inner shelf sites compared with middle to outer shelf sites. If the atmospheric contribution to upper water column current variability is relatively constant across the shelf, then the Gulf Stream must comprise the other fraction of the total current fluctuation at middle and outer shelf sites.

Pietrafesa and Janowitz [1980] found that at both the shelf break (sites H, J in Figure 1) and midshelf (sites E, C in Figure 1), a northeasterly alongshore current occasionally will diminish in magnitude or possibly reverse direction with a decrease in temperature. Typically, the current returns to a northeasterly set and temperature increases within several days. The rise in temperature usually peaks at values characteristic of the Gulf Stream front. The cross-shelf current changes from positive (offshore) to negative (onshore) to positive again during this period and

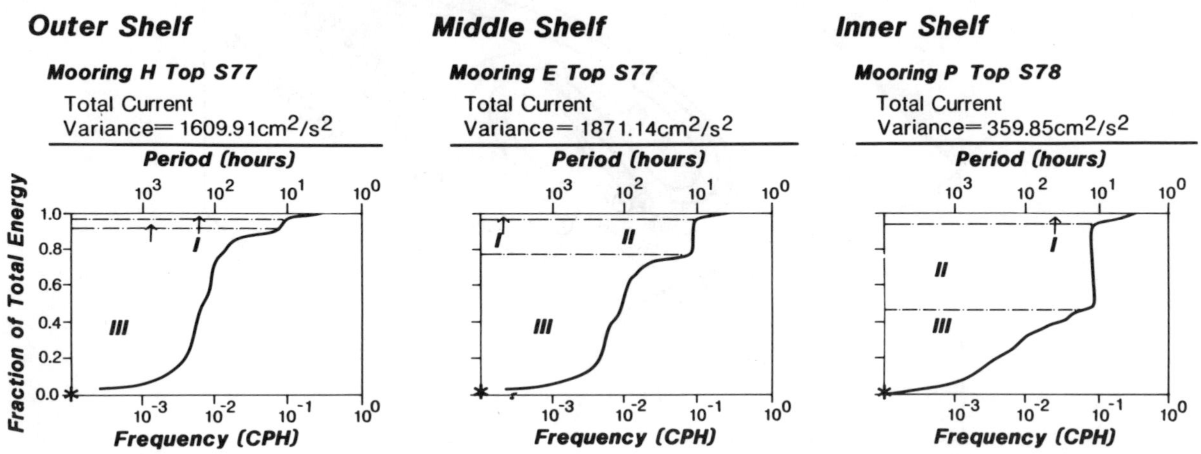

Fig. 7. Current fluctuation energy distributions for inner, middle, and outer shelf regions, showing the fraction of total energy contributed from (I) short-period (less than 1 day) processes, (II) semidiurnal tidal processes, and (III) long-period atmospheric and oceanic random processes.

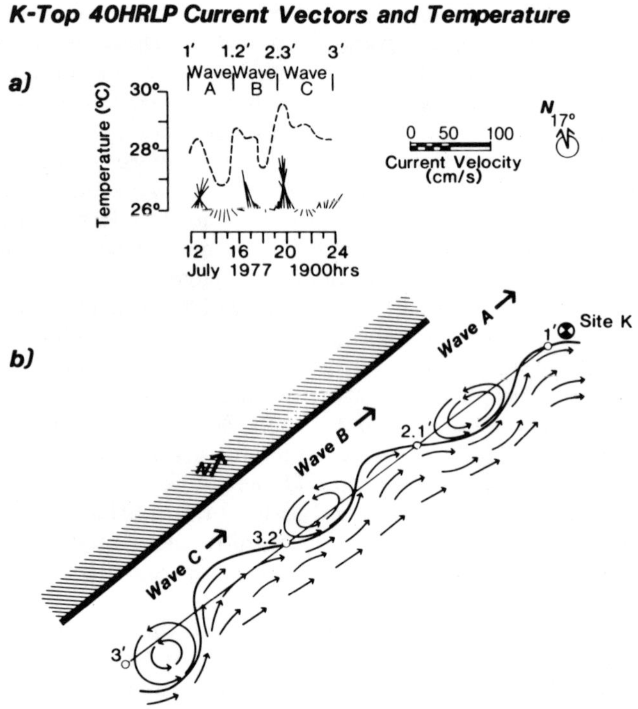

Fig. 8. (*a*) Current and temperature data at mooring K-top during passage of Gulf Stream frontal waves. (*b*) Schematic streamline pattern of three consecutive Gulf Stream frontal meanders moving past mooring K.

precedes changes in alongshore currents. The drop in temperature related to the decrease in the alongshore current could be caused by frontal waters that have moved offshore, allowing a horizontal advection of cooler shelf/slope waters, or from an upwelling of cool, deep shelf/slope waters. Salinity data indicate that the converging water masses are of Gulf Stream origin. These types of summertime variations in currents and temperatures, typical of middle to outer shelf waters, are manifestations of the Gulf Stream lateral meanders [*Webster*, 1961*a*, *b*] that occur about every 6–7 days with no apparent seasonal preference.

Site K (Figure 1) surface current and temperature data collected during the summer of 1977, when divided into three time series of July 12–16, 16–19, and 19–23 show that three northeasterly propagating frontal waves passed the mooring (Figure 8*a*). The three are noted as waves A, B, and C corresponding to lines 1-1', 2-2', and 3-3' in Figure 8*b*. As the crest of wave A approaches the current meter site (point 1 in Figure 8*b* and day 12 in Figure 8*a*), northeasterly flow increases and temperature rises. As wave A moves by the mooring, the trough approaches the site (day 13), and alongshore flow decreases in magnitude and finally reverses in direction as the temperature drops (day 15). Wave B moves past K along line 2-2' between days 16 and 19, and wave C moves past K between days 19 and 22.5 along line 3-3'. The streamlines of flow within the crests and troughs of the waves are as shown in Figure 8*b*. As the axis of the Gulf Stream front shifts farther onshore, the current vector moves right to left while the tempera-

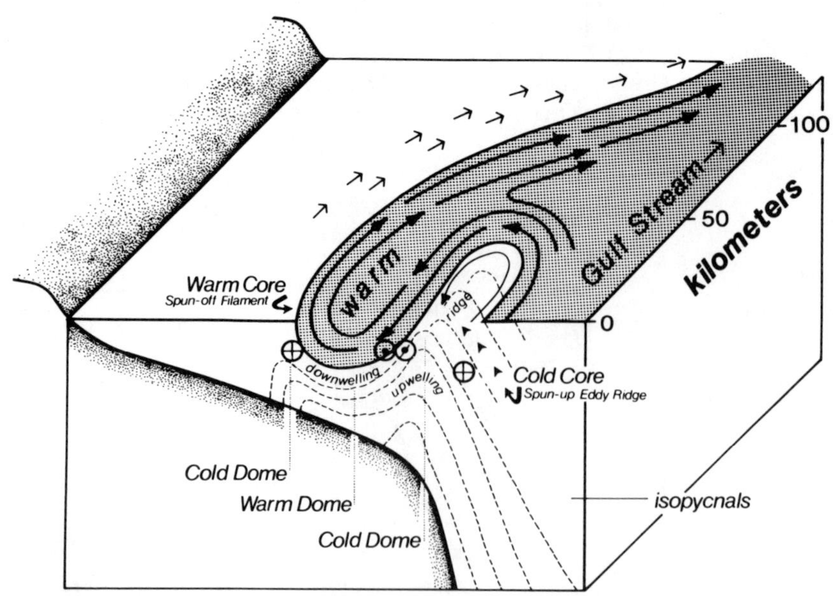

Fig. 9. Conceptual Gulf Stream frontal filament and offshore eddy ridge based on Charleston current meter mooring and VHRR data. Note clockwise flow around warm core filament and counterclockwise flow in cold core eddy ridge.

ture rises and then falls, as suggested by wave B. Wave B probably had an amplitude similar to wave A. The amplitude of wave C, on the other hand, may have been much larger than the amplitudes of A or B, or alternatively wave C passed along a skewed line. Satellite very high resolution radiometer (VHRR) imagery (R. Perchall, personal communication, 1981) shows five meanders of the Gulf Stream front into Raleigh Bay, over site K, during this period.

This conceptual model of a Gulf Stream frontal meander passing a moored current meter suggests a manifestation of outer shelf upwelling. As the crest of a wave (meander) passes a site, the front moves offshore and the trough of the wave appears. The offshore movement of the front effects a one-sided divergence, which is compensated for by horizontal onshore flow or by upwelling of deeper water. Upwelling seems to be principally responsible for supplying water to the system. In addition, since the trough of the wave is rotating cyclonically, upwelling results from the positive vertical vorticity generated.

Gulf Stream frontal filaments are most likely meanders or waves that grow in amplitude and fold back around the cool pool sitting in the wave trough [*Pietrafesa and Janowitz*, 1980; *Sun*, 1982]. As the wave crest folds back toward the coast, flow is to the north on the shoreward side and to the south on the seaward side of the filament, resulting in a rapid doming up of isopycnals (Figure 9). As the filament advects downstream along the front, a series of shelf break/slope cold eddies forms an upwelled ridge. Such filaments can be seen in CZCS chlorophyll and VHRR temperature imagery [*McClain et al.*, 1984].

Summary

Physical processes in the Carolina Capes region of the SAB are primarily driven by synoptic scale (2 days to 2 weeks) wind and Gulf Stream events. Bottom topography, with prominent shoals extending to the shelf break, also plays an important role in the circulation. Topography-enhanced upwelling occurs at the shelf break. Wind stress is the dominant forcing mechanism in the inner and middle shelf regions, while the outer shelf is dominated by Gulf Stream frontal events. Throughout the region, alongshore current velocities are greater than cross-shelf velocities. Middle and outer shelf alongshore currents show a limited response to seasonal forcing mechanisms, which is not apparent on the inner shelf.

Acknowledgments. Funding for this work was provided by the Department of Energy under contract AS09-F6EY00902; the University of North Carolina Sea Grant, NOAA, Department of Commerce under contract 04-6-158-44054; the State of North Carolina, Department of Administration; and the Minerals Management Service (formerly Bureau of Land Management) under contract 209-04.

References

Anderson, W., J. Gehringer, and E. Cohen, Physical oceanographic, biological and chemical data—South Atlantic coast of the United States, *U.S. Fish Wildl. Serv. Spec. Sci. Rep. Fish.*, *178*, 1956*a*.

Anderson, W., J. Gehringer, and E. Cohen, Physical oceanographic, biological and chemical data—South Atlantic coast of the United States, *U.S. Fish Wildl. Serv. Spec. Sci. Rep. Fish.*, *198*, 1956*b*.

Anderson, W., J. Gehringer, and E. Cohen, Physical oceanographic, biological and chemical data—South Atlantic coast of the United States, *U.S. Fish Wildl. Serv. Spec. Sci. Rep. Fish.*, *210*, 1957*a*.

Anderson, W., J. Gehringer, and E. Cohen, Physical oceanographic, biological and chemical data—South Atlantic coast of the United States, *U.S. Fish Wildl. Serv. Spec. Sci. Rep. Fish.*, *234*, 1957*b*.

Anderson, W., J. Gehringer, and E. Cohen, Physical oceanographic, biological and chemical data—South Atlantic coast of the United States, *U.S. Fish Wildl. Serv. Spec. Sci. Rep. Fish.*, *248*, 1958*a*.

Anderson, W., J. Gehringer, and E. Cohen, Physical oceanographic, biological and chemical data—South Atlantic coast of the United States, *U.S. Fish Wildl. Serv. Spec. Sci. Rep. Fish.*, *265*, 1958*b*.

Anderson, W., J. Gehringer, and E. Cohen, Physical oceanographic, biological and chemical data—South Atlantic coast of the United States, *U.S. Fish Wildl. Serv. Spec. Sci. Rep. Fish.*, *278*, 1959*a*.

Anderson, W., J. Gehringer, and E. Cohen, Physical oceanographic, biological and chemical data—South Atlantic coast of the United States, *U.S. Fish Wildl. Serv. Spec. Sci. Rep. Fish.*, *303*, 1959*b*.

Anderson, W., J. Gehringer, and E. Cohen, Physical oceanographic, biological and chemical data—South Atlantic coast of the United States, *U.S. Fish Wildl. Serv. Spec. Sci. Rep. Fish.*, *313*, 1959*c*.

Blanton, J. O., Exchange of Gulf Stream water with North Carolina shelf water in Onslow Bay during stratified conditions, *Deep Sea Res.*, *18*, 167–178, 1971.

Blanton, J. O., L. P. Atkinson, L. J. Pietrafesa, and T. N. Lee, The intrusion of Gulf Stream water across the continental shelf due to topographically-induced upwelling, *Deep Sea Res. Part A*, *28*, 393–405, 1981.

Brink, K. H., A comparison of long coastal trapped wave theory with observations off Peru, *J. Phys. Oceanogr.*, *12*, 897–913, 1982.

Brooks, D. A., and J. M. Bane, Jr., Gulf Stream deflection by a bottom feature off Charleston, South Carolina, *Science*, *201*, 1225–1226, 1978.

Bumpus, D. F., The circulation over the continental shelf south of Cape Hatteras, *EOS Trans. A.G.U.*, *36*, 601–611, 1955.

Bumpus, D. F., A description of the circulation on the continental shelf of the east coast of the U.S., *Prog. Oceanogr.*, *6*, 111–157, 1973.

Bumpus, D. F., and E. L. Pierce, The hydrography and the distribution of chaetognaths over the continental shelf off North Carolina, *Deep Sea Res., 3*, suppl., 92–109, 1956.

Chao, S. Y., and G. S. Janowitz, The effect of a localized topographic irregularity on the flow of a boundary current along the continental margin, *J. Phys. Oceanogr., 9*, 900–910, 1979.

Chao, S. Y., and L. J. Pietrafesa, The subtidal response of sealevel to atmospheric forcing in the Carolina Capes, *J. Phys. Oceanogr., 10*, 1246–1255, 1980.

Curtin, T. B., Oceanographic field observations off North Carolina–Winter survey—2–12 February 1978, *Rep. 79-3*, 120 pp., Dep. of Mar. Sci. and Eng., N. C. State Univ., Raleigh, 1979*a*.

Curtin, T. B., Oceanographic field observations off North Carolina—Spring survey—12–22 March 1978, *Rep. 79-4*, 168 pp., Dep. of Mar. Sci. and Eng., N. C. State Univ., Raleigh, 1979*b*.

Jacobsen, J. P., Physical Oceanography, in *A Socio-Economic Environmental Baseline Summary for the South Atlantic Region Between Cape Hatteras, North Carolina, and Cape Canaveral, Florida.* 247 pp., Virginia Institute of Marine Science, Gloucester Point, 1974.

Janowitz, G. S., and L. J. Pietrafesa, A model and observations of time dependent upwelling over the mid-shelf and slope, *J. Phys. Oceanogr., 10*, 1574–1583, 1980.

Janowitz, G. S., and L. J. Pietrafesa, The effects of alongshore variation in bottom topography on a boundary current—Topographically induced upwelling, *Cont. Shelf Res., 1*, 123–141, 1982.

Kirshen, P. H. (Ed.), *Summary and analysis of physical oceanographic and meteorological information on the continental shelf and Blake Plateau from Cape Hatteras to Cape Canaveral,* report, vol. 2, pp. 19–213, Dep. of the Inter., Washington, D.C., 1979.

Lee, T. N., and L. P. Atkinson, Low-frequency current and temperature variability from Gulf Stream frontal eddies and atmospheric forcing along the southeast U.S. outer continental shelf, *J. Geophys. Res., 88*, 4541–4567, 1983.

McClain, C. R., L. J. Pietrafesa, and J. A. Yoder, Observations of Gulf Stream–induced and wind–driven upwelling in the Georgia Bight using ocean color and infrared imagery, *J. Geophys. Res., 89*, 3705–3723, 1984.

Pietrafesa, L. J., Continental shelf processes affecting the oceanography of the South Atlantic Bight, progress report, 765 pp., U.S. Dep. of Energy, Washington, D.C., 1981.

Pietrafesa, L. J., Survey of a Gulf Stream frontal filament, *Geophys. Res. Lett., 10*, 203–206, 1983.

Pietrafesa, L. J., and G. S. Janowitz, On the effects of buoyancy flux on continental shelf circulation, *J. Phys. Oceanogr., 9*, 911–918, 1979*a*.

Pietrafesa, L. J., and G. S. Janowitz, A note on the identification of a Gulf Stream spin-off eddy from Eulerian data, *Geophys. Res. Lett., 6*, 549–552, 1979*b*.

Pietrafesa, L. J., and G. S. Janowitz, On the dynamics of the Gulf Stream front in the Carolina Capes, in *Proceedings of the Second International Symposium on Stratified Flows*, pp. 184–197, Tapir Publications, Trondheim, Norway, 1980.

Pietrafesa, L. J., J. O. Blanton, and L. P. Atkinson, Evidence for deflection of the Gulf Stream at the Charleston Rise, *Gulfstream, 4*(9), 3 and 6–7, 1978.

Pietrafesa, L. J., S. Y. Chao, and G. S. Janowitz, The variability of sea level in the Carolina Capes, *Sea Grant Work. Pap. 81-11*, 39 pp., Univ. of N. C., Chapel Hill, 1981.

Rooney, D. M., G. S. Janowitz, and L. J. Pietrafesa, A simple model of deflection of the Gulf Stream by the Charleston Rise, *Gulfstream, 4*(11), 3 and 6–7, 1978.

Singer, J. J., L. P. Atkinson, J. O. Blanton, and J. A. Yoder, Cape Romain and the Charleston Bump: Historical and recent hydrographic observations, *J. Geophys. Res., 88*, 4685–4697, 1983.

Stefansson, U., L. P. Atkinson, and D. F. Bumpus, Seasonal studies of hydrographic properties and circulation of the North Carolina shelf and slope waters, *Deep Sea Res., 18*, 383–420, 1971.

Sun, L. C., On the dynamic variability of the Gulf Stream front, Ph.D. dissertation, 144 pp., N. C. State Univ., Raleigh, 1982.

Thompson, J. D., and J. J. O'Brien, Time-dependent coastal upwelling, *J. Phys. Oceanogr., 3* 33–46, 1973.

Webster, F., A description of Gulf Stream meanders off Onslow Bay, *Deep Sea Res., 8*, 130–143, 1961*a*.

Webster, F., The effects of meanders on the kinetic energy balance of the Gulf Stream, *Tellus, 13*, 392–401, 1961*b*.

Weisberg, R. H., and L. J. Pietrafesa, Kinematics and correlation of the surface wind field in the South Atlantic Bight, *J. Geophys. Res., 88*, 4593–4610, 1983.

Wittman, P. A., L. J. Pietrafesa, and G. S. Janowitz, The thermal transition of the Carolina Capes shelf, *Cont. Shelf Res.*, 1985.

Shelf Circulation From Cape Canaveral to Cape Fear During Winter

T. N. Lee, V. Kourafalou, J. D. Wang, and W. J. Ho

Rosenstiel School of Marine and Atmospheric Science, University of Miami, Florida

J. O. Blanton and L. P. Atkinson

Skidaway Institute of Oceanography, Savannah, Georgia

L. J. Pietrafesa

Department of Marine, Earth, and Atmospheric Science, North Carolina State University, Raleigh

Subtidal current and sea level response to wind and Gulf Stream forcing are investigated for the South Atlantic Bight shelf during winter conditions. Low-frequency flow variability in the outer shelf results primarily from wavelike meanders and eddies in the Gulf Stream front that occur in a 2-day to 2-week period band. Current meter derived vertically integrated momentum balances indicated that these large-amplitude flow events are in approximate geostrophic balance with baroclinic pressure gradients induced by northward propagating Gulf Stream disturbances. Low-frequency flow at midshelf is primarily a local Ekman response to wind forcing. Cross-shelf momentum balance for the total water column is between the alongshelf geostrophic current and the cross-shelf barotropic pressure gradient resulting from wind-induced sea level changes at the coast. This balance holds for both mean and fluctuating parts of the flow, with the alongshelf barotropic current lagging sea level by 6 to 12 hours and alongshelf wind by 12 to 24 hours. The alongshelf balance of momentum is between the Coriolis force, alongshelf pressure gradient and alongshelf wind stress for the mean flow, with additional contribution from the local alongshelf acceleration of the water column and alongshelf bottom stress for the fluctuating flow. Near the transition from midshelf to outershelf flow regimes, which occurs at about the 40-m isobath, there is a significant contribution to mean and fluctuating alongshelf momentum from the divergence of cross-shelf transport of alongshelf momentum. An alongshelf slope of Gulf Stream origin of the order of -10^{-7} appears to make a significant contribution to the observed mean northward flow over the shelf. The inner shelf is characterized by a band of low-salinity water from river runoff, which is vertically stratified throughout the year and forms a surface front at the eastern edge of the band. The alongshelf component of the flow is highly correlated with local wind stress, and the vertical distribution of the current is influenced by an internal (baroclinic) pressure gradient deriving from the river runoff. Alongshelf currents tend to lag wind-forcing events by about 3 hours and lead coastal sea level changes by about 15 hours at periods of 3 to 5 days. Mean volume transports for the innershelf and midshelf regions combined are approximately 20×10^4 m^3 s^{-1} northward, which indicates a shelf residence time of 3 months. A vertically integrated two-dimensional model has been adapted for the study of the wind-driven flow on the South Atlantic Bight (SAB) shelf during the winter season. Wind data are used as an input to the model, and current data are used to verify the model results. Initially, a constant wind field in time and space is applied over the whole domain; the predicted flow pattern shows good agreement with observations. Model results using observed spatially and temporally varying wind fields are also in good agreement with midshelf observations. At the shelf break the comparison is rather poor owing to the strong influence of the Gulf Stream in the outer shelf data, which was neglected in the model. Water particle trajectories resulting from wind forcing, tidal forcing, and the combined effect of an alongshelf slope seem to have a significant effect on the net displacement of water particles as well as on the alongshelf volume transport. Comparison of momentum balances from model results and observations indicates that low-frequency current variability at midshelf during winter is primarily an Ekman response to local wind forcing.

Introduction

Circulation on the southeast U.S. continental shelf during winter is investigated using data collected as part of a Department of Energy and Minerals Management Services supported interdisciplinary study of the region. The purpose of this paper is to describe and quantify the subtidal frequency response of the shelf waters to wind and Gulf Stream forcing using observational and modeling techniques.

Topography of the South Atlantic Bight (SAB) shelf between Cape Canaveral, Florida, and Cape Fear, North Carolina, is relatively uncomplicated. Isobaths tend to parallel the curved coastline, diverging north of Cape Canaveral as the shelf broadens and converging south of Cape Romain as the shelf narrows (Figure 1). The shelf break occurs at about 75-m depth. Shelf widths are greatest (120 km) off Brunswick and Savannah, Georgia, decreasing to a minimum of about 50 km off Cape Canaveral. Shelf length from Cape Canaveral to Cape Fear is approximately 700 km. The northern boundary of the region is formed by a shoal extending seaward of Cape Fear. East of Savannah a topographic ridge known as the Charleston Bump [*Brooks and Bane,* 1978] protrudes beneath the Gulf Stream and has a dramatic effect on the position of the Gulf Stream and outer shelf flow variability in the northeast portion of the study area.

The SAB shelf can be subdivided into three distinct flow regimes [*Atkinson et al.,* 1983]. Inner shelf waters (0- to 20-m isobaths) are characterized by an alongshelf band of low-salinity stratified waters near the coast formed from river runoff [*Blanton,* 1980, 1981; *Blanton and Atkinson,* 1978]. Local wind forcing influences the flow and sea level variability in this region [*Weber and Blanton,* 1980; *Bumpus,* 1973], and there is a southward baroclinic current present in autumn [*Blanton,* 1981].

In the midshelf zone (21–40 m), subtidal frequency flow is primarily dependent on local wind forcing [*Lee and Brooks,* 1979; *Klinck et al.,* 1981; *Lee and Atkinson,* 1983] through an Ekman frictional equilibrium process similar to that of the Middle Atlantic Bight [*Beardsley and Butman,* 1974; *Scott and Csanady,* 1976; *Beardsley and Winant,* 1979] and of the Pacific northwest shelf [*Hickey and Hamilton,* 1980; *Hickey,* 1981]. Stratification changes seasonally: vertically well mixed conditions characterize fall and winter, while vertical stratification prevails during spring and summer [*Atkinson et al.,* 1983].

The outer shelf (41–75 m) is strongly influenced by Gulf Stream frontal disturbances such as northward propagating wavelike meanders and cold cyclonic frontal eddies [*Lee and Brooks,* 1979; *Lee et al.,* 1981; *Lee and Atkinson,* 1983]. The western edge of the Gulf Stream generally lies within ±15 km of the shelf break in the region south of 32°N [*Bane and Brooks,* 1979]. Between 32° and 33°N the Charleston Bump induces an offshore meander of the Gulf Stream [*Brooks and Bane,* 1978; *Pietrafesa et al.,* 1978; *Legeckis,* 1979]. Downstream of the Charleston Bump, enlarged wavelike meanders can displace the Gulf Stream front up to 100 km from the shelf break [*Legeckis,* 1979; *Bane and Brooks,* 1979]. These enlarged meanders are kinematically similar to the smaller frontal eddies off the Georgia shelf and suggest a dynamic connection [*Brooks and Bane,* 1981; *Bane et al.,* 1981; *Lee and Atkinson,* 1983].

Methods

During the winter/spring of 1980 a multiinstitutional study of physical, chemical and biological processes on the southeast U.S. shelf was undertaken (Georgia Bight Experiment (GABEX) I) [cf. *Lee and Atkinson,* 1983]. As part of this study an array of subsurface current meter moorings was deployed on the shelf between Cape Canaveral, Florida, and Cape Romain, South Carolina, from February 16 to July 2, 1980 (Figure 1). Local wind data were obtained from an NDBC buoy and a fixed oceanographic tower off Savannah, Georgia. Additional wind and sea level data were obtained from coastal stations. All time series were filtered with a 40-hour low-passed (HLP) Lanczos filter to produce low-frequency ($f < 0.6$ cpd) records. Current and wind data were rotated into an isobath coordinate system (u positive offshore and v positive alongshelf toward the north).

Hydrographic and biological sampling was conducted in the vicinity of the current meter array. Two ships were involved: *Eastward* and *Columbus Iselin.* The *Eastward* was used to investigate shelf processes, while the *Columbus Iselin* was used to locate and track Gulf Stream frontal disturbances with the aid of a telefax system for receiving satellite sea surface temperature (SST) images from the NOAA-NESS field station in Miami. The *Columbus Iselin* also received a detailed SST map every 3 days from a U.S. Coast Guard aircraft equipped with an airborne radiation thermometer.

Outer Shelf

Time Domain

The primary difference between the outer shelf region of the SAB and that of other continental shelves is the commanding presence of the nearby Gulf Stream. The Gulf

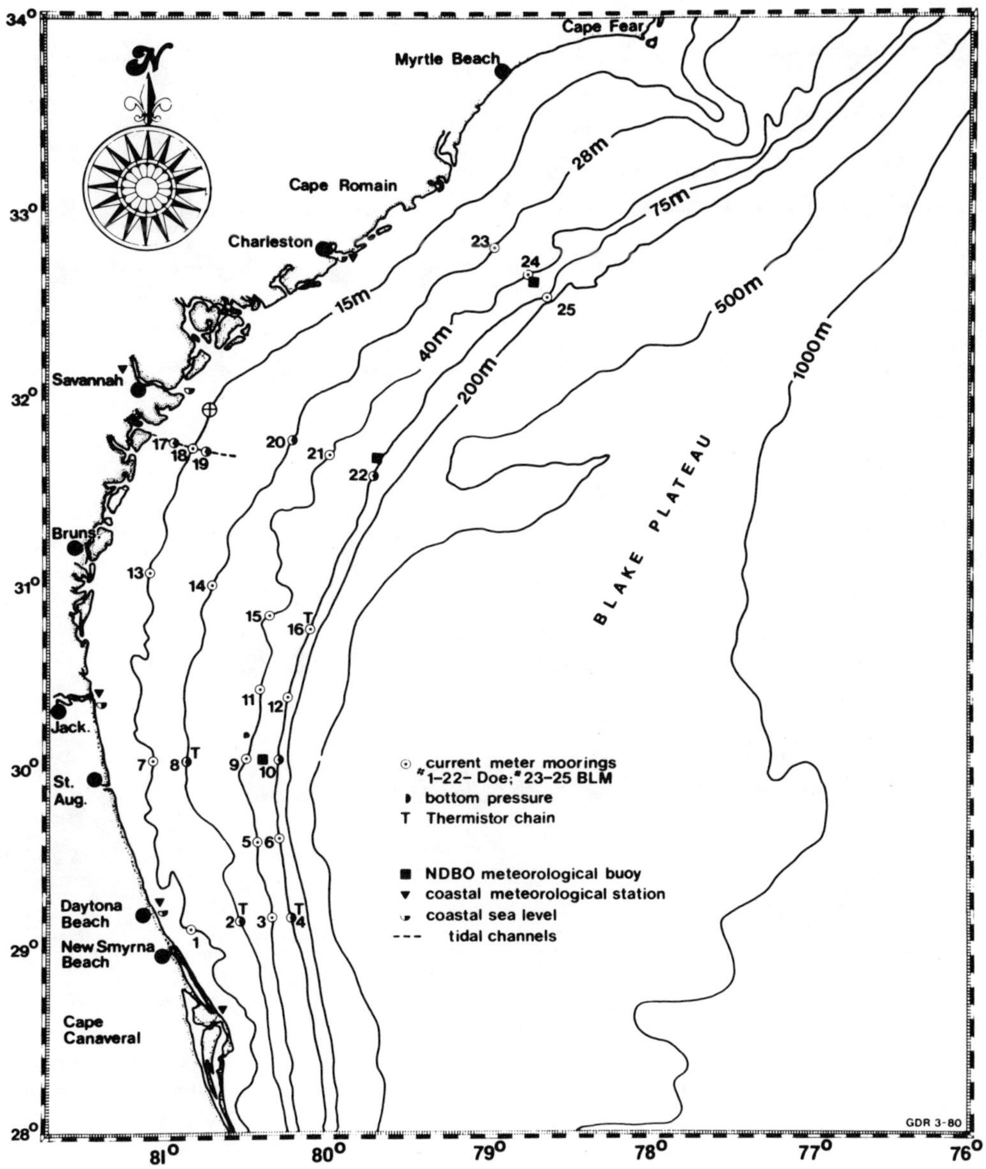

Fig. 1. GABEX I subsurface current meter array, February 17 to June 26, 1980.

Stream influence is clearly indicated in the 40-HLP filtered current meter data from the shelf break (Figure 2). In the southern part of the array (moorings 4, 6, and 10), current speeds were often greater than 100 cm s^{-1}, and strong northward flows persisted for most of the measurement period. The persistent northward flow did not occur at the shelf break off Savannah, Georgia (mooring 22), and the number of flow reversals increased. At the Cape Romain transect, prolonged southward flow events at the shelf break had little visual similarity to currents off Savannah.

Subtidal frequency current and temperature fluctuations at the shelf break were strongly coupled throughout the water column at 30°N (Figure 3). Onshore flow events led decreasing northward currents and decreasing temperatures. Thus cold anomalies occurred simultaneously with cyclonic perturbations of the basic northward flow. Approximately 20 cold cyclonic disturbances were observed at the shelf break during the 4-month experiment, giving an average recurrence period of about 6 days (Figure 3; events 1 to 14 are examples). Vertical shears at the shelf break ranged from about 2×10^{-3} s^{-1} to 1×10^{-2} s^{-1} in the fluctuations and 1×10^{-2} s^{-1} in the mean, indicating a significant baroclinic component in both the fluctuating and mean flows. The baroclinicity was greatest at times of maximum northward flow, corresponding to times when the Gulf Stream front was located near the shelf break.

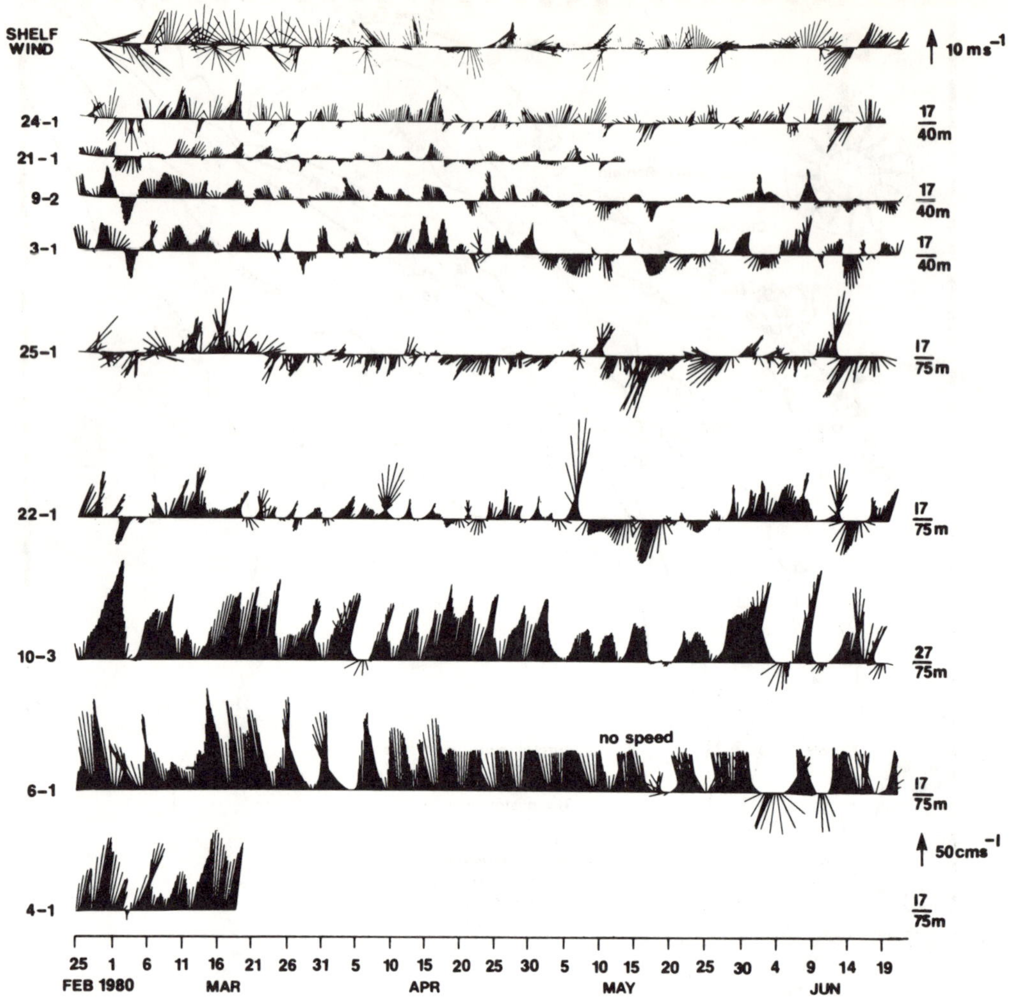

Fig. 2. Time series of 6-hour rotated 40 HLP wind and current vectors from the upper layer for February 25 to June 21, 1980. Current meter identification is given on the left; instrument and water depths are given on the right with scale arrows. See Figure 1 for locations.

Movement of Gulf Stream disturbances along the shelf break can be seen by pattern recognition of events in the 75-m current meter data. Event identification is enhanced by band pass filtering the current meter data to remove fluctuations with periods less than 40 hours and greater than 2 weeks and plotting the data according to the mooring separation distance (Figure 4). Cold cyclonic perturbations traveled northward along the shelf break at about 55 cm s^{-1} and were not visually correlated with either wind or coastal sea level events. The movement is clearest in the lower layer due to the greater number of available instruments but is also apparent in the upper layer. Events 10 to 14 traveled northward at nearly a constant speed. Event 12 was recognizable over alongshelf distances of up to 425 km. These perturbations are more pronounced at the Cape Romain transect, especially the cross-shelf component, which was larger in magnitude than the alongshelf component in the upper layer at mooring 25. This mooring location is downstream of the Charleston Bump in a region of meander amplification [*Bane*, 1983]. Downstream of the Charleston Bump satellite IR images indicate the mean northward propagation speed of meanders as 46 cm s^{-1} [*Legeckis*, 1979].

Frequency Domain

Low-frequency currents from the shelf break at 30°N had well-defined energy peaks at periods of 5 to 6 days for u and 7 to 9 days for v (Figure 5). The velocity components were coherent at these periods with u leading v by about 50° to 70°, which is indicative of propagating cyclonic fluctuations. Coherence was also significant at periods of 2 to 3 days, but the energy levels were considerably reduced. Current meter cospectra of u versus v and u versus T (Figure 5) show well-defined peaks at the 7-day period, indi-

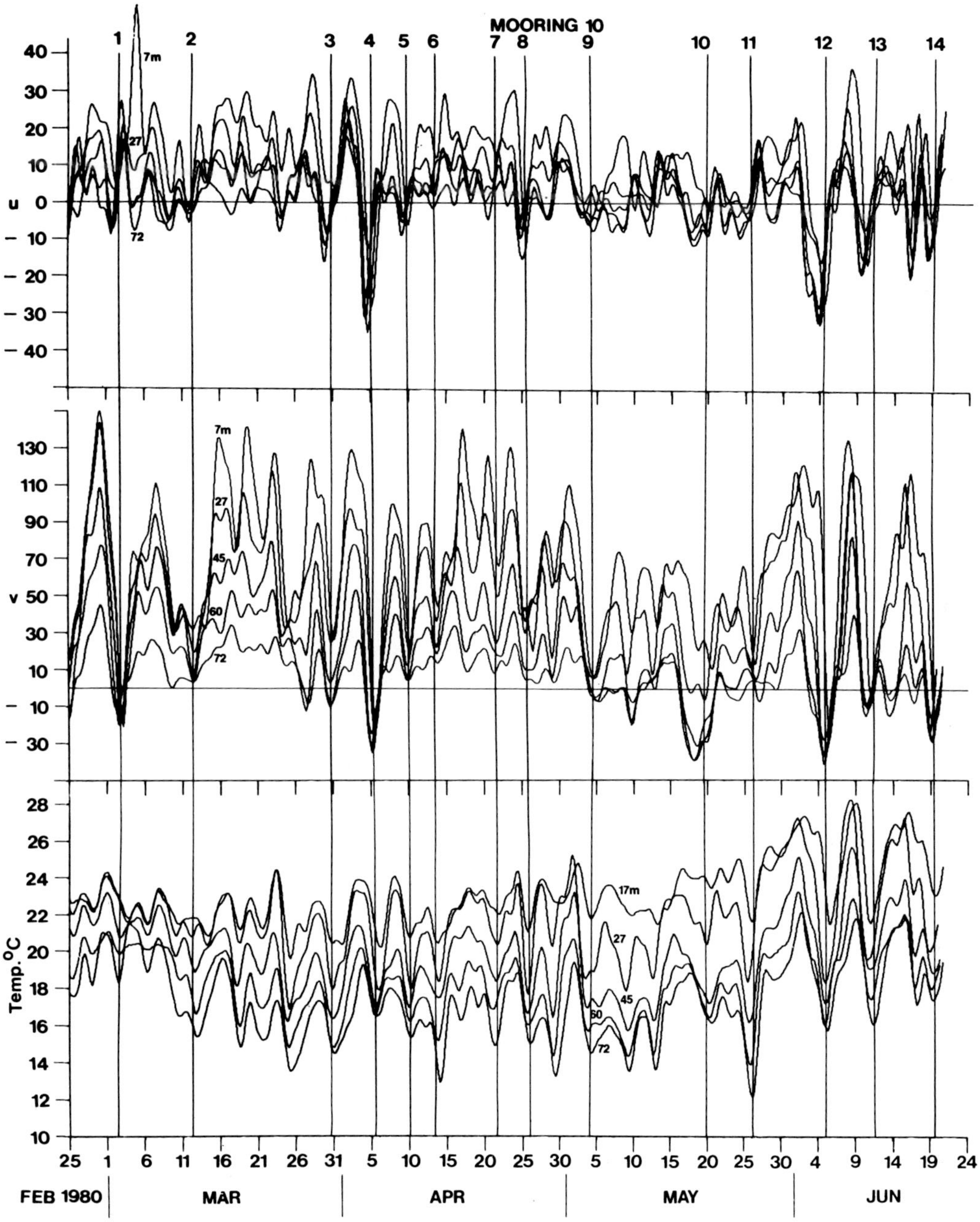

Fig. 3. Time series of 6-hour rotated 40 HLP *u*, *v*, and *T* from mooring 10. Vertical lines are for event identification.

cating an offshore transport of northward momentum and heat on weekly time scales associated with the passage of Gulf Stream disturbances. Coherence between shelf wind components was low in the 5- to 9-day band where high coherence occurred between current components.

Coherence and phase relationships between alongshelf pairs of current meters at the shelf break were determined by *Lee and Atkinson* [1983]. Coherence decreased rapidly with separation distance, thus defining an alongshelf coherence scale of approximately 100 km for *u*, *v*, and *T* (Figure 6, shown for *T* only). Phase estimates show a near-linear increase of negative phase with separation for distances of less than the coherence scale. Negative phase lags indicate northward propagation, and the scatter of points tends to fall around a line equivalent to a constant phase speed of 55 cm s^{-1}. Similar coherence scales and

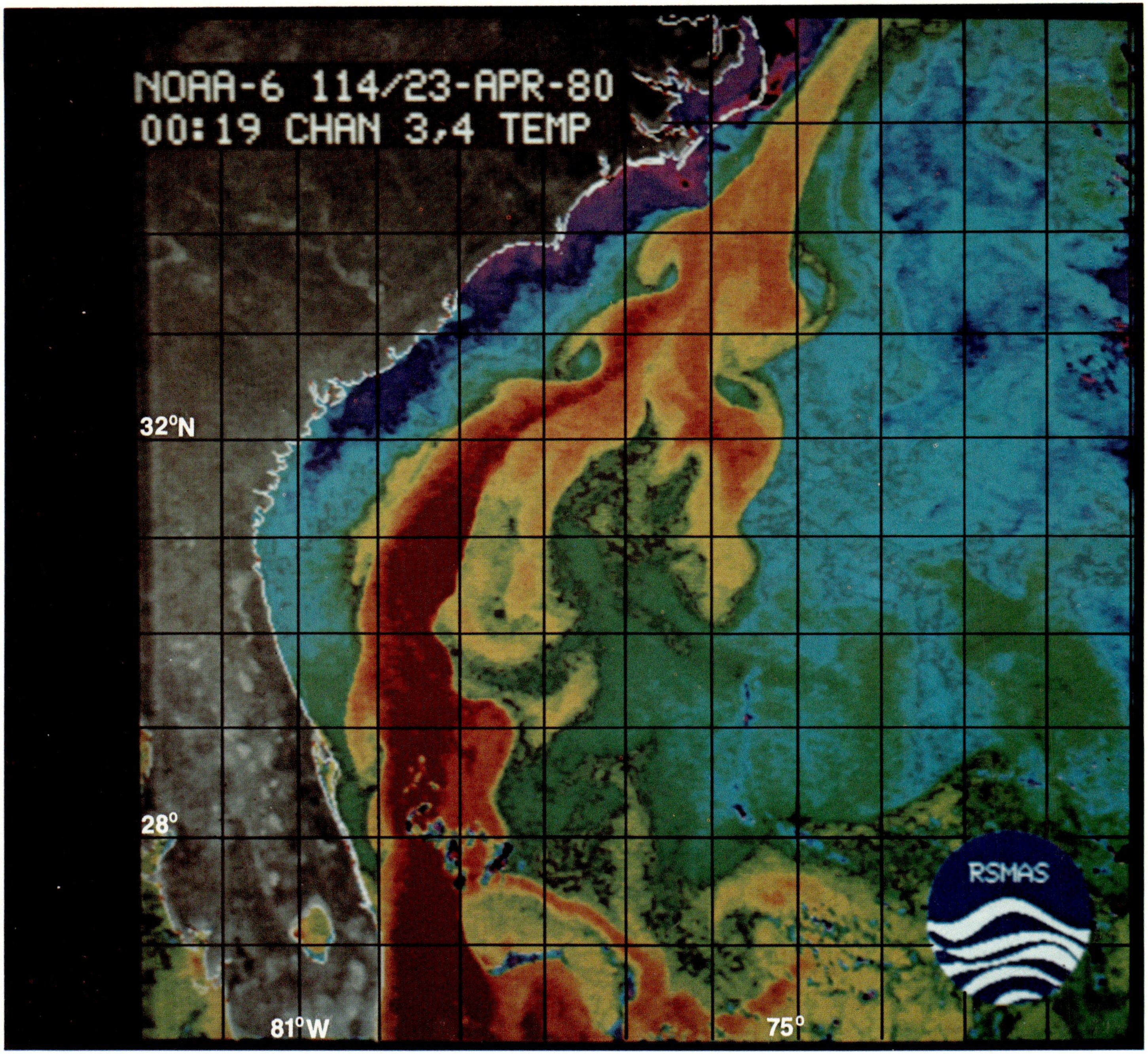

Plate 1. Satellite VHRR thermal image of the Gulf Stream at 0019 UT on April 23, 1980 (prepared by Otis Brown and Bob Evans of the University of Miami). Gulf Stream is shown by the red hues. Blue indicates colder water.

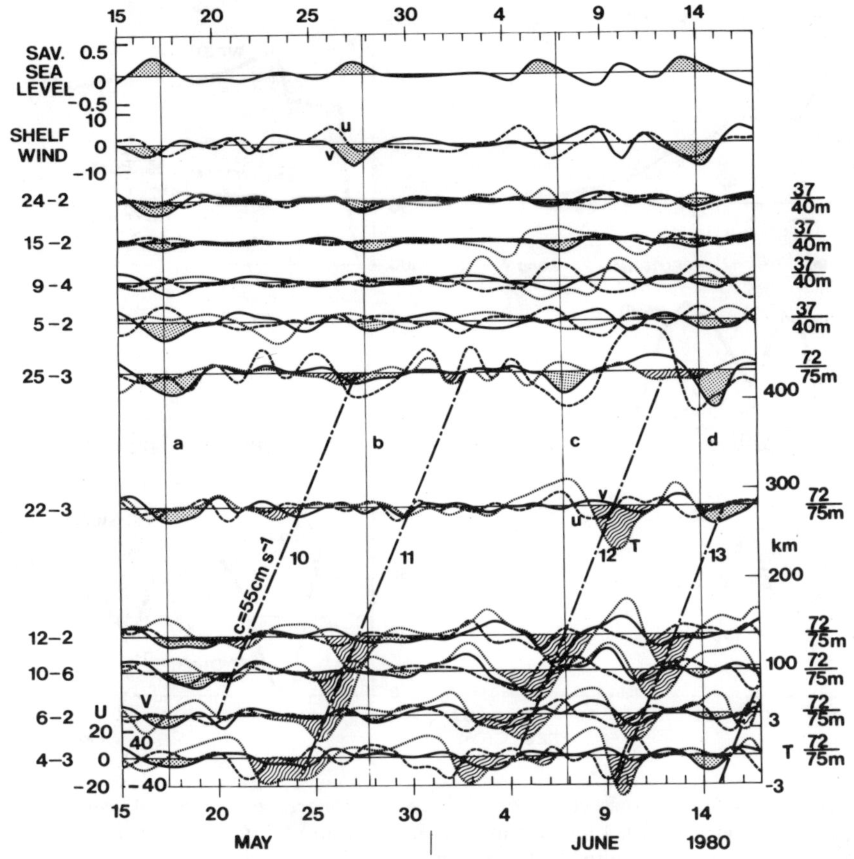

Fig. 4. Time series of 6-hour rotated, band-pass-filtered lower-layer current components (centimeters per second) u (dashed lines), v (solid lines) and T (dotted lines); shelf wind components (meters per second) u (dashed lines), v (solid lines) for the period May 15 to June 17. Propagating events (10–13) shown by slanted lines with phase speed c = 55 cm s^{-1} connecting cold anomalies at the shelf break (wavy shading). Nonpropagating events (a–d) shown by vertical lines connecting southward wind and current events (dotted shading) and coastal sea level setup (dotted shading).

phase speeds were found for u and v as well [*Lee and Atkinson,* 1983].

Mean Conditions

Mean flows over the outer shelf appear to be closely related to the position of the Gulf Stream (Figure 7). South of the Savannah transect the western edge of the Gulf Stream lies close to the shelf break, and northward mean currents of about 60 cm s^{-1} result in the upper layer. Onshore/offshore displacements of the Gulf Stream front are generally less than 15 km in this region [*Bane and Brooks,* 1979]. At the southern transect there was a mean onshore flow near the bottom at the shelf break and offshore flow at the four mooring sites to the north, indicating the possibility of a net water movement onto the shelf south of the array and off the shelf to the north.

Mean flows were generally weak at the Savannah transect for both the 4-month and monthly means, indicating that the Gulf Stream front was probably displaced further offshore at this position. Seaward of the Savannah transect the Charleston Bump protrudes onto the Blake Plateau and into the path of the Gulf Stream and could cause an offshore displacement (see, for example, *Brooks and Bane* [1978], *Pietrafesa et al.* [1978], and *Legeckis* [1979]). The Savannah area appears to be a transition region between strong northerly and southerly mean flow regimes.

At Cape Romain, which is downstream of the Charleston Bump, the mean flow at the shelf break was significantly southward with an offshore component in the lower layer and onshore in the upper layer. Previous investigators have shown the existence of a quasi-persistent offshore displacement of the Gulf Stream in the vicinity of the Charleston Bump [*Pietrafesa et al.,* 1978; *Brooks and Bane,* 1978; *Legeckis,* 1979; *Olson et al.,* 1983; *Bane,* 1983; *Singer et al.,* 1983]. *Bane* [1983] used available satellite imagery to show that the seaward deflection of the Gulf Stream occurred about 70% of the time. Bane concluded that the deflection was due to a bottom steering effect of the Charleston Bump

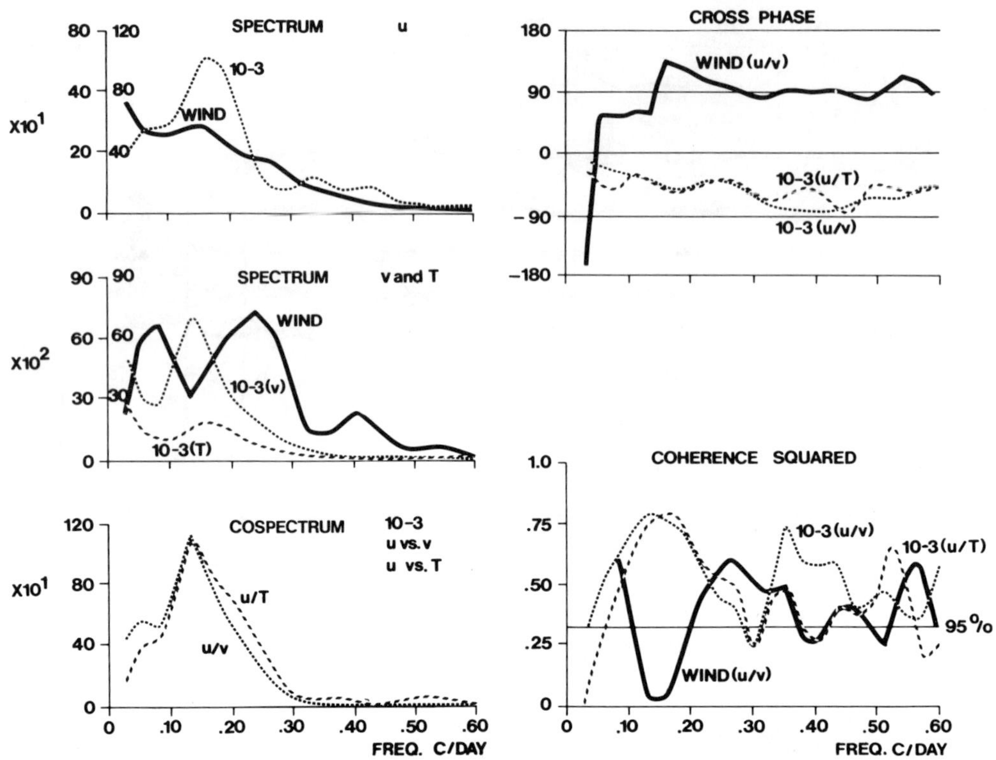

Fig. 5. Spectra (cm² s⁻² cpd⁻¹), cospectrum, cross phase, and coherence squared between u and v 40 HLP rotated velocity components of buoy winds, 10-3 currents and temperature (T) for GABEX I. Degrees of freedom, 16.8; number of lags, 74; effective bandwidth, 0.036 cpd. Spectra scale on left is for currents and right is for wind.

but with considerable shorter-term variability due to growing and propagating shorter-period Gulf Stream meanders. Both *Bane* [1983] and *Singer et al.* [1983] have shown the existence of a dome-shaped volume of cold water located immediately downstream of the Charleston Bump between the displaced Gulf Stream and the shelf, which is believed to be maintained by upwelling. *Singer et al.* [1983] found that 75% of the doming in hydrographic data occurred within the 200- and 400-m isobaths. Geostrophy suggests that a cyclonic circulation should exist around such a domed feature. Since mooring 25 was located on the western side of the cold dome, the observed prolonged southward flow events (Figures 2 and 7) may well have been produced by this cyclonic gyre.

Gulf Stream Frontal Eddies

Shipboard CTD and biological observations were made in the region of the current meter array and the adjacent Gulf Stream front during April 10 to 26, 1980. Current and temperature records from the outer shelf indicate that a succession of four cold cyclonic perturbations occurred during the shipboard sampling period (Figure 3, events 5, 6, 7, and 8). From the shipboard observations, the anomalously cool waters consisted of newly upwelled, nutrient-enriched North Atlantic Central Water [*Dunstan and Atkinson,* 1976; *Yoder et al.,* 1981].

The formation, growth, and northward movement of event 8 were documented with shipboard hydrography and remote thermal imagery from April 21 to 27. On April 22 and 23 the event was seen in satellite thermal imagery as well-defined frontal "shingle" [*Von Arx et al.,* 1955] seaward of Cape Canaveral (Plate 1) and by shipboard SST off St. Augustine and Jacksonville, Florida on April 25 and 26 (Figure 8). The shingle pattern has been found to be the characteristic SST signature of cyclonic cold core frontal eddies in this region [*Lee et al.,* 1981; *Lee and Atkinson,* 1983]. The disturbance was sampled hydrographically in detail during a 20-hour period on April 24 and 25 and again over a 37-hour period on April 25 and 26 (Figures 8 and 9).

The northward speed of the event was estimated using remote imagery, shipboard mapping, and current meter records, all of which indicated a speed of about 55 cm s⁻¹. A warm filament of 23° to 24°C Gulf Stream water approximately 15 to 20 m in depth extended southward from the front around a cold core with a minimum surface temperature of 22°C. The alongshelf dimension of the feature was approximately 130 km, and the cross-shelf distance

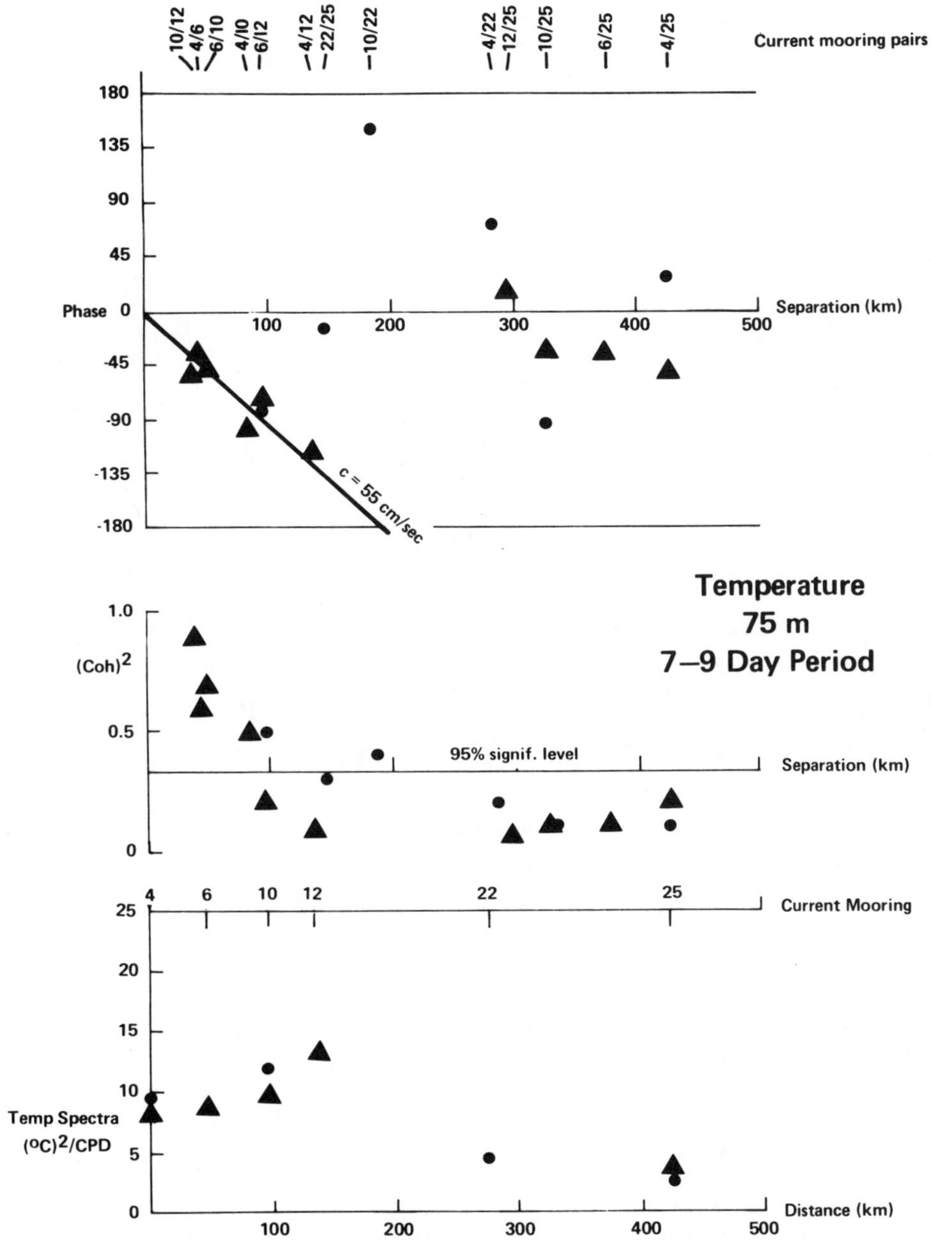

Fig. 6. Alongshelf coherence squared, phase, and energy density for the 40 HLP temperature from current meter pairs at shelf break in 7- to 9-day period band; dots are for upper layer and triangles are for lower layer.

from western edge of the filament to the center of the cold core was about 25 km. A typical temperature, density, and nitrate section through the center of the feature is shown in Figure 9. Uplifted temperatures in the cold dome extended across the outer shelf beneath the warm filament in a cold subsurface intrusion of nutrient-enriched waters. Upwelling velocities responsible for the uplifted temperature structure were estimated at 10^{-2} cm s^{-1}. Euphotic zone nitrate concentrations were 5 μmol L^{-1} beneath the warm filament and 10 μmol L^{-1} near the shelf break.

Midshelf

Time Domain

Subtidal frequency current records from the 28- and 40-m isobaths at 30°N are shown in Figure 10 along with shelf

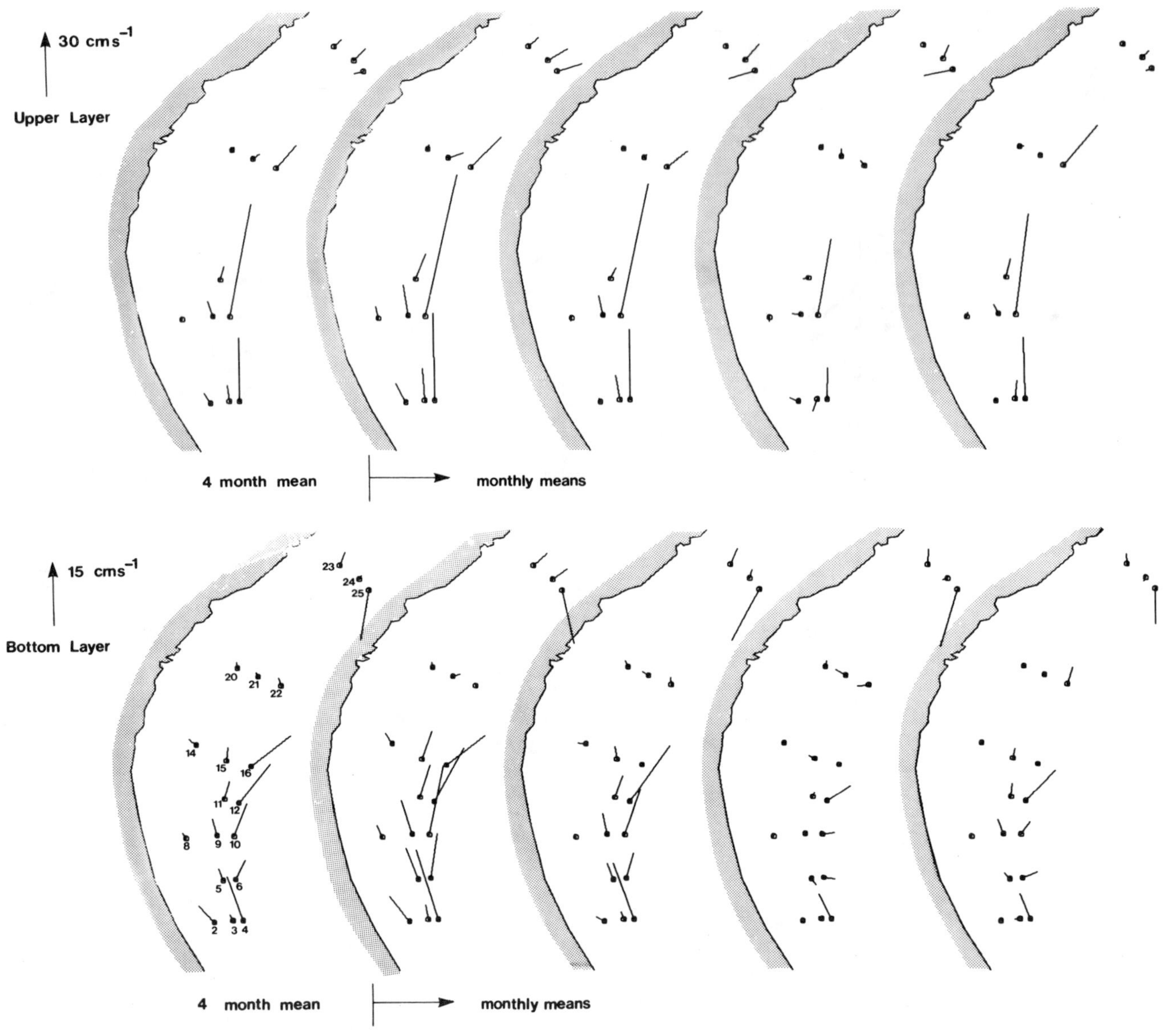

Fig. 7. Mean flows at the 28-, 40-, and 75-m isobaths during GABEX I. Top panel is for upper layer (17 m) using a scale of 30 cm s^{-1}. Lower panel is for lower layer (3 m above bottom) using a scale of 15 cm s^{-1}. First scene is the 4-month averages; remaining scenes are for consecutive monthly averages.

break currents, shelf wind, and coastal synthetic subsurface pressure (SSP) from the Savannah River NOS tide gauge. The SSP record was constructed by adding atmospheric pressure to the coastal sea level data to make it comparable to bottom pressure data. Low-fequency currents at the 28- and 40-m isobaths appear to be highly correlated between stations and with local wind and coastal SSP but poorly correlated to events at the shelf break. Current fluctuations show high coherence over the vertical arrays (Figures 10 and 11) with little decrease in amplitude, indicating a predominantly barotropic character. The correlation between currents and temperature does not appear to be as strong as at the shelf break. The signature of the cool cyclonic perturbations is not as clear at 40 m as at the shelf break, but it is still recognizable for many of the events (5, 6, 7, 8, 11, 12, 13, and 14 of Figure 11).

Temperature oscillations at 40 m had smaller amplitudes and less vertical correlation than at the shelf break. On several occasions temperature jumps occurred as the water column suddenly warmed, which is indicative of advection of a warm front past the mooring. Northward current speeds also increased at these times.

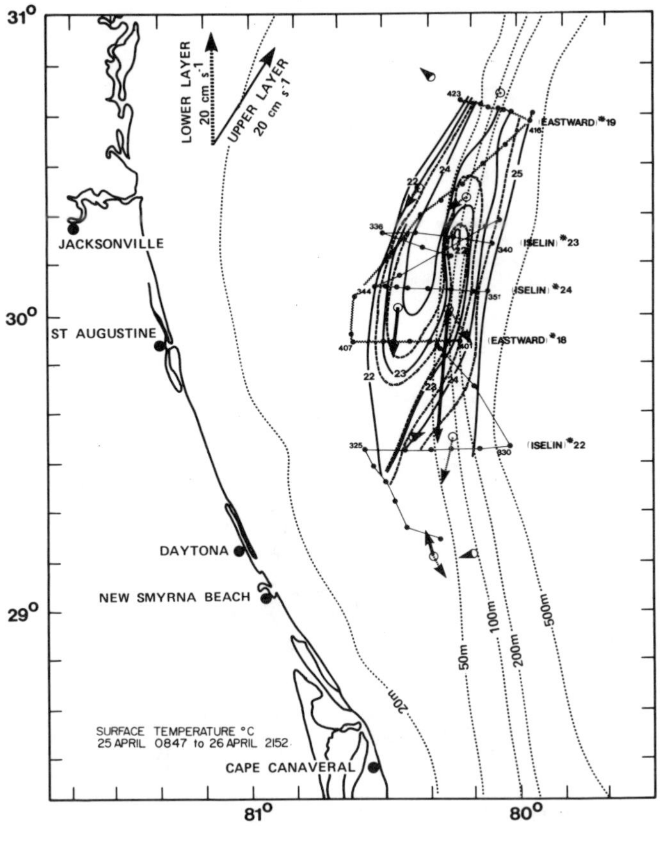

Fig. 8. Ship-measured surface temperature (degrees Centigrade) on April 25 (0847 hours) to April 26 (2152 hours). Straight lines with dots show *Iselin* ship track and stations. Wavy lines and dots show *Eastward* track and stations. Hydrographic sections are labeled with ship name and end point station numbers. Daily averaged currents on April 26 are shown with arrows.

A seasonal increase in stratification was apparent at the shelf break (Figure 2), but it is more evident at 40 m (Figure 11). Stratification was weak in February and increased during spring and into early summer. An isothermal bottom layer of at least 13-m thickness existed for most of the mooring period, presumably due to tidal current stirring.

Frequency Domain

Spectra of currents from the 40-m isobath (not shown) have energy peaks at 5- to 6-day periods for u and 7- to 9-day periods for v, as was found at the shelf break, but they are not as well resolved and the coherence between u and v is weak. The only significant coherence between u and v in the upper layer (7 m) is at 2- and 4-day periods. In the lower layer the velocity components were rectilinear for periods greater than 2 days. Coherence was high for periods greater than 9 days but only marginally significant at periods of 1.8, 2.5, and 3.3 to 4 days. Shelf winds also had energetic coherent fluctuations between u and v at the same periods as the coherent lower layer motions at the 40-m isobath (1.8, 2.5, 3 to 4, and 12 days). Neither the currents at 40 m nor the winds showed significant coherence between u and v in the 5- to 9-day period band where the coherence was highest at the shelf break (Figure 5).

Coherence and phase relationships between alongshelf currents and alongshelf winds are shown in Figure 12; alongshelf currents versus coastal sea level are shown in Figure 13. The only significant coherence between currents at the shelf break with local wind and sea level occurred for near-bottom alongshelf flow at the 9-day period. Alongshelf flow in the upper layer at the 40-m isobath had significant coherence with local alongshelf winds at periods of 4 and 7 to 12 days and even higher coherence with sea level at periods of 7 to 12 days. Near-bottom alongshelf currents at 40 m were very strongly correlated with local wind and sea level at periods of 3 to 4 and 7 to 12 days. The coherence was generally higher between currents and sea level at these periods. Lower-layer velocity fluctuations lagged the wind by about 12 to 18 hours and sea level by about 6 to 12 hours (Figure 10). Also, sea level was highly coherent with alongshelf winds at periods of 10 days and with cross-shelf winds at periods of 2 to 7 days and appeared to lag wind events by about 6 hours.

At the 40-m isobath, coherence of alongshelf current fluctuations was high over alongshelf distances of up to 425 km with small and nearly constant phase lags, which is indicative of a local response to coherent large-scale wind forcing (Figure 14). The cross-shelf velocity components and temperature did not show any consistent alongshelf coherent features at the 40-m isobath, so they are not presented.

Mean Conditions

Mean flows for the February to June period at the 28- and 40-m isobaths were northward at all mooring sites, with an onshore component in the southern part and offshore component in the northern part (Figure 7). There were no appreciable differences with depth, indicating a dominant barotropic mode [*Lee and Atkinson*, 1983]. Thus there appears to have been a net onshore flow north of Cape Canaveral, followed by strictly northward alongshelf flow and then offshore flow in the vicinity of Savannah and Cape Romain.

Inner Shelf

Time Domain

The shallowness of the inner shelf (0–20 m) complicates conceptualization. The surface and bottom boundary layers merge and the alongshelf component of flow is highly correlated with local wind stress [*Schwing et al.*, 1983]. During winter, vertical density stratification is relatively weak offshore of the inner shelf frontal zone, and the correlation between wind and current at the 15-m isobath is very strong (Figure 15).

The water is vertically stratified within the frontal zone practically all year as a result of freshwater discharge from

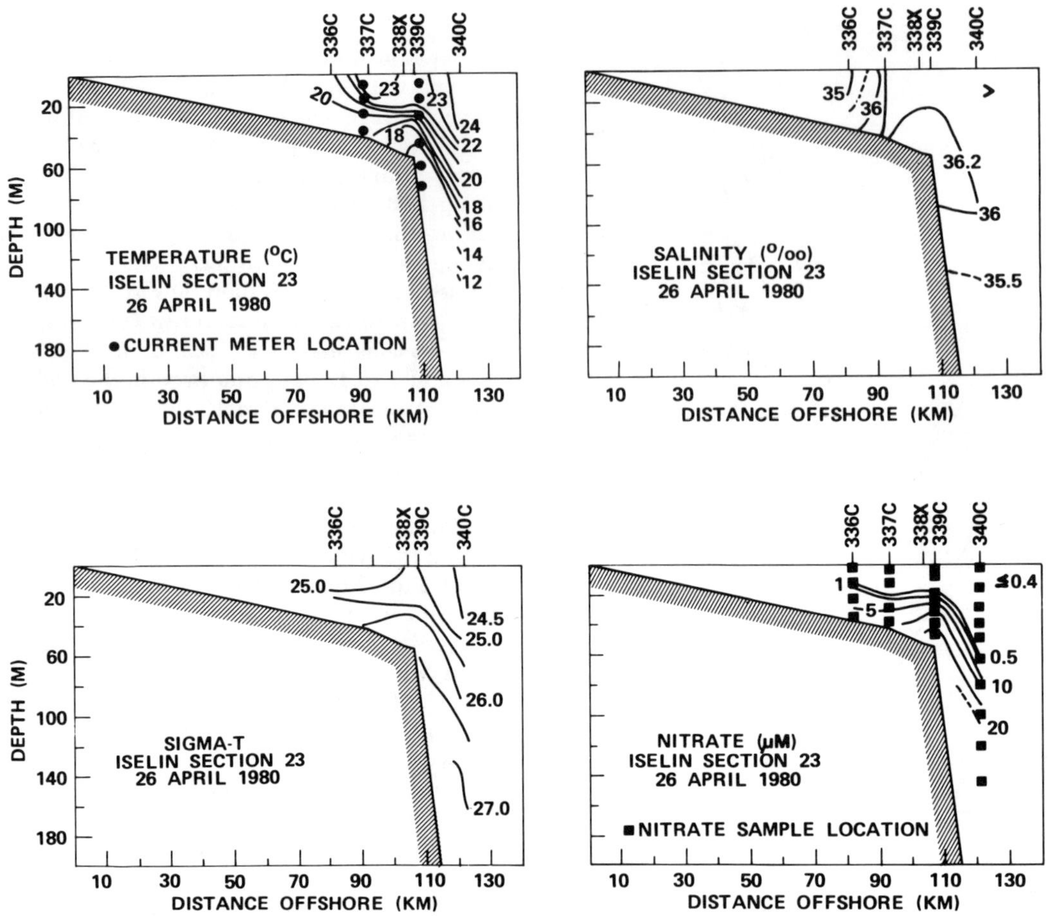

Fig. 9. Temperature (degrees Centigrade), salinity (per mill), density (σ_t), and nitrate (μM) sections across a Gulf Stream frontal eddy on April 26 (0028 to 0757 hours). Stations are numbered along the top from *Iselin* section 23 (Figure 8).

several rivers. This introduces an internal (baroclinic) pressure gradient which influences the vertical distribution of currents. There exist two snapshots of this environment, one of which is described by *Blanton* [1981].

During autumn the southwestward wind stress yields an onshore Ekman transport. The offshore pressure gradient and southward transport are in geostrophic balance [*Blanton,* 1981]. Density surfaces in the frontal zone slope upward and offshore by 10 m for every 20 or 30 km. The coastal currents during autumn are driven southward not only by wind stress, but are reinforced by southward geostrophic flow (Figure 16). Even so, present evidence suggests that the southward flow is slower than one would predict simply from thermal wind calculations [*Blanton,* 1981]. The strength is apparently modified by the presence of barotropic northward flow as noted at midshelf (see the discussion on momentum balance).

During spring the wind stress is northeastward, and the Ekman transport is offshore. The inner shelf density field has a relatively small slope of 0.1 to 0.3 m per 1000 m, and a two-layer system exists on the inner shelf (Figure 16). The circulation is offshore above and onshore below the pycnocline. This circulation prevails during much of spring and early summer [*Blanton and Atkinson,* 1983], and it is primarily responsible for the dispersal of river discharges stored in the upper layers of the inner shelf frontal zone.

Frequency Domain

Schwing et al. [1983] have analyzed currents in 10-m water depth off the South Carolina coast together with local winds and sea level. The variance in all data is concentrated principally in two frequency bands: 2 to 12 days (synoptic band) and 12 to 25 hours (tidal and daily bands). This is in agreement with findings on the middle shelf. Alongshelf currents and winds are highly coherent in the synoptic band with currents lagging winds by about 3 hours. Coastal sea level is also coherent with alongshelf current. This differs with midshelf results in one respect: coastal

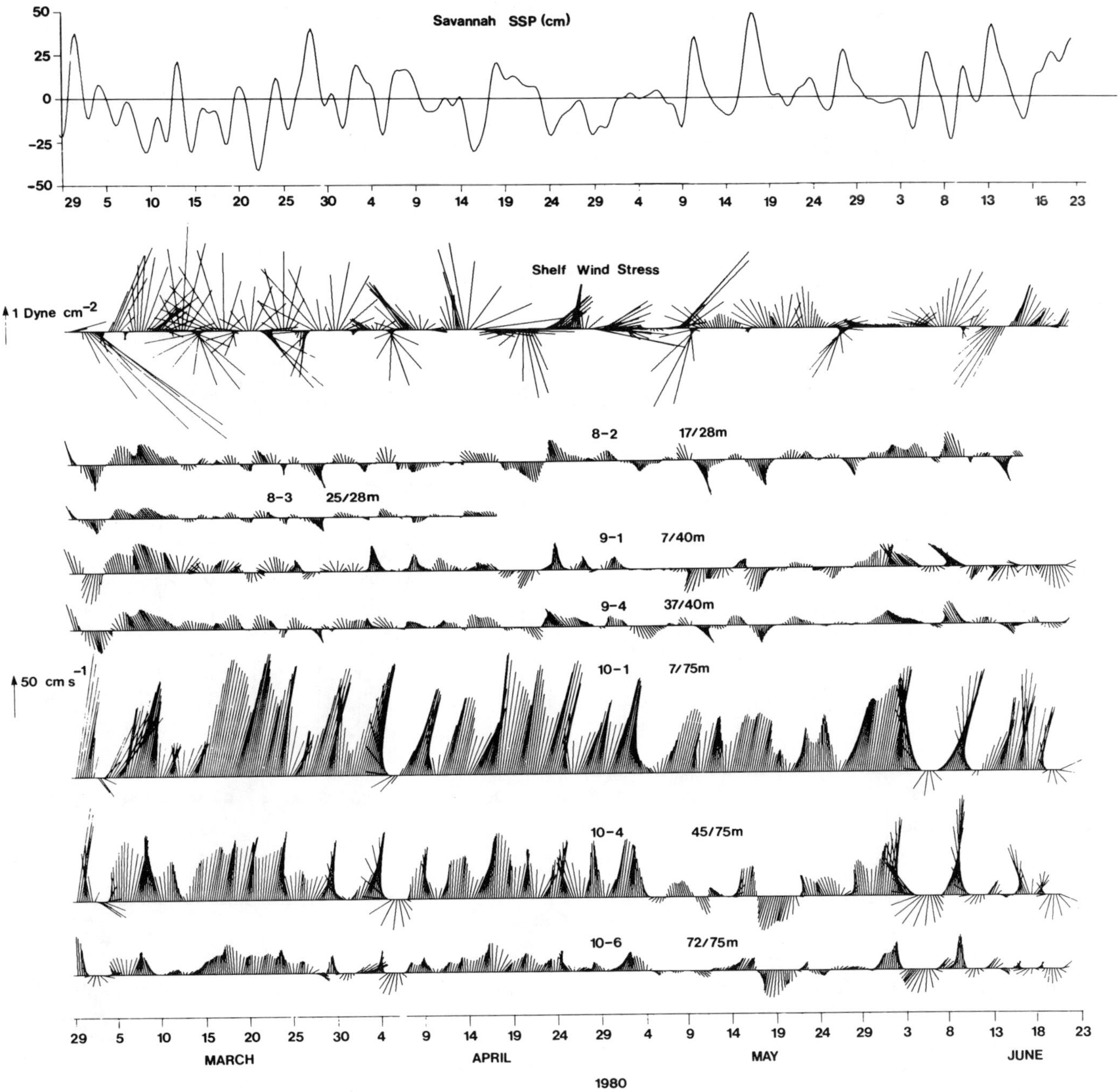

Fig. 10. Coastal synthetic subsurface pressure (SSP) from Savannah River; shelf wind stress and current vectors from the 75-, 40-, and 28-m isobaths at 30°N during GABEX I. All data filtered with a 40-hour low-pass filter and subsampled every 6 hours. Instrument and water depths shown next to current meter identification.

sea level lags alongshelf current by about 15 hours at periods of about 3 to 5 days. This has suggested to *Schwing et al.* [1983] that the inner shelf current may drive sea level changes.

Similar results were obtained at the platform offshore Savannah (Figure 17). When the principal axis of velocity (3 m off bottom) is correlated with sea level at the Savannah River entrance, we find statistically significant lags of 8 to 15 hours between peaks/troughs in velocity and troughs/peaks in sea level. Apparently, coastal currents

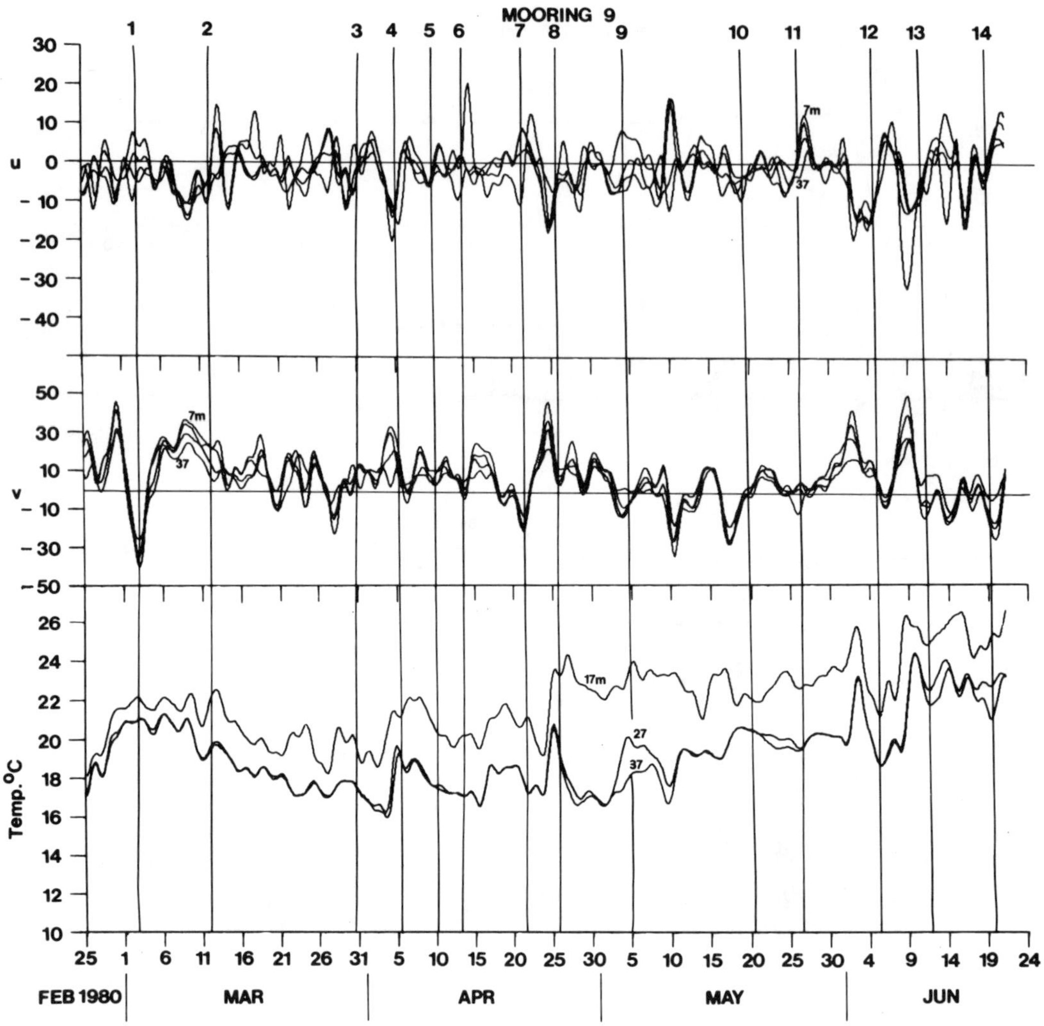

Fig. 11. Time series of 6-hour rotated 40 HLP *u, v,* and *T* from mooring 9. Vertical lines are for event identification.

must transport water away from or toward the coast for several hours before a change in sea level is registered at the coast.

Wind-Driven Circulation

For a northward alongshelf wind, simple Ekman dynamics predict an offshore transport in the surface friction layer that will cause a setdown of coastal sea level, which in turn will maintain a northward geostrophic interior flow, $V_g = (g/f)(\partial\eta/\partial x)$, where η is sea surface elevation [*Beardsley and Butman,* 1974; *Lee and Brooks,* 1979]. The opposite occurs for southward winds. *Scott and Csanady* [1976] have shown that for the Middle Atlantic Bight a frictional equilibrium resulted for the balance of alongshelf momentum for the entire water column, with alongshelf wind stress being balanced by an alongshelf sea level slope plus bottom friction. Parameter $\bar{\tau}_{ys} = gh\ (\partial\eta/\partial y) + \bar{\tau}_{yb}$, where $\bar{\tau}_{ys}$ is the event-averaged surface stress in the y direction and $\bar{\tau}_{yb}$ is the alongshelf bottom stress given by $\bar{\tau}_{yb} = r\bar{v}_b$, where r is a bottom resistance coefficient and $\bar{v}_b$ is the event-averaged near bottom velocity. The alongshelf slope was believed to result from offshore influences [*Csanady,* 1978; *Beardsley and Winant,* 1979]. *Tebeau and Lee* [1979] found a similar balance for midshelf flow measurements in the SAB during winter conditions. They reported a resistance coefficient of 0.1 cm s^{-1}, whereas *Scott and Csanady* [1976] found 0.16 cm s^{-1} for the Middle Atlantic Bight. *Tebeau and Lee* [1979] estimated a mean alongshelf slope of $\partial\eta/\partial y = -1.7 \times 10^{-7}$, which is almost exactly the value *Sturges* [1974] found for the region using geostrophic and steric leveling in the Gulf Stream. *Blanton* [1981] calcu-

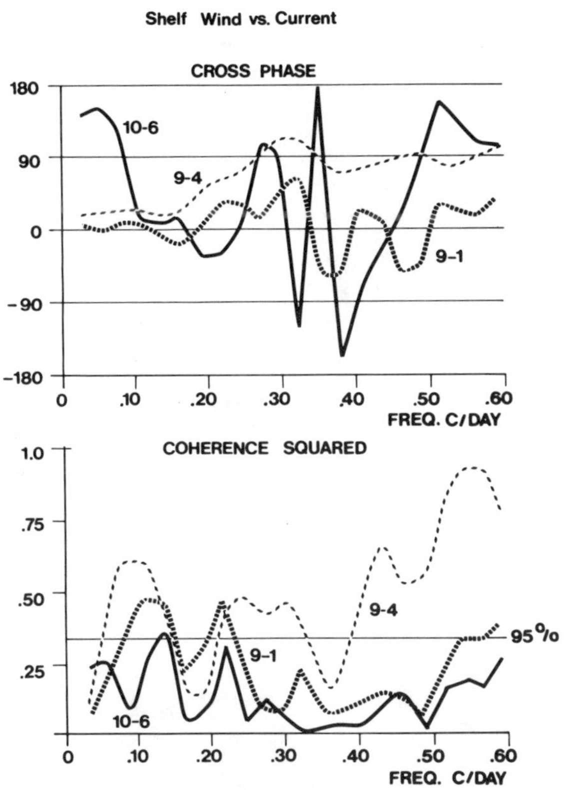

Fig. 12. Cross phase and coherence squared between 40 HLP rotated buoy winds (v component) and currents (v component) from 9-1, 9-4, and 10-6. Degrees of freedom, 16.8; number of lags, 74; effective bandwidth, 0.036 cpd.

lated $\partial\eta/\partial y = -2.9 \times 10^{-7}$ for midshelf regions, which supports a weak barotropic current northward in 20 m of water.

Evidence supporting a simple Ekman response to wind forcing is clearly indicated in current, coastal SSP, and bottom pressure data (Figure 18). The SSP and pressure data are demeaned time series. Vertically averaged alongshelf currents are shown for moorings 8, 9, and 10, together with shelf wind stress. Bottom pressure and coastal SSP fluctuations indicate high coherence between all stations with small phase lags. These fluctuations are also highly correlated with wind events in the 2- to 14-day period band, indicating that bottom pressure variations in the synoptic band are predominantly produced by changes in sea level rather than density (barotropic rather than baroclinic). Some exceptions to this may occur at times at the shelf break (station 22). Low-frequency variations in cross-shelf and alongshelf sea level slopes (interpreted from bottom pressure differences) appear to result largely from amplitude differences between stations. Greatest amplitudes (±10 to 30 cm) occurred at the coast near Savannah, Georgia, where the shelf is widest, and decreased offshore to ±7 to 20 cm at the 15-m isobath (stations 1 and 17) and ±3 to 7 cm at the shelf break (station 22).

Sea level slopes and barotropic flows at the 28- and 40-m isobaths appear to be a response to local wind forcing. Northward winds (event lines in Figure 18) that are coherent over the shelf domain cause a greater drop in sea level near the coast than at the shelf edge. This results in a positive cross-shelf slope that is highly coherent with and leads the northward barotropic flow by 6 to 12 hours. The barotropic flow response occurs nearly simultaneously over the 425-km array separation. The opposite occurs for southward winds. *Hickey* [1981] found for the Pacific northwest shelf that subtidal fluctuations of alongshelf current, wind, and subsurface pressure were all coherent over an alongshelf separation of 480 km. She concluded also that the current fluctuations were a forced response to local winds at each station and the high alongshelf coherence was due to the large spatial coherence scale of the wind forcing and not to freely propagating, topographically trapped waves.

As a consequence of the greater sea level fluctuations near Savannah, a northward wind can also cause a negative alongshelf slope in the region south of Savannah that

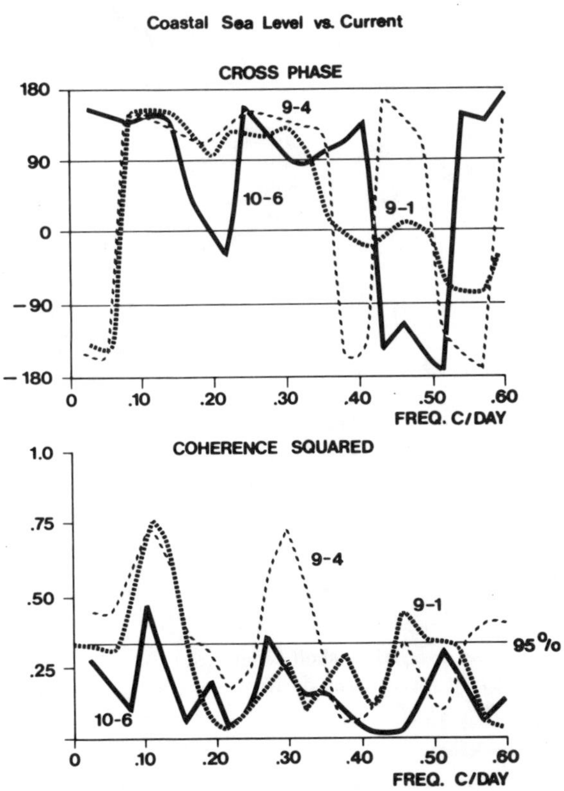

Fig. 13. Cross phase and coherence squared between 40 HLP Savannah sea level and currents (v component) from 9-1, 9-4, and 10-6. Degrees of freedom 16.8; number of lags, 74; effective bandwidth, 0.036 cpd.

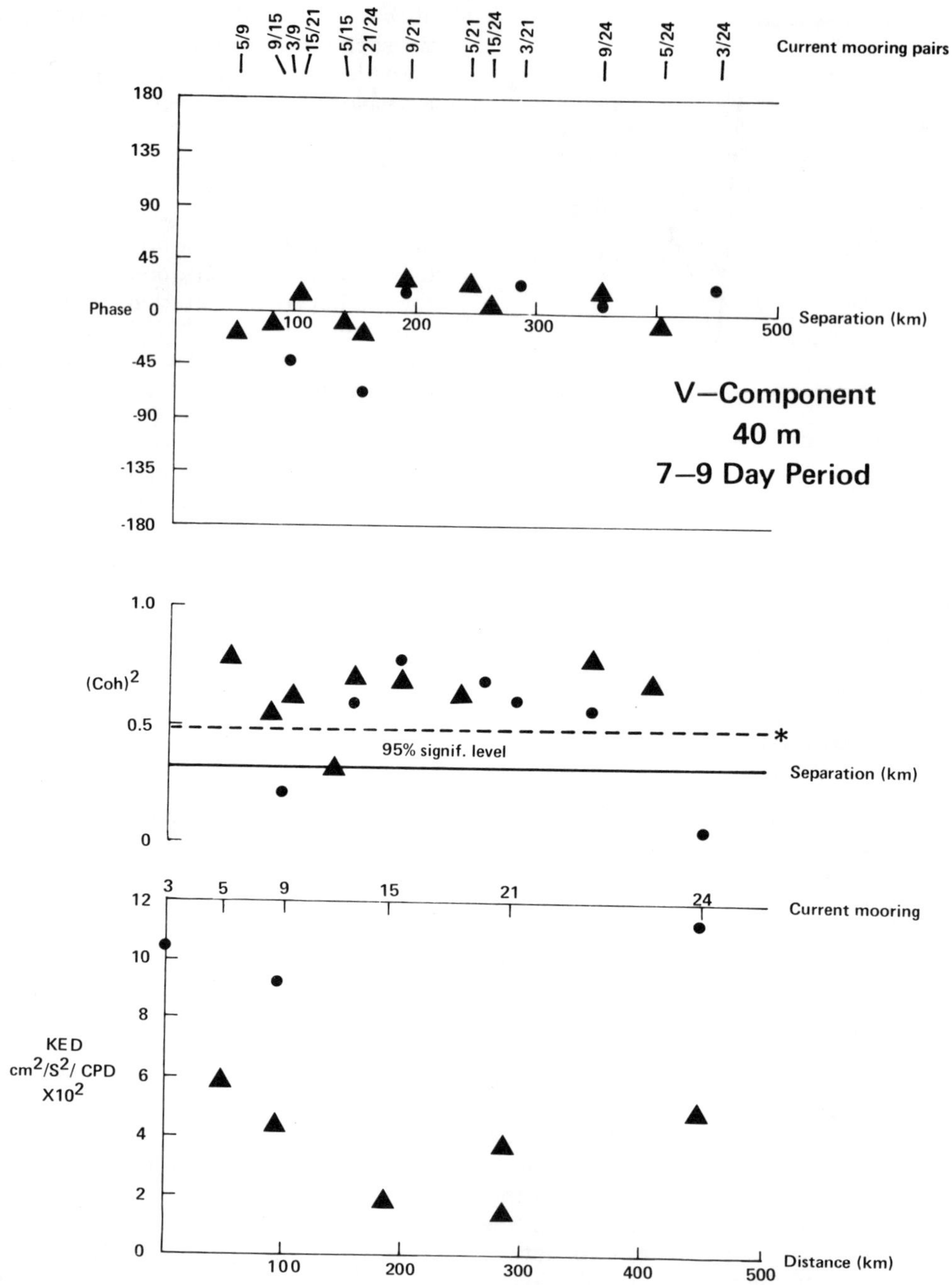

Fig. 14. Alongshelf coherence squared, phase, and energy density for the 40 HLP rotated v component from current meter pairs at 40-m isobath in 7- to 9-day period band; dots are for upper layer and triangles for lower layer.

can further enhance the northward barotropic flow [*Scott and Csanady,* 1976]. Again, the opposite effect would occur for southward winds.

Figure 18 also indicates a long-period oscillation in alongshelf sea level slope at the 15-m isobath that may be related to either wind or density effects. During March the mean winds were northeastward and the alongshelf slope was mostly negative at the 15-m isobath. Since sea level setdown would be greater at Savannah for these winds, a negative slope would result. However, the alongshelf slope was mostly positive in April and May, which is when we expect the lowest salinity water on the inner shelf due to

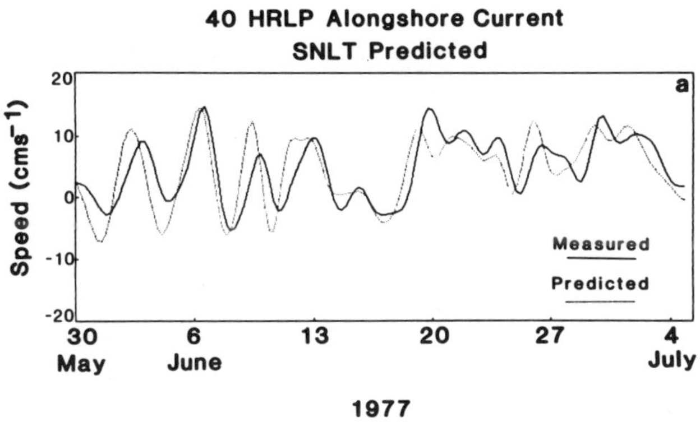

Fig. 15. Alongshore current and wind speed (40 HLP) at Savannah Navigational Light Tower (SNLT). See Figure 16 for location.

river discharges [*Blanton and Atkinson,* 1983]. With lower density in the vicinity of the Savannah River outflow, sea level could stand higher off Savannah relative to New Smyrna Beach. This trend is also apparent in the cross-shelf sea level slope off Savannah and may also be an effect of the freshwater discharge accumulating primarily in the nearshore zone [*Atkinson et al.,* 1978].

Momentum Balances

Vertically integrated momentum balances (Tables 1 and 2, Figures 19 and 20) give further support to local Ekman dynamics accounting for the subtidal current variability at midshelf. For an explanation of terms and methods of estimation using the GABEX I array data, see *Lee et al.* [1984]. The mean sea level slope terms were derived from the balance of the other terms. The baroclinic terms were not computed for the 40-m isobath. Cross-shelf momentum balances at both the 28- and 40-m isobaths indicate that for mean conditions the Coriolis force is in approximate geostrophic balance with the cross-shelf barotropic pressure gradient. The fluctuating flow has a lesser but still

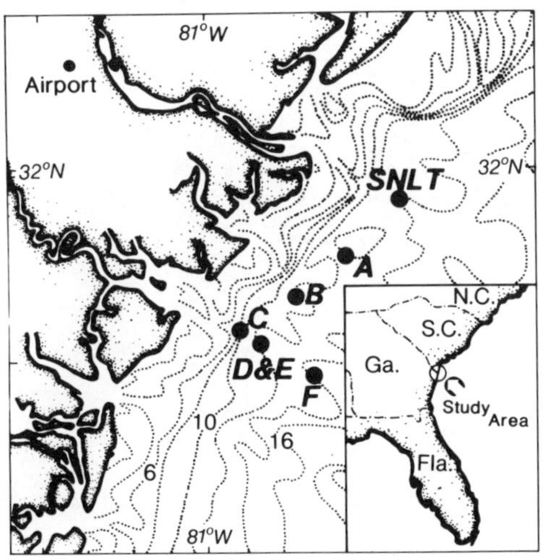

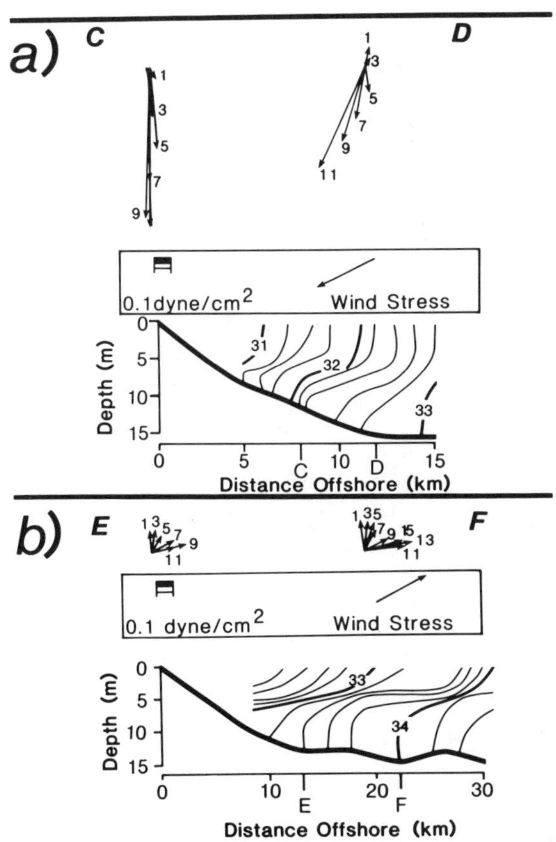

Fig. 16. Average currents for four tidal cycles for two opposing wind regimes. Vertical sections show salinity distribution at beginning of experiment. Numbers adjacent to vectors denote distance in meters above bottom. (*a*) November 7–9, 1979; (*b*) April 2–4, 1981.

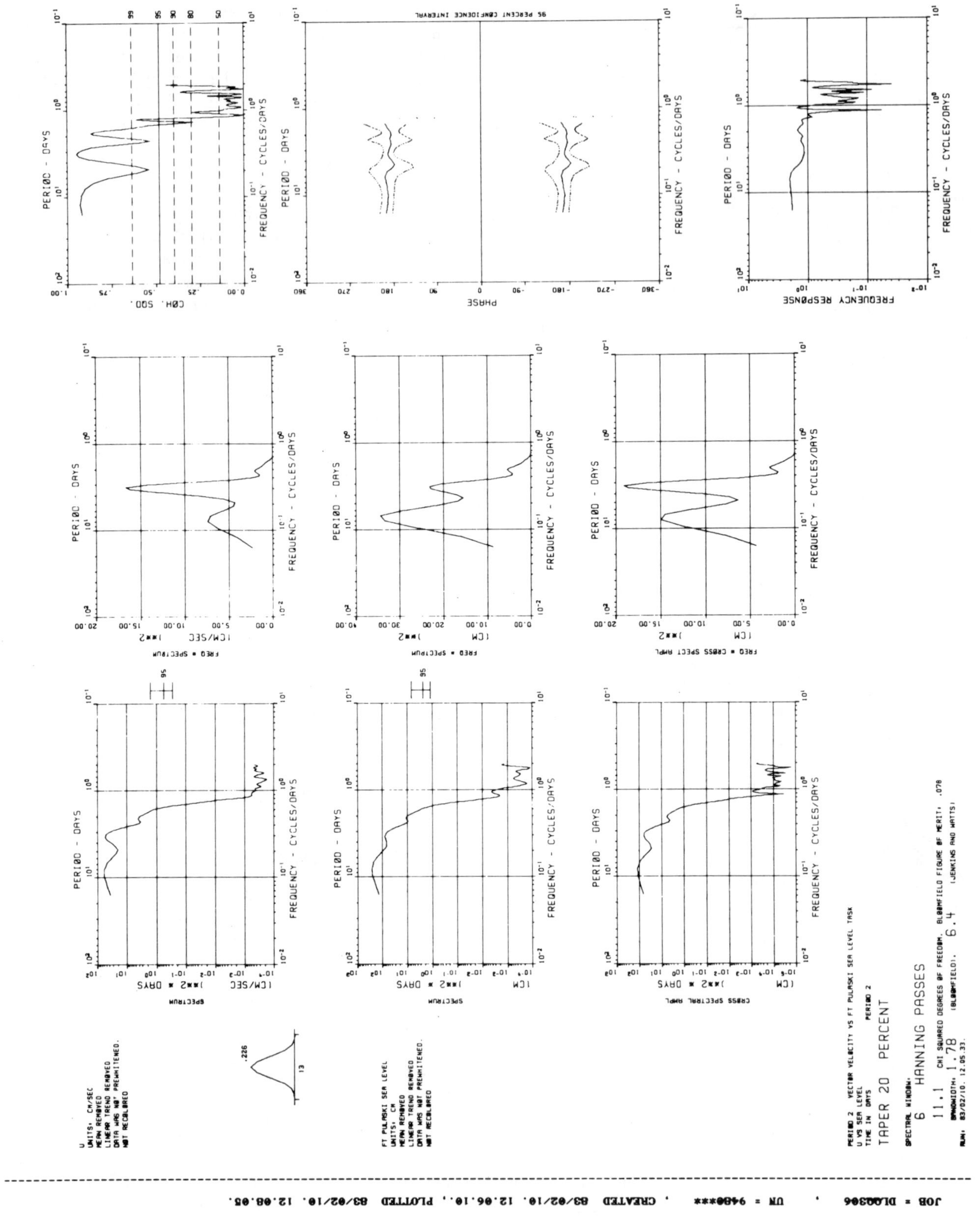

Fig. 17. Cross-correlation matrix between principal axis of current 3 m above bottom at SNLT and coastal sea level at Savannah River mouth.

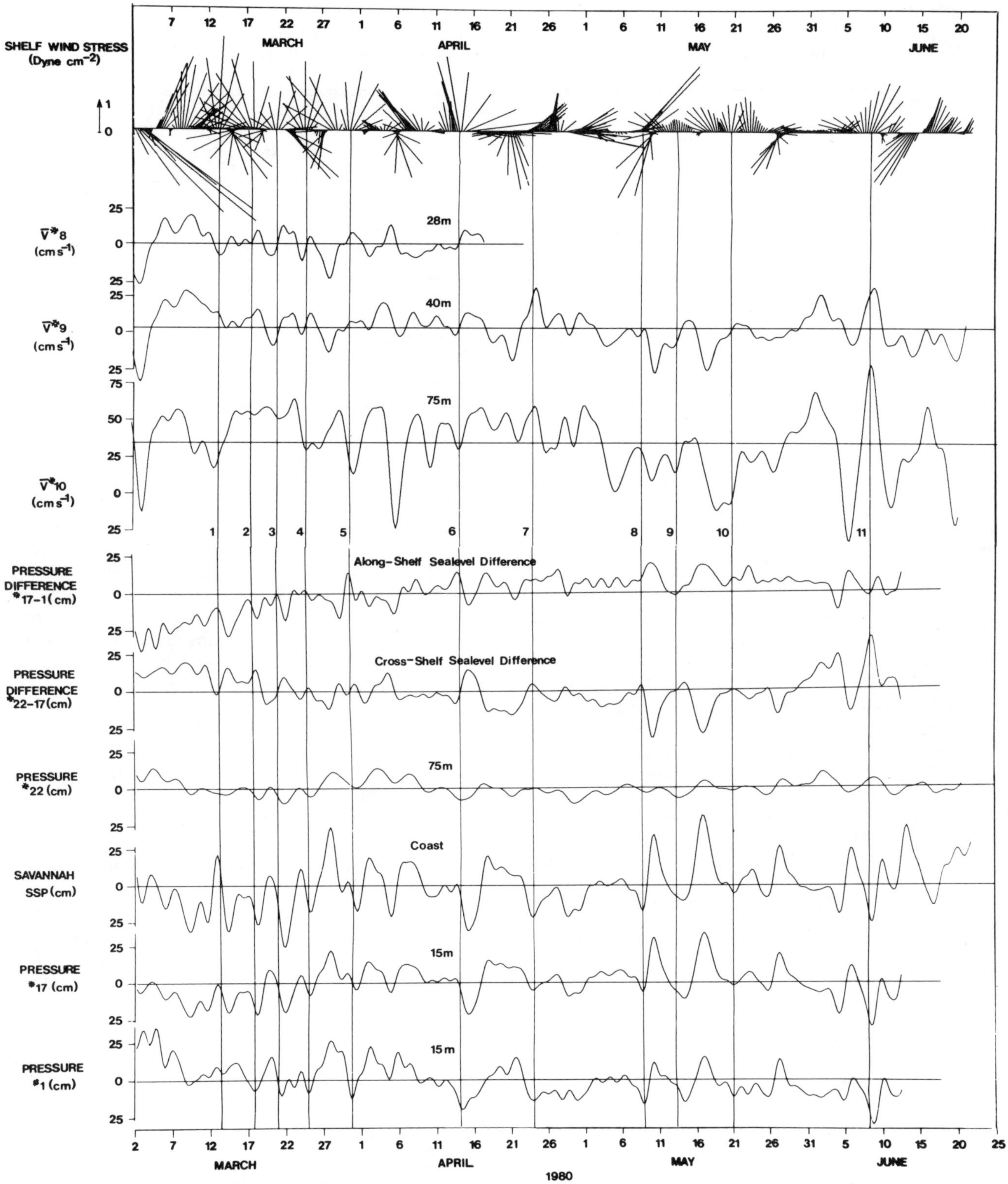

Fig. 18. Shelf wind stress, vertically averaged current from the 28-, 40-, and 75-m isobaths at 30°N, bottom pressure at the 15-m and 75-m isobaths (sites 1, 17, and 22), coastal SSP, and cross-shelf and alongshelf sea level differences (surface slopes) during GABEX I. Wind and currents are subtidal 40 HLP filtered records. All bottom pressure and coastal SSP are band passed filtered from 40 hours to 2 weeks.

TABLE 1. Term Estimates of the Vertically Integrated Momentum Equations for the 28-m Isobath

	Cross Shelf, $\times 10^{-4}$ cm s^{-2}								
	$\frac{\partial \bar{u}}{\partial t}$	$+\frac{\partial \bar{u}^2}{\partial x}$	$+\frac{\partial \bar{u}\bar{v}}{\partial y}$	$-f\bar{v}$	$=$	$-g\frac{\partial \eta}{\partial x}$	$-\frac{g}{H\rho_o}\int_{-h}^{o}\frac{\partial \alpha}{\partial x}dz$	$+\frac{\tau_{xs}}{\rho_o H}$	$-\frac{\tau_{xb}}{\rho_o H}$
Mean	−0.03	+ 0.05	+ 0.01	− *1.0*	=	*−1.6*	+ 0.3	+ 0.3	+ 0.05
Standard deviation	±0.6	± 0.2	± 0.04	± *7.0*	=	*±7.0*	+ (0.2 − 2.8)	± *2.5*	± 0.3
	Alongshelf, $\times 10^{-4}$ cm s^{-2}								
	$\frac{\partial \bar{v}}{\partial t}$	$+\frac{\partial \bar{v}^2}{\partial y}$	$+\frac{\partial \bar{u}\bar{v}}{\partial x}$	$+f\bar{u}$	$=$	$-g\frac{\partial \eta}{\partial y}$	$-\frac{g}{H\rho_o}\int_{-h}^{o}\frac{\partial \alpha}{\partial y}dz$	$+\frac{\tau_{ys}}{\rho_o H}$	$-\frac{\tau_{yb}}{\rho_o H}$
Mean	0.03	− 0.001	− 0.04	− *0.6*	=	*−1.4*		+ *0.8*	+ 0.3
Standard deviation	*±1.2*	± 0.06	± 0.2	± *2.7*	=	*±3.6*	+ (0.1 − 1.9)	± *3.2*	± *0.6*

The larger term estimates are italicized.

significant contribution from the cross-shelf wind stress. The mean alongshelf momentum balance at 28 m is between the Coriolis term, alongshelf surface slope, and alongshelf wind stress. The fluctuation balance has additional contributions from local alongshelf accelerations of total water column momentum and bottom stress. Estimates of the baroclinic term from hydrographic data indicate that at times this term may also contribute to both the cross-shelf and alongshelf fluctuating balances.

At the 40-m isobath, which is close to the transition between midshelf and outer shelf flow regimes, the mean alongshelf momentum balance is between the nonlinear momentum advection term, Coriolis force, alongshelf surface slope, and alongshelf surface and bottom stresses. An alongshelf slope of the order of -10^{-7} is needed to balance the mean alongshelf momentum, which is close to the alongshelf slope values reported by *Sturges* [1974] and *Tebeau and Lee* [1979].

The mean advection of water column momentum, $\langle \partial \bar{u}\bar{v}/\partial x \rangle$, at 40 m is large and positive, which indicates that there was a net offshore advection of northward momentum. The northward passage of Gulf Stream frontal eddies produces a net offshore flux of northward momentum $(+\langle u'v' \rangle)$ at this location, which transfers kinetic energy from the eddies to the mean flow of the Gulf Stream [*Lee and Atkinson,* 1983]. Thus the outer shelf strip between 40- and 75-m water depths at 30°N appears to be a region where the Gulf Stream imparts mean northward momentum to the shelf through the negative alongshelf surface slope and the net onshore flow just north of Cape Canaveral. This momentum transfer is locally compensated by an offshore advection of northward momentum arising from the passage of Gulf Stream frontal eddies.

Momentum balances similar to those discussed above have been made over very short periods for the inner shelf. The balances were derived from anchored ship observations covering four tidal cycles for the two cases described previously (Figure 16). The variations about the mean are

TABLE 2. Term Estimates of the Vertically Integrated Momentum Equations for the 40-m Isobath

	Cross Shelf, $\times 10^{-4}$ cm s^{-2}								
	$\frac{\partial \bar{u}}{\partial t}$	$+\frac{\partial \bar{u}^2}{\partial x}$	$+\frac{\partial \bar{u}\bar{v}}{\partial y}$	$-f\bar{v}$	$=$	$-g\frac{\partial \eta}{\partial x}$	$-\frac{g}{H\rho_o}\int_{-h}^{o}\frac{\partial \alpha}{\partial x}dz$	$+\frac{\tau_{xs}}{\rho_o H}$	$-\frac{\tau_{xb}}{\rho_o H}$
Mean	−0.003	− 0.7	+ 0.03	− *6.0*	=	*−6.4*		+ 0.2	+ 0.2
Standard deviation	±0.6	± 0.8	± 0.06	± *8.6*	=	*±7.4*		± *1.8*	± 0.4
	Alongshelf, $\times 10^{-4}$ cm s^{-2}								
	$\frac{\partial \bar{v}}{\partial t}$	$+\frac{\partial \bar{v}^2}{\partial y}$	$+\frac{\partial \bar{u}\bar{v}}{\partial x}$	$+f\bar{u}$	$=$	$-g\frac{\partial \eta}{\partial y}$	$-\frac{g}{H\rho_o}\int_{-h}^{o}\frac{\partial \alpha}{\partial y}dz$	$+\frac{\tau_{ys}}{\rho_o H}$	$-\frac{\tau_{yb}}{\rho_o H}$
Mean	0.1	− 0.08	+ *2.8*	− 1.7	=	*1.0*		+ *0.5*	− *0.4*
Standard deviation	*±1.3*	± 0.1	+ *2.4*	+ *3.0*	=	*+3.6*		± *2.2*	± *1.0*

The larger term estimates are italicized.

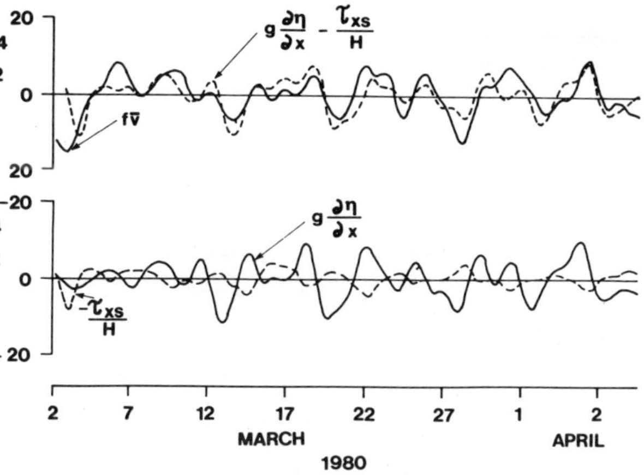

Fig. 19. Vertically averaged cross-shelf momentum balance 28-m isobath: moorings 8, 9, 20, GABEX I.

primarily caused by tidal fluctuations [*Blanton,* 1981].

Some of the cross-front terms in the momentum balances are clearly more important than others (Table 3). During autumn the Coriolis term was balanced by a combination of the barotropic and and baroclinic pressure gradient and the cross-front wind stress. *Blanton* [1981] calculated a balance that omitted the baroclinic term. As Table 3 shows, this term is important because the cross-front pressure gradient (due to the steep slope of the isopycnals) was particularly strong for this event. Sea level would drop about 0.6 cm in the offshore direction for the autumn case.

Cross-front balances for the spring case involved a different set of terms. First, the small frontal slope induced a negligible baroclinic pressure gradient. Thus in the upper layer the Coriolis term was primarily balanced by cross-front wind stress and the tidal stress term. In the lower layer the balance was simply between the Coriolis term and the tidal stress term.

This result, taken at face value, suggests that a steady northward current is induced by tidal fluctuations and would be part of the background alongfront flow. Clearly, we do not have sufficient data to make a strong case. Knowledge of alongshore velocity gradients is essential because there are large gradients in the tidal stress terms near the coast [*Blanton,* 1980]. More definite balances would require data from long-term moorings on the inner shelf similar to those that are now available for the midshelf and outer shelf (Tables 1 and 2).

In general, no clear balance emerges for alongfront momentum with the possible exception of the upper layer flow in spring (Table 3*b*), when Ekman balance predominated.

Volume Transport

Alongshelf volume transports between the 15- and 70-m isobaths were computed for the 30°N transect using current meter data from moorings 8 and 9 (Figure 1). Details of the transport calculations can be found in the work of *Kourafalou et al.* [1983]. Maximum alongshelf volume transport was about 85 $\times$ 10^4 m^3 s^{-1} during an intense wind event in early March. Typical transport fluctuations in the 2- to 14-day period band had amplitudes of 30 to 40 $\times$ 10^4 m^3 s^{-1} and were highly coherent with wind events through the Ekman frictional response mechanism discussed earlier. Mean transport was toward the north at 16 $\times$ 10^4 m^3 s^{-1} for the total 40-day period. If the inner shelf waters between 0- and 15-m depths were included in this estimate using the northward mean flow from site 8 of 1.4 cm s^{-1}, then the total mean alongshelf transport would be increased by only 0.2 $\times$ 10^4 m^3 s^{-1}. Therefore the volume transports calculated from data of moorings 8 and 9 should be reasonably good estimates of the total alongshelf transport at 30°N.

Mean alongshelf volume transport can be used to estimate shelf residence time, provided the 41-day average is typical of long-term mean conditions. The volume of shelf water between Cape Canaveral and Cape Fear between the coast and the 60-m isobath was estimated by *Atkinson et al.* [1983] as 1540 km^3. If the mean alongshelf transport is relatively constant, then it would take approximately three months to transport a volume equivalent to the shelf volume through the domain. This residence time compares with a 2-month estimate based on volume transport in the Onslow Bay [*Blanton and Pietrafesa,* 1978]. Previous estimates using salt conservation have also predicted shelf residence times of about 3 months [*Atkinson et al.,* 1978, 1983].

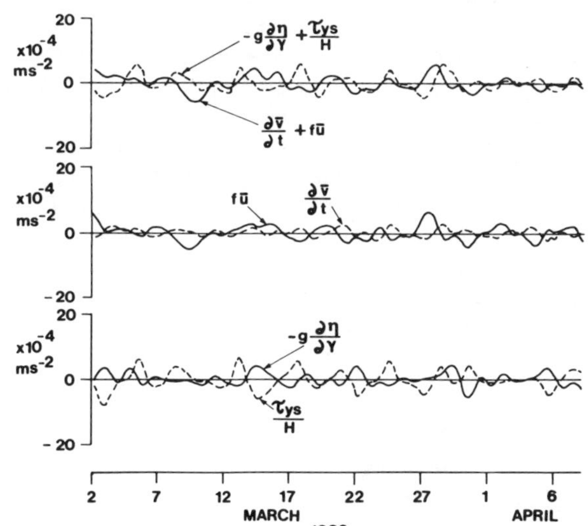

Fig. 20. Vertically averaged alongshelf momentum balance 28-m isobath: moorings 8, 9, 20, GABEX I.

TABLE 3. Term Estimates of the Vertically Integrated Momentum Balances for the Inner Shelf Frontal Zone

Cross Front, $\times 10^{-4}$ cm s^{-2} (Vertically Integrated)

Date	$\frac{\partial \bar{u}}{\partial t}$	$+\frac{\partial \bar{u}\bar{u}}{\partial x}$	$-f\bar{v}$	$= -g\frac{\partial \eta}{\partial x}$	$-\frac{g}{\rho H}\int_{-h}^{o}\frac{\partial \alpha}{\partial x}dz$	$+\frac{\tau_{sx}}{\rho H}$	$-\frac{\tau_{bx}}{\rho H}$	$-\frac{\partial\langle u'u'\rangle}{\partial x}$
Nov. 7–9, 1979	−0.24	+ ~0	+ *3.40*	= *19.9**	− *13.2*	− 1.40	− 0.5	− 1.60

Alongfront, $\times 10^{-4}$ cm s^{-2} (Vertically Integrated)

Date	$\frac{\partial \bar{v}}{\partial t}$	$+\frac{\partial \bar{u}\bar{v}}{\partial x}$	$+f\bar{u}$	$= -g\frac{\partial \eta}{\partial y}$	$+\frac{\tau_{sy}}{\rho H}$	$-\frac{\tau_{by}}{\rho H}$	$-\frac{\partial\langle u'v'\rangle}{\partial x}$
Nov. 7–9, 1979	*−0.44*	+ 0.17	+ *2.14*	= *2.97*	− *2.90*	+ *0.80*	+ *1.00*

Cross Front, $\times 10^{-4}$ cm s^{-2} (Two Layers)

	Date	$\frac{\partial \bar{u}}{\partial t}$	$+\frac{\partial \bar{u}\bar{u}}{\partial x}$	$-f\bar{v}$	$= -g\frac{\partial \eta}{\partial x}$	$-\frac{g}{\rho H}\int_{-h}^{o}\frac{\partial \alpha}{\partial x}dz$	$+\frac{\tau_{sx}}{\rho H}$	$-\frac{\tau_{bx}}{\rho H}$	$-\frac{\partial\langle u'u'\rangle}{\partial x}$
Upper	April 2–4, 1981	−0.37	+ 0.06	− *1.99*	= −0.15		+ *1.95*		− *4.10*
Lower	April 2–4, 1981	0.25	+ ~0	− *2.58*	= 0.28	~ 0		+ 0.06	− *2.67*

Alongfront, $\times 10^{-4}$ cm s^{-2} (Two Layers)

	Date	$\frac{\partial \bar{v}}{\partial t}$	$+\frac{\partial \bar{u}\bar{v}}{\partial x}$	$+f\bar{u}$	$= -g\frac{\partial \eta}{\partial y}$	$+\frac{\tau_{sy}}{\rho H}$	$-\frac{\tau_{by}}{\rho H}$	$-\frac{\partial\langle u'v'\rangle}{\partial x}$
Upper	April 2–4, 1981	−0.21	+ 0.07	+ *1.83*	= −0.52	+ *2.20*		+ 0.014
Lower	April 2–4, 1981	*−0.52*	− 0.01	− *0.32*	= *−0.74*		− *0.49*	+ *0.38*

**Blanton* [1981] reported a value $g\,(\partial\eta/\partial x) = 5.5 \times 10^{-4}$ cm s^{-2}. Because of a mistake in the algebraic sign, the correct value should be -6.9×10^{-4}. The earlier study neglected $\partial u/\partial x$.

The larger term estimates are italicized. No alongfront gradients were measured. In all cases the barotropic pressure gradient term was calculated.

Modeling Wind-Driven Flow

As was pointed out previously, winter flow on the shelf between Cape Canaveral and Cape Fear is mainly driven by tides and wind, while significant Gulf Stream effects are observed primarily on the outer shelf. During the winter months (November–April) the shelf is vertically well mixed, and horizontal density gradients are small. During the rest of the year the thermocline stratification is strong [*Stefansson et al.*, 1971]. Current meter measurements from the SAB midshelf region showed no significant change in the currents with depth during winter, indicating Ekman turning was negligible over the observed portion of the water column [*Kourafalou et al.*, 1983].

These findings encourage the adaptation of a one-layer two-dimensional numerical model for the study of wind-driven flow in the SAB during the winter season.

Model Development

The formulation of the model has been described in detail by *Wang and Connor* [1975] and *Wang* [1978]. The complete Reynolds equations and the continuity equation are integrated over the water depth to eliminate dependence on the vertical coordinate, z. The hydrostatic assumption and Boussinesq approximation are introduced, leading to the shallow water long-wave equations:

$$\frac{\partial \eta}{\partial t} + \frac{\partial q_x}{\partial x} + \frac{\partial q_x}{\partial y} = 0 \tag{1}$$

$$\frac{\partial q_x}{\partial t} + \frac{\partial u q_x}{\partial x} + \frac{\partial v q_x}{\partial y} = -gH\frac{\partial \eta}{\partial x} + \frac{1}{\rho}(\tau_{xs} - \tau_{xb}) + fq_y + \frac{\partial}{\partial x}\left(E_{xx}\frac{\partial q_x}{\partial x}\right) + \frac{\partial}{\partial y}\left[E_{xy}\left(\frac{\partial q_x}{\partial y} + \frac{\partial q_y}{\partial x}\right)\right] \tag{2}$$

$$\frac{\partial q_y}{\partial t} + \frac{\partial u q_y}{\partial x} + \frac{\partial v q_y}{\partial y} = -gH\frac{\partial \eta}{\partial y} + \frac{1}{\rho}(\tau_{ys} - \tau_{yb}) - fq_x + \frac{\partial}{\partial x}\left[E_{xy}\left(\frac{\partial q_x}{\partial y} + \frac{\partial q_y}{\partial_x}\right)\right] + \frac{\partial}{\partial y}\left(E_{yy}\frac{\partial q_y}{\partial y}\right) \tag{3}$$

where (x,y) are horizontal coordinates, t is time, η is surface elevation around mean sea level (msl), (q_x,q_y) are volume transports per unit width and time in x and y directions, (u,v) are vertically averaged velocities, g is gravity, $H = h + \eta$ and is total water depth, h is depth from a level water surface, $\vec{\tau}_s$ is wind stress and is equal to $\rho_{air}C_D\bar{U}\bar{U}$, $\vec{\tau}_b$ is bottom stress and is equal to $\rho C\bar{u}\bar{u}$, f is the Coriolis parameter, (E_{xx}, E_{xy}, E_{yz}) are eddy viscosities, U is wind speed, and C_D and C are surface and bottom drag coefficients, respectively. For details of model adaptation to the SAB and values of drag coefficients, refer to *Kourafalou et al.* [1983].

A numerical finite element method in space is coupled with a simple finite difference scheme in time to integrate

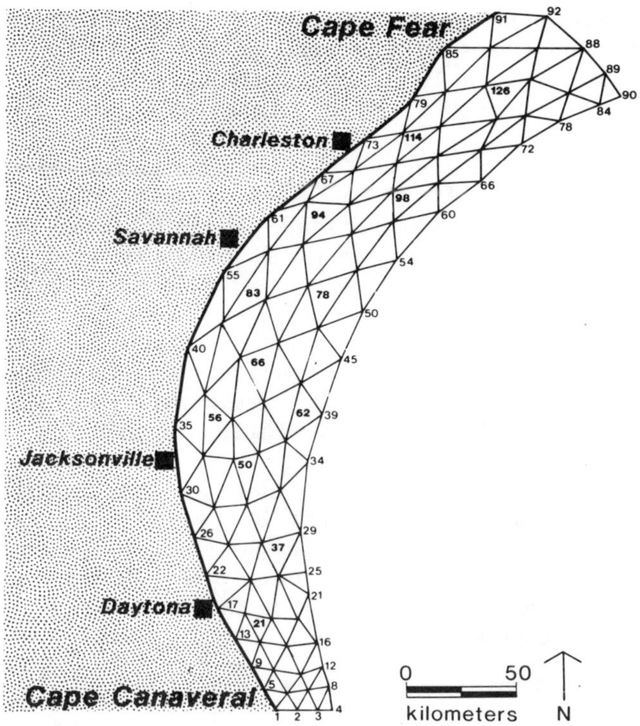

Fig. 21. Finite element grid of model domain; the SAB is divided into 144 linear triangular elements defined by 93 node points.

the equations. The model is predictive (prognostic) in that, apart from lateral boundary conditions, only the bottom friction and wind stress coefficients need be prescribed. The boundary conditions for this model are of two types: prescribed water surface elevations at the ocean boundary, and a free slip condition and no flow through the land boundary.

The SAB is divided into 144 triangular elements and 93 nodes, as shown in Figure 21. The grid varies in size according to the bathymetry and the shape of the shoreline, with an average node separation of about 40 km. The ocean boundary is taken at the 75-m isobath, which is also the location of the shelf break.

Analysis of wind data [*Weisberg and Pietrafesa,* 1983] shows high coherence of the wind field measured at different locations over the SAB shelf. This coherent wind field is advantageous for the modeling of the flow, because it permits linear interpolation between coastal and shelf wind stations in order to prescribe the wind field over the whole shelf domain. The interpolation scheme uses three wind stations: Daytona Beach and Charleston at the coast and the offshore meteorological buoy 41005 (Figure 1). The wind observations from the coastal station were increased by a factor of 2.3 in order to represent the wind above the water. This factor was determined by comparing wind data from the Savannah coastal station and the Savannah lighthouse station 15 km offshore (Figure 1).

Wind-Driven Flow

An 8-day wind event (March 16–23, 1980) was chosen for modeling of the wind-driven flow. Figure 22 shows the wind measured at the three stations, and Figures 23 and 24 show the flow that the model predicts under the influence of this wind field on March 18 (following northward winds) and March 19 (following southward winds). The arrows indicate the predicted vertically averaged current velocity vectors in centimeters per second, while the thicker vectors represent the observed vertically averaged currents from the 28-, 40-, and 75-m isobaths.

At the beginning of the modeled period, winds were northwestward and appear to have generated a northward current pattern (Figure 23). As the wind turned southeastward, both measured and predicted currents turned to the right in order to adjust to the change in the wind pattern and reach a steady southward flow (Figure 24). It generally took 12 to 18 hours to complete a current flow reversal (both measured and predicted) following a reversal of wind direction.

Modeled current vectors appear to be in good agreement both in speed and direction with the observed currents, especially when the flow has steadied under sufficiently steady wind patterns. However, the comparison of the current vectors at the 75-m isobath (moorings 22 and 25) is less satisfactory owing to the influence of Gulf Stream forcing on the measurements, which causes measured currents to be stronger than predicted. A more detailed comparison of modeled currents and observations can be found in the work of *Kourafalou et al.* [1984].

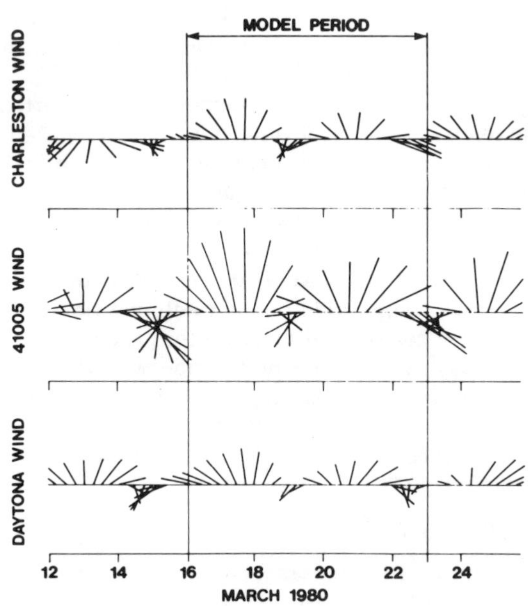

Fig. 22. Observed winds March 16–23, 1980, wind event.

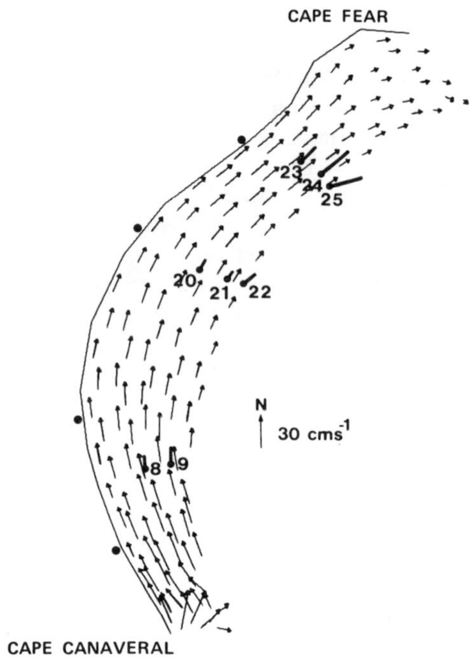

Fig. 23. Comparison of observed (thick lines) and modeled (arrows) vertically averaged velocities. Plot 5: March 18, 1980, at 0600 hours.

Particle Trajectories

The effect of winds, tides, and alongshelf slope forcing is examined by comparing particle trajectories computed with wind forcing only, and with tidal forcing combined with an alongshelf slope. The particles were initially located at the same positions as shown in Figure 21. The model was run with tidal-forcing conditions and a mean alongshelf sea level slope of -1.67×10^{-7} [*Tebeau and Lee,* 1979]. Figure 25 shows the net displacement of each particle under the combined effects of wind, tides, and alongshelf slope. A tendency to follow the isobaths is obvious, while the net displacement appears greater at deeper depths, a trend that was predicted by *Csanady* [1978] as a consequence of the alongshelf surface slope.

The net particle displacement over 8 days for wind forcing only was comparable to the combined tide and slope case. A characteristic net distance that the particles traveled in 8 days in the midshelf region was 70 km, which indicates an average Lagrangian speed of 8 cm s^{-1} with approximately 3 to 4 cm s^{-1} due to wind forcing, 2 to 3 cm s^{-1} due to tides, and about 2 cm s^{-1} due to the applied alongshelf slope.

Volume Transport

The alongshelf volume transport due to wind forcing only was computed from the model results for the cross-shelf section through moorings 8 and 9 (Figure 26) and compared with the transport estimated from the measured currents (dotted and solid lines, respectively, in Figure 26). The agreement between the transport curves is remarkable with maximum transport values reaching 40×10^4 m^3 s^{-1} (measured) and 36×10^4 m^3 s^{-1} (model), while the minimum values were -22×10^4 m^3 s^{-1} and -15×10^4 m^3 s^{-1}, respectively.

The effect of alongshelf slope on volume transport was also computed. Hourly alongshelf transports were computed with the model over a tidal period for the cross-shelf section at Savannah where the shelf width is widest (120 km) using tidal forcing for the cases of alongshelf slope of 0 and alongshelf slope of -1.67×10^{-7} [*Kourafalou et al.,* 1984]. Comparison indicates that the effect of the alongshelf slope on the alongshelf volume transport is of the order of 20×10^4 m^3 s^{-1} at this section and is comparable in magnitude to the alongshelf tidal transport. The Lagrangian analyses (trajectories) indicated that the combination of tidal transport plus alongshelf slope transport over an 8-day period appears to be of comparable magnitude to the net wind-driven transport for the 8 days. Since tidal and alongshelf slope forcing are persistent, as opposed to the variable nature of the wind, they may have greater influence on maintaining mean flows.

Predicted Sea Level Fluctuations

Model-predicted sea elevation contours are also shown for northward and southward wind events. In Figure 27 (March 18, 1800 UT), a drop in coastal sea level of approximately 14 cm occurred near Savannah in response to

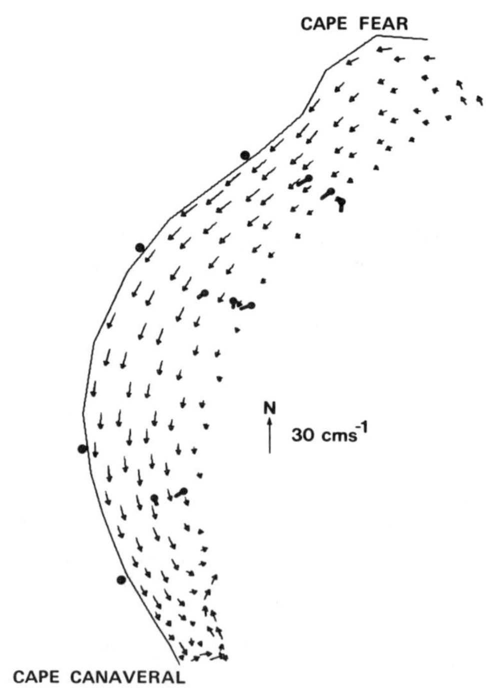

Fig. 24. Comparison of observed (thick lines) and modeled (arrows) vertically averaged velocities. Plot 8: March 19, 1980, at 1800 hours.

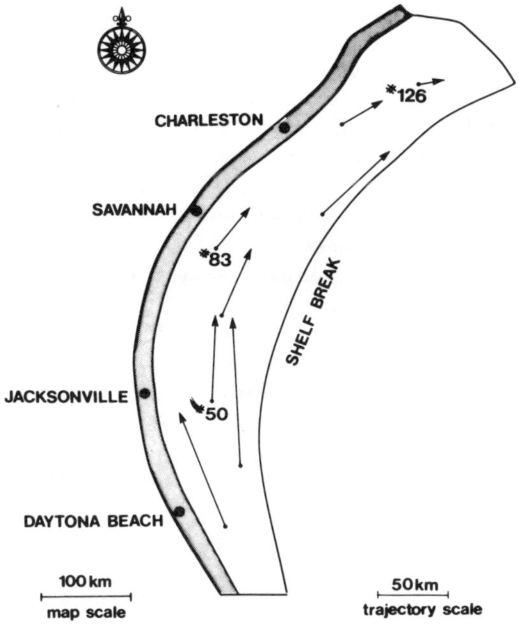

Fig. 25. Net particle motion for an 8-day period: March 16 to 24, 1980.

northward winds and appears to have produced the northward flow pattern shown in Figure 23. In Figure 28 (March 20, 0600 UT) a coastal sea level rise of about 12 cm due to the southward wind appears to be the driving force for the southward flow pattern shown in Figure 24. These predicted patterns are in close agreement with the measured sea level fluctuations shown in Figure 26. Alongshelf transport variations appear to lag changes in cross-shelf slope by 6 to 12 hours and alongshelf wind reversals by 12 to 18 hours.

Momentum Balances

The model was used to evaluate vertically integrated momentum balances at midshelf and to compare these to the calculations made from current meter data and presented in Tables 1 and 2. A nearly steady state situation of northward flow was chosen (March 21, 1980), and the terms were computed at midshelf for 30°N (Table 4).

The cross-shelf momentum balance is clearly geostrophic, while the alongshelf momentum balance is between the Coriolis term, the alongshelf surface slope, and wind stress. Bottom friction and local acceleration terms are smaller but possibly not negligible. The term estimates from current meter measurements are standard deviations from 41-day records over the period March 2 to April 11, 1980, and represent typical amplitudes of the fluctuations. Although the model terms were estimated for a single time step representing a steady state northward wind case, similar results are obtained from the current meter derived estimates both in the magnitude of term estimates and in the resulting balances. The cross-shelf fluctuation balance is again primarily barotropic-geostrophic, with a lesser contribution from the cross-shelf wind stress. The cross-shelf wind is small in the model estimates because a northward wind case was used in the evaluation. The alongshelf

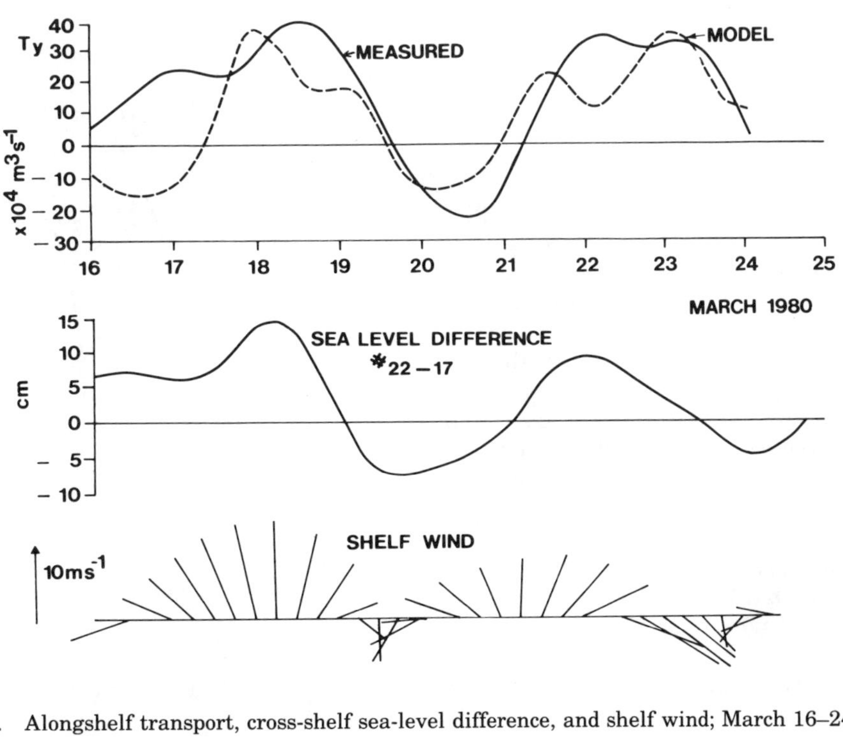

Fig. 26. Alongshelf transport, cross-shelf sea-level difference, and shelf wind; March 16–24, 1980.

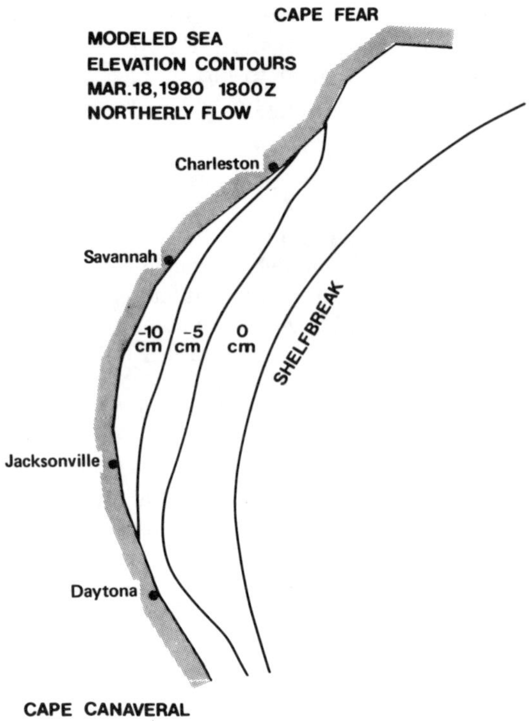

Fig. 27. Predicted sea elevation contours; March 18, 1980, at 1800 hours (northward flow).

fluctuation balance involves the Coriolis, barotropic pressure gradient, wind stress, bottom stress, and local acceleration terms, which are all of comparable magnitude, as in the model findings.

Conclusions

Analysis of currents on the SAB shelf during winter conditions lends further support to the concept of separation of flow regimes by processes controlling low-frequency variability.

Subtidal frequency current and temperature variability on the outer shelf (depths of >40 m) result largely from northward propagating Gulf Stream frontal disturbances. Current meter, hydrographic, and satellite observations indicate that these disturbances are cold cyclonic eddies embedded in the Gulf Stream front. These disturbances propagated northward at speeds of 50 to 70 cm s^{-1}, producing coherent fluctuations of cross-shelf and alongshelf velocity components and temperature throughout the water column in the 5- to 9-day period band and had alongshelf coherence scales of 100 km.

The SST signatures of frontal eddies consist of shingle patterns [*Von Arx et al.,* 1955] which develop as warm filaments of near-surface Gulf Stream water 10 to 15 m deep, pulled out of the front and wrapped around the eddy's cold core. Presumably this is due to the cyclonic eddy circulation, in much the same manner as streamers associated with warm and cold core rings [*Vukovich and Grissman,* 1978]. The eddy circulation appears to be in nearly geostrophic equilibrium with the uplifted density surfaces in the cold core. Estimates of cross-shelf momentum balances for the total water column indicate that both the mean and fluctuating parts of the alongshelf flow within frontal eddies are in approximate geostrophic balance with the cross-shelf baroclinic pressure gradient [*Lee et al.,* 1984].

The formation process of frontal eddies is not well understood. The shingle signature observed in satellite imagery appears to be connected to growing wavelike meanders of the Gulf Stream front [*Legeckis,* 1975, 1979; *Stumpf and Rao,* 1975]. Eddies can develop in just a few days and possibly dissipate just as fast. Theoretical investigations predict that both barotropic [*Niiler and Mysak,* 1971] and baroclinic [*Orlanski,* 1969; *Orlanski and Cox,* 1973] instabilities can exist in the frontal region with wave properties for the fastest growing waves that match reasonably well with satellite observations, i.e., wavelengths of 100 to 200 km and periods of 10 days. Our observations show that cyclonic eddies travel in conjunction with the offshore meander portion of the waves. Similar findings were reported from the Florida Straits [*Lee and Mayer,* 1977] and off Onslow Bay, North Carolina [*Bane et al.,* 1981]. Frontal eddies do not appear to be directly or simply connected to wind forcing. However, wind events may act to trigger a frontal instability which could grow into an eddy.

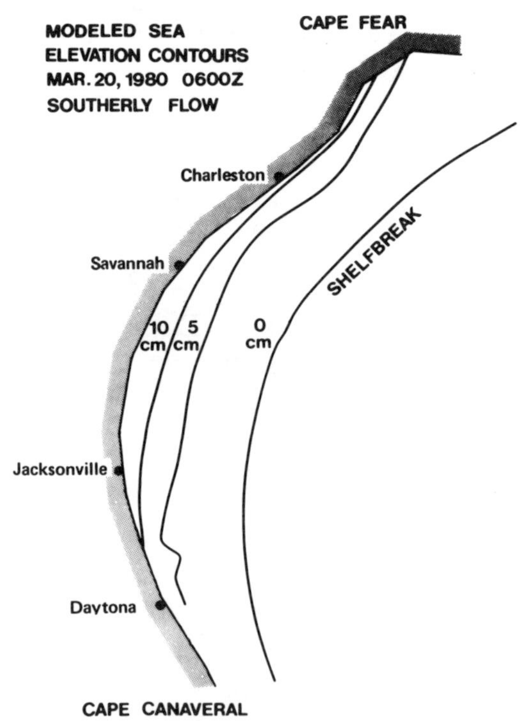

Fig. 28. Predicted sea elevation contours; March 20, 1980, at 0600 hours (southward flow).

TABLE 4. Model and Current Meter Derived Estimates of the Vertically Integrated Momentum Balances for the 28-m Isobath During a Northward Wind Event

	Cross Shelf, $\times 10^{-4}$ cm s^{-2}							
	$\frac{\partial \bar{u}}{\partial t}$	$+\frac{\partial \bar{u}^2}{\partial x}$	$+\frac{\partial \bar{u}\bar{v}}{\partial y}$	$-f\bar{v}$	$=$	$-g\frac{\partial \eta}{\partial x}$	$+\frac{\tau_{xs}}{\rho H}$	$-\frac{\tau_{xb}}{\rho H}$
Model	−0.3	+ 0.01	− 0.02	− *6.6*	=	*−7.6*	+ 0.3	− 0.15
Measurements	±0.6	± 0.2	± 0.04	± *7.0*	=	*±7.0*	± *2.5*	± 0.3
	Alongshelf, $\times 10^{-4}$ cm s^{-2}							
	$\frac{\partial \bar{v}}{\partial t}$	$+\frac{\partial \bar{v}^2}{\partial y}$	$+\frac{\partial \bar{u}\bar{v}}{\partial x}$	$+f\bar{u}$	$=$	$-g\frac{\partial \eta}{\partial y}$	$+\frac{\tau_{ys}}{\rho H}$	$-\frac{\tau_{yb}}{\rho H}$
Model	*−0.4*	+ 0.1	− 0.04	+ *2.1*	=	*2.0*	+ *1.2*	− *0.5*
Measurements	± *1.3*	± 0.06	± 0.2	± *2.6*	=	*±3.6*	± *3.2*	± *0.6*

Measurement values are current meter standard deviations for the 41-day period from March 20 to April 11, 1980. Italicized values are those with the largest magnitudes.

Dissipation of frontal eddies appears to occur through a rapid elongation process brought about by the large horizontal shear across the Gulf Stream front. The kinematic and dynamic properties of the enlarged frontal disturbances that form downstream of the Charleston Bump appear to be consistent with the eddies we observed upstream of the bump [*Bane et al.,* 1981; *Brooks and Bane,* 1981]. Occasionally, shingles are observed in satellite imagery that remain recognizable as they propagate from the upstream to downstream region with considerable enlargement occurring downstream of the bump [*Lee et al.,* 1981]. However, tracking of events in satellite imagery requires about one week of consecutive cloud-free days, which usually only occurs in April. Our shelf break current meter records also indicate that frontal disturbances can maintain continuity between regions and produce larger fluctuations downstream of the Charleston Bump.

Frontal eddies have considerable influence on primary production on the outer shelf. Upwelling in the cold core, together with onshore flow in the cyclonic circulation, transports deeper nutrient-enriched Gulf Stream waters into the euphotic zone for phytoplankton uptake. Upwelling velocities are estimated at 8.5 m d^{-1}. Rapid utilization of newly upwelled nutrients results in elongated patches of high chlorophyll that propagate with the cold core and have similar dimensions [*Yoder et al.,* 1981]. Since the demise of frontal eddies appears due to a shear-induced dissipation process rather than a sinking of isopycnal surfaces, the upwelled nutrients should remain on the outer shelf, possibly causing chlorophyll bands in the Gulf Stream front.

Low-frequency variability in the midshelf (20- to 40-m isobaths) of the SAB during winter appears to follow simple, locally forced Ekman dynamics. Coherent alongshelf winds over the shelf domain generate offshore/onshore transports in the surface Ekman layer that cause sea level to rise or fall at the coast and thus produce cross-shelf surface slopes that are in approximate geostrophic balance with an alongshelf barotropic interior flow. Sea level response to alongshelf wind forcing is greatest near the coast in the vicinity of Savannah, Georgia, where the shelf is widest. This response produces cross-shelf and alongshelf slopes in the 2- to 14-day period synoptic weather band that lag the wind events by 6 to 12 hours. Alongshelf barotropic currents, with near-zero phase lags over downstream coherence length scales greater than 400 km, lag the cross-shelf slopes by 6 to 12 hours. There is little indication in this region of southward propagating continental shelf waves.

The forced wave response to local winds appears to be a characteristic property of winter shelf dynamics on a wide variety of continental shelves and generally accounts for a large fraction of the low-frequency flow variability, which is contained in a quasi-barotropic, first-mode empirical orthogonal function. This type of locally forced response has been observed in the Middle Atlantic Bight [*Beardsley and Butman,* 1974; *Scott and Csanady,* 1976; *Csanady,* 1976, 1978; *Beardsley and Winant,* 1979; *Beardsley and Boicourt,* 1981], the west Florida shelf [*Mitchum and Sturges,* 1983], and the Pacific northwest shelf [*Hickey and Hamilton,* 1980; *Hickey,* 1981; *Smith,* 1981]. Free coastal trapped waves that are generated by remote forcing have been found to make significant contributions to low-frequency flow variability on the Peru shelf [*Smith,* 1978; *Allen and Romea,* 1980; *Allen and Smith,* 1981; *Romea and Smith,* 1983]. There is also evidence in the Middle Atlantic Bight for a combined wave response including both a forced wave and a free coastal trapped wave with about equal energy levels [*Wang,* 1979; *Noble and Butman,* 1979; *Ou et al.,* 1981].

Vertically integrated momentum balances give further support for local Ekman response to wind forcing in the SAB. Cross-shelf momentum balances at both the 28- and 40-m isobaths indicate that the mean alongshelf flow is in geostrophic balance with the cross-shelf barotropic pressure gradient. The fluctuating flow has an additional contribution from the cross-shelf wind. The momentum balance in the alongshelf direction is between the Coriolis force, alongshelf slope, and alongshelf wind stress at the 28-m isobath for mean conditions, with an additional contribution from the local acceleration for the fluctuating part. At the 40-m isobath, which is close to the transition between the midshelf and outer shelf flow regimes, the mean alongshelf momentum balance is between nonlinear momentum advection, Coriolis force, alongshelf slope, and alongshelf surface and bottom stresses. An alongshelf slope of the order of -10^{-7} is needed to balance the mean alongshelf momentum. *Sturges* [1974] found a similar slope in the Gulf Stream adjacent to the SAB shelf from both steric and geostrophic leveling, indicating a Gulf Stream origin for the slope. Apparently, the Gulf Stream imparts northward momentum to the shelf through the alongshelf surface slope and a net onshore flow in the vicinity of Cape Canaveral. The momentum gained is compensated for locally at 30°N by an offshore momentum advection of Gulf Stream frontal eddies and over the shelf length by a net offshore transport in the vicinity of the Charleston Bump.

The mean alongshelf volume transport in the SAB during winter was northward at approximately 20×10^4 m^3 s^{-1}. This transport appears to result from a combination of wind, tidal residual, and alongshelf slope forcing as determined from observations and model results. Apparently, the mean northward transport was sustained by onshore flow near Cape Canaveral and offshore flow in the vicinity of the Charleston Bump. The shelf residence time estimated from alongshelf transport is 3 months.

Wind-driven flow in the SAB can be satisfactorily hindcast during the winter with a vertically integrated two-dimensional model. The wind-induced flow patterns hindcast from an observed time-varying wind event were in good agreement with measured currents at midshelf and inner shelf locations. Model results were not as satisfactory on the outer shelf due to the neglect of effects of the Gulf Stream.

An alongshelf slope of Gulf Stream origin of the order of -10^{-7} (sloping downward toward the north) has a significant effect on shelf net particle trajectories and volume transports. The alongshelf volume transport due to the slope is of comparable magnitude to the net alongshelf tidal transport, while the combination of alongshelf slope and net tidal transports is comparable to wind-driven alongshelf transports.

Model results further indicate that local Ekman dynamics explain the observed winter shelf water response to wind forcing in the midshelf to inner shelf of the SAB. Sea level setup or setdown typically lags the alongshelf wind by 6–12 hours followed 6–12 hours later by the establishment of a near-uniform alongshelf flow. Model estimates of vertically integrated momentum balances at midshelf agree closely with estimates derived from current meter observations.

Acknowledgments. We thank J. Green and G. Rosiello for their technical assistance. Funding for this study was provided by the Department of Energy under contract DE-AS05-76EV05163 and the Minerals Management Service (formerly Bureau of Land Management) under prime contract AAT851-CT1-25 to Science Applications, Inc., of Raleigh, North Carolina, subcontract 11-82029-11. Funding for data from the 15- and 28-m isobaths collected by L. J. Pietrafesa of North Carolina State University was provided by Department of Energy contract DOE-AS09-74-00902.

References

Allen, J. S., and R. D. Romea, On coastal trapped waves at low latitudes in a stratified ocean, *J. Fluid Mech., 98,* 535–585, 1980.

Allen, J. S., and R. L. Smith, On the dynamics of wind-driven shelf currents, *Philos. Trans. R. Soc. London Ser. A, 302,* 617–634, 1981.

Atkinson, L., J. O. Blanton, and E. Haines, Shelf flushing rates based on distribution of salinity and fresh water in the Georgia Bight, *Estuarine Coastal Mar. Sci.,* 1, 464–472, 1978.

Atkinson, L., T. N. Lee, J. Blanton, and W. S. Chandler, Climatology of the southeastern U.S. continental shelf waters, *J. Geophys. Res., 88,* 4705–4718, 1983.

Bane, J. M., Jr., Initial observations of the subsurface structure and short-term variability of the seaward deflection of the Gulf Stream off Charleston, South Carolina, *J. Geophys. Res., 88,* 4673–4684, 1983.

Bane, J. M., Jr., and D. A. Brooks, Gulf Stream meanders along the continental margin from Florida Straits to Cape Hatteras, *Geophys. Res. Lett., 6,* 280–282, 1979.

Bane, J. M., Jr., D. A. Brooks, and K. R. Lorenson, Synoptic observations of the three-dimensional structure, propagation, and evolution of Gulf Stream meanders along the Carolina continental margin, *J. Geophys. Res., 86,* 6411–6425, 1981.

Beardsley, R. C., and W. C. Boicourt, On estuarine and continental shelf circulation in the Middle Atlantic Bight, in *Evolution of Physical Oceanography,* edited by B. A. Warren and C. Wunsch, pp. 198–233, MIT Press, Cambridge, Mass., 1981.

Beardsley, R. C., and B. Butman, Circulation on the New England continental shelf: Response to strong winter storms, *Geophys. Res. Lett., 1,* 181–184, 1974.

Beardsley, R. C., and C. D. Winant, On the mean circulation in the Mid-Atlantic Bight, *J. Phys. Oceanogr., 9,* 612–619, 1979.

Blanton, J. O., The transport of freshwater off a multi-inlet coastline, in *Modeling of Estuaries and Wetland Processes,* edited by P. Hamilton and K. B. MacDonald, pp. 49–64, Plenum, New York, 1980.

Blanton, J. O., Ocean currents along a nearshore frontal

zone on the continental shelf of the southeastern United States, *J. Phys. Oceanogr., 11,* 1627–1637, 1981.
Blanton, J. O., and L. P. Atkinson, Physical transfer processes between Georgia tidal inlets and nearshore waters, in *Estuarine Interactions,* edited by M. Wiley, pp. 514–532, Academic, Orlando, Fla., 1978.
Blanton, J. O., and L. P. Atkinson, Transport and fate of river discharge on the continental shelf of the southeastern United States, *J. Geophys. Res., 88,* 4730–4739, 1983.
Blanton, J. O., and L. J. Pietrafesa, Flushing of the continental shelf south of Cape Hatteras by the Gulf Stream, *Geophys. Res. Lett., 5,* 495–498, 1978.
Brooks, D. A., and J. M. Bane, Jr., Gulf Stream deflection by a bottom feature off Charleston, SC, *Science, 201,* 1225–1226, 1978.
Brooks, D. A., and J. M. Bane, Jr., Gulf Stream fluctuations and meanders over the Onslow Bay upper continental slope, *J. Phys. Oceanogr., 11,* 247–256, 1981.
Bumpus, D. F., A description of the circulation on the continental shelf of the east coast of the United States, *Prog. Oceanogr., 6,* 111–156, 1973.
Csanady, G. T., Mean circulation in shallow seas, *J. Geophys. Res., 81,* 5389–5399, 1976.
Csanady, G. T., The arrested topographic wave, *J. Phys. Oceanogr., 8,* 47–62, 1978.
Dunstan, W. M., and L. P. Atkinson, Sources of new nitrogen for the South Atlantic Bight, in *Estuarine Processes,* edited by M. Wiley, pp. 69–78, Academic, Orlando, Fla., 1976.
Hickey, B. M., Alongshore coherence on the Pacific northwest continental shelf (January–April 1975), *J. Phys. Oceanogr., 11,* 822–835, 1981.
Hickey, B. M., and P. Hamilton, A spin-up model as a diagnostic tool for interpretation of current and density measurements on the continental shelf of the Pacific northwest, *J. Phys. Oceanogr., 10,* 12–24, 1980.
Klinck, J. M., L. J. Pietrafesa, and G. S. Janowitz, Continental shelf circulation induced by a moving, localized wind stress, *J. Phys. Oceanogr., 11,* 836–848, 1981.
Kourafalou, V., J. D. Wang, T. N. Lee, and L. J. Pietrafesa, Modeling of winter circulation on the southeast U.S. continental shelf, *Tech Rep. 83009,* 97 pp., Univ. of Miami, Miami, Fla., 1983.
Kourafalou, V., J. D. Wang, and T. N. Lee, Circulation on the continental shelf of the southeastern United States, 3, Modeling the winter wind-driven flow, *J. Phys. Oceanogr., 14,* 1022–1031, 1984.
Lee, T. N., and L. P. Atkinson, Low-frequency current and temperature variability from Gulf Stream frontal eddies and atmospheric forcing along the southeast U.S. outer continental shelf, *J. Geophys. Res., 88,* 4541–4568, 1983.
Lee, T. N., and D. Brooks, Initial observations of current, temperature and coastal sea level response to atmospheric and Gulf Stream forcing, *Geophys. Res. Lett., 6,* 321–324, 1979.
Lee, T. N., and D. Mayer, Low frequency current variability and spin-off eddies on the shelf off southeast Florida, *J. Mar. Res., 35,* 193–220, 1977.
Lee, T. N., L. P. Atkinson, and R. Legeckis, Observations of a Gulf Stream frontal eddy on the Georgia continental shelf, April 1977, *Deep Sea Res., 28A,* 347–378, 1981.
Lee, T. N., W.-J. Ho, V. Kourafalou, and J. D. Wang, Circulation on the continental shelf of the southeastern United States, 1, Subtidal response to wind and Gulf Stream forcing during winter, *J. Phys. Oceanogr., 14,* 1001–1012, 1984.
Legeckis, R., Applications of synchronous meteorological satellite data to the study of time dependent sea surface temperature changes along the boundary of the Gulf Stream, *Geophys. Res. Lett., 2,* 435–438, 1975.
Legeckis, R., Satellite observations of the influence of bottom topography on the seaward deflection of the Gulf Stream off Charleston, SC, *J. Phys. Oceanogr., 9,* 483–497, 1979.
Mitchum, G. T., and W. Sturges, Wind-driven currents on the west Florida shelf, *J. Phys. Oceanogr., 12,* 1310–1317, 1982.
Niiler, P. P., and L. A. Mysak, Barotropic waves along an eastern continental shelf, *Geophys. Astrophys. Fluid Dyn., 2,* 273–278, 1971.
Noble, M., and B. Butman, Low-frequency wind-induced sea level oscillations along the east coast of North America, *J. Geophys. Res., 84,* 3227–3236, 1979.
Olson, D. B., O. B. Brown, and S. R. Emmerson, Gulf Stream frontal statistics from Florida Straits to Cape Hatteras derived from satellite and historical data, *J. Geophys. Res., 88,* 4569–4578, 1983.
Orlanski, I., The influence of bottom topography on the stability of jets in a baroclinic fluid, *J. Atmos. Sci., 26,* 1216–1232, 1969.
Orlanski, I., and M. D. Cox, Baroclinic instability in ocean currents, *Geophys. Astrophys. Fluid Dyn., 4,* 297–332, 1973.
Ou, H. W., R. C. Beardsley, D. Mayer, W. C. Boicourt, and B. Butman, An analysis of subtidal current fluctuations in the Middle Atlantic Bight, *J. Phys. Oceanogr., 11,* 1383–1392, 1981.
Pietrafesa, L. J., J. O. Blanton, and L. P. Atkinson, Evidence for deflection of the Gulf Stream by the Charleston Rise, *Gulf Stream, 4,* 3–7, 1978.
Romea, R. D., and R. L. Smith, Further evidence for coastal trapped waves along the Peru coast, submitted to *J. Phys. Oceanogr.,* 1984.
Schwing, F. B., B. J. Kjerfve, and J. E. Sneed, Nearshore currents on the South Carolina continental shelf, *J. Geophys. Res., 88,* 4719–4729, 1983.
Scott, J. T., and G. T. Csanady, Nearshore currents off Long Island, *J. Geophys. Res., 81,* 5401–5409, 1976.
Singer, J. J., L. Atkinson, J. O. Blanton, and Y. A. Yoder, Cape Romain and the Charleston Bump: Historical and

recent hydrographical observations, *J. Geophys. Res., 88,* 4685–4698, 1983.
Smith, R. L., Poleward propagating perturbations in currents and sea level along the Peru coast, *J. Geophys. Res., 83,* 6083–6092, 1978.
Smith, R. L., A comparison of the structure and variability of the flow field in three coastal upwelling regions: Oregon, northwest Africa and Peru, in *Coastal Upwelling, Coastal and Estuarine Ser.,* vol. 1, edited by F. A. Richards, pp. 107–118, AGU, Washington, D. C., 1981.
Stefansson, V., L. P. Atkinson, and D. F. Bumpus, Hydrographic properties and circulation of the North Carolina shelf and slope waters, *Deep Sea Res., 18,* 383–420, 1971.
Stumpf, H. G., and P. K. Rao, Evolution of Gulf Stream eddies as seen in satellite infrared imagery, *J. Phys. Oceanogr., 5,* 388–393, 1975.
Sturges, W., Sea level slope along continental boundaries, *J. Geophys. Res., 79,* 825–830, 1974.
Tebeau, P. A., and T. N. Lee, Wind induced circulation on the Georgia shelf, *Tech. Rep. 79003,* 177 pp., Univ. of Miami, Miami, Fla., 1979.
Von Arx, W. S., D. F. Bumpus, and W. S. Richardson, On the fine structure of the Gulf Stream front, *Deep Sea Res., 3,* 46–65, 1955.
Vukovich, F. M., and B. W. Grissman, Further studies of a cold eddy on the eastern side of the Gulf Stream using satellite data and ship data, *J. Phys. Oceanogr., 8,* 838–843, 1978.
Wang, D.-P., Low-frequency sea level variability on the Middle Atlantic Bight, *J. Mar. Res., 37,* 683–697, 1979.
Wang, J. D., Real-time flow in unstratified shallow water, *J. Waterway Port Coastal Ocean Div. Am Soc. Civ. Eng., WW1,* 53–68, 1978.
Wang, J. D., and J. J. Connor, Mathematical modeling of near coastal circulation, *Tech. Rep. 200,* 272 pp., R. M. Parsons Lab., Dep. of Civ. Eng., Mass. Inst. of Technol., Cambridge, MA, 1975.
Weber, A. H., and J. O. Blanton, Monthly mean wind fields for the South Atlantic Bight, *J. Phys. Oceanogr., 10,* 1256–1263, 1980.
Weisberg, R. H., and L. J. Pietrafesa, Kinematics and correlation of the surface wind field in the South Atlantic Bight, *J. Geophys. Res., 88,* 4593–4610, 1983.
Yoder, J. A., L. P. Atkinson, T. N. Lee, H. H. Kim, and C. K. McClain, Role of Gulf Stream frontal eddies in forming phytoplankton patches on the outer southeastern shelf, *Limnol. Oceanogr., 26,* 1103–1110, 1981.

The Tidal Regime in the South Atlantic Bight

LEONARD J. PIETRAFESA,[1] J. O. BLANTON,[2] J. D. WANG,[3] V. KOURAFALOU,[3] T. N. LEE,[3] AND K. A. BUSH[2]

Tidal currents, particularly at the M_2 frequency, dominate South Atlantic Bight inner shelf and midshelf dynamics, with M_2 tides accounting for almost 80% of the total kinetic energy there. Maximum energy occurs offshore of Savannah, where the shelf is widest. Outer shelf dynamics are less influenced by tides and more dependent on Gulf Stream interaction. There appear to be some seasonal differences in the tidal currents. A simple vertically integrated two-dimensional tidal model was used to reproduce tidal characteristics in the midshelf region during well-mixed winter conditions.

Introduction

Tides and tidal currents in coastal waters play important roles in local circulation, mixing processes, and sediment transport. This is particularly true in the South Atlantic Bight (SAB) where tidal motions account for 75–90% of the total energy of inner shelf (depths less than 20 m) waters throughout the year. The purpose of this chapter is to examine the tidal regime of the SAB as evinced by data collected since 1975. In addition, a brief review of tidal dynamics is included, as is a description of a tidal flow model with application to the SAB.

The periodic rise and fall of the sea surface known as the tide results from the gravitational forces of the earth-moon-sun system. Although tides, because of their importance to navigation, have long been observed and recorded, the first theories of tidal dynamics were not developed until the late seventeenth century. *Newton* [1687] presented the first theory of oceanic response to tide-generating forces, accounting for some of the chief characteristics of the tides. *Laplace* [1775] was the first to present a tidal theory that identified and separated tidal constituents and that included the horizontal components of the Coriolis acceleration in the equations of motion.

An important step in the historic development of tidal dynamics was the canal theory of tides, which describes oscillations in narrow channels of constant depth and vertical sides. *Airy* [1842] is considered the founder of the canal theory of tides, while *Sterneck* [1915] provided a numerical integration method for dealing with basins of complex shapes. *Sverdrup* [1926] included the Coriolis force in the canal theory equations, while *Fleming* [1938] included both rotation and a horizontal pressure gradient. *Fjeldstad* [1936] looked at the influence of friction on tidal currents. *Proudman* [1953] condensed, applied, and explained existing tidal theory.

Redfield [1958] hypothesized that the M_2 tide, the primary lunar semidurnal tidal constituent along the eastern United States continental shelf, is a cooscillation generated by the ocean tide at the continental slope and is dynamically similar to tidal cooscillations occurring in narrow channels. The variations in tidal intervals and elevations seen along the coast can then be explained in terms of the change in velocity of the tidal wave as it crosses the shallow shelf and by its reflection at the coast. Concepts derived from the canal theory of tides were used with coastal tide data to predict tidal properties across the continental shelf. *Platzman* [1971] presented the basic mathematics and physics of ocean tides. *Mofjeld* [1976] provided a review of tidal currents, including discussions of frictional and rotational effects. *Pietrafesa* [1983] contrasted U.S. western and eastern coastal ocean tides. *Clarke and Battisti* [1981] developed a continental margin theory, which includes alongshore pressure gradients and rotation, to explain the effects of smooth continental shelves on barotropic tides. They addressed the question of the amplification of semidiurnal tides on wide continental shelves and also derived a simple method, neglecting friction, for estimating tides on continental shelves given coastal tidal data. *Battisti and Clarke* [1982] presented a simple model to predict barotropic tidal currents on continental shelves and used this model to describe the M_2 tidal currents on the east coast of the United States. The model includes linearized bottom friction and variable topography.

Review of Tidal Dynamics

Tides in coastal regions can be treated as long (shallow water) waves propagating shoreward from the open ocean and being partially reflected seaward by the shallow bottom of the continental margin. The propagating incident and reflected waves are called Sverdrup waves, while the superposition of incident and reflected waves is called a

[1]Department of Marine, Earth, and Atmospheric Sciences, North Carolina State University, Raleigh.
[2]Skidaway Institute of Oceanography, Savannah, Georgia.
[3]Rosenstiel School of Marine and Atmospheric Science, University of Miami, Florida.

Poincaré wave. The hydrodynamical equations of motion for tides consider the dynamical effects of gravity acting on the pressure field through changes in the mass distribution and the effects of the earth's rotation. Mass distribution is manifested by changes in the sea surface elevation. The simplified equations of motion for free barotropic long waves, which describe tidal motion in domains small enough to ignore the earth's curvature, are

$$\frac{\partial u}{\partial t} - fv = -g\frac{\partial \eta}{\partial x} \tag{1}$$

$$\frac{\partial v}{\partial t} + fu = -g\frac{\partial \eta}{\partial y} \tag{2}$$

$$\frac{\partial \eta}{\partial t} + \frac{\partial (uh)}{\partial x} + \frac{\partial (vh)}{\partial y} = 0 \tag{3}$$

where u, v are the horizontal components of velocity in the x and y directions, respectively, f is the Coriolis parameter, t is time, g is gravitational acceleration, η is sea surface elevation, and h is water depth. The coordinate system is right handed with x positive offshore and y positive poleward. Equation (3) is the vertically integrated continuity equation. The Sverdrup wave is one solution to these equations and can be regarded as an elemental propagator of tidal signals [*Platzman*, 1971]. Sverdrup waves of a given frequency σ, where $\sigma = \pm[f^2 + (k^2 + l^2)gh]^{1/2}$, cannot exist poleward of the latitude where $\sigma = f$. Here k and l are horizontal wave numbers. According to this criterion, the limiting latitudes for diurnal tidal constituents are typically less than 30°, while the limiting latitudes for semidiurnal tidal constituents are typically greater than 70°. Crests and troughs of Sverdrup waves are horizontal, and particle motion is elliptical with clockwise rotation in the nothern hemisphere.

In the case of reflection of Sverdrup waves from a coast or bottom boundary, the superposition of incident and reflected waves results in a Poincaré wave. Poincaré wave fronts are perpendicular to the coast with characteristic nodal lines perpendicular to the wave fronts. Water parcel trajectories are elliptical with major axes in the direction of propagation and with clockwise rotation. The ratio v/u of minor to major axis, in the absence of friction, is approximately equal to f/ω, where f is the Coriolis parameter and ω is the frequency of the tidal constituent. This means that at higher latitudes tidal ellipses become more circular as v/u increases.

If bottom friction is included, the equations of motion become

$$\left(\frac{\partial}{\partial t} + \frac{r}{h}\right)u - fv = -g\frac{\partial \eta}{\partial x} \tag{4}$$

$$\left(\frac{\partial}{\partial t} + \frac{r}{h}\right)v + fu = -g\frac{\partial n}{\partial y} \tag{5}$$

$$\frac{\partial \eta}{\partial t} + \frac{\partial (uh)}{\partial x} + \frac{\partial (vh)}{\partial y} = 0 \tag{6}$$

where r is a bottom frictional stress coefficient and the other terms are as previously defined. *Battisti and Clarke* [1982] presented such a formulation of the tidal equations. They assumed that cross-shelf variations in bottom topography occur over much smaller length scales than do alongshore variations and that bottom stress can be approximated as a linear function of the root-mean-square velocity. Bottom friction causes the ratio of v/u to be less than the local ratio of f/ω and also results in a reduction of near-bottom current amplitudes and a flattening of the tidal ellipses.

In the vertical, tidal currents at any depth reach a maximum when frictional forces just balance the effective pressure gradient, which is the resultant of the pressure gradient and the tide-generating (gravitational) forces. It is assumed that the bottom frictional forces diminish with distance above the bottom, so that bottom currents reach their maxima earlier than do near-surface currents. Consequently, there is a vertical phase lag in tidal current velocities. The equation of motion for a tidal current given by *Proudman* [1953] is

$$\frac{\partial u}{\partial t} - A\frac{\partial^2 u}{\partial z^2} = Fe^{-i\omega t} \tag{7}$$

where $A(\partial^2 u/\partial z^2)$ is a friction term with constant eddy viscosity A and $Fe^{-i\omega t}$ is an oscillating pressure gradient due to the tidal wave. The pressure gradient is constant with depth, but the friction term decreases with increasing distance above the bottom. The tidal current reaches a maximum when $\partial u/\partial t = 0$.

Kundu et al. [1981] derived a relatively simple solution to the problem of tidal flow over variable bathymetry by assuming that flow above the frictional bottom boundary layer describes a known tidal ellipse. Their theory also predicts that current elipses become narrower, that major axes veer counterclockwise as distance above the bottom decreases, and that bottom currents lead surface currents by approximately 1 hour.

Tides in the SAB

Kinematical and Dynamical Properties

The principal tidal harmonic constituents in the SAB are listed in Table 1. Although there are nine semidurnal and six diurnal constituents, the M_2 partial tide, with a period of 12.42 solar hours, dominates SAB tidal dynamics [*Haight,* 1942; *Redfield,* 1958; *Clarke and Battisti,* 1981; *Battisti and Clarke,* 1982; *Pietrafesa,* 1983]. The North Atlantic Ocean tide is in fact chiefly semidiurnal, since the dimensions of the North Atlantic basin are such that its natural period of resonance is close to the M_2 period [*Redfield,* 1980].

Tebeau and Lee [1979] and *Lee and Brooks* [1979] found that tidal currents accounted for 80–90% of the total cross-shelf and about 20–40% of the alongshelf current varia-

TABLE 1. Principal Tidal Harmonic Constituents in the South Atlantic Bight

Name of Partial Tide	Symbol	Period	Predicted Tidal Potential Function Gradient Coefficient Ratio (Normalized on the M_2 Partial Tide)	Measured Energy Density Ratio (Normalized on the M_2 Partial Tide for Instrument 8 (2))
Luni-solar semidiurnal	K_2	11.97	0.127	0.038
Principal solar	S_2	12.00	0.466	0.037
Larger solar elliptic	T_2	12.01	0.027	0.038
Smaller lunar elliptic	L_2	12.19	0.028	0.010
Principal lunar	M_2	12.42	1.000	1.000
Larger lunar evectional	ω_2	12.63	0.036	0.080
Larger lunar elliptic	N_2	12.66	0.192	0.071
Variational	ν_2	12.87	0.031	0.006
Lunar elliptic second order	$2N_2$	12.91	0.025	0.005
Smaller lunar elliptic	J_1	23.10	0.033	0.010
Luni-solar diurnal	K_1	23.93	0.584	0.051
Principal solar diurnal	P_1	24.07	0.194	0.048
Smaller lunar elliptic	M_1	24.84	0.033	0.009
Principal lunar diurnal	O_1	25.82	0.415	0.007
Larger lunar elliptic	Q_1	26.87	0.079	0.004

bility in the midshelf region of the SAB. Kinetic energy distribution in the SAB is summarized in Figure 1, which shows normalized energy spectra for near-surface inner shelf, midshelf, and outer shelf regions seaward of Savannah and indicates the fraction of the total kinetic energy occurring within different frequency ranges. Energy at semidiurnal frequencies dominates the inner shelf and midshelf regions, accounting for almost 80% of the total kinetic energy there. At the outer shelf, semidiurnal energy accounts for less than 30% of the total energy. Thus the M_2 tide is the predominant contributor to the overall energetics of the inner shelf and midshelf regions, while outer shelf dynamics are probably more dependent on Gulf Stream interactions.

M_2 tidal characteristics in the SAB were investigated using data collected during 1975–1981. Figure 2 is an SAB area map showing locations of current meter moorings from which data are available. Each mooring had instruments at 17 m below the surface and 3 m above the bottom. Sampling intervals of 10 or 20 min were sufficient to resolve the principal tidal constituents. In addition to these current records, several stations along the Georgia coast provided nearshore tidal current data. These coastal stations, shown in Figure 3, cover an alongshore distance of about 25 km and an offshore distance of about 12 km near the Savannah River inlet. Long-term tidal records at two depths are available from the Savannah Navigational Light Tower (SNLT), while short-term records lasting four tidal cycles are available from the other coastal stations.

Calculations of M_2 kinetic energy for all the locations shown in Figure 2 indicate that maximum energy occurs at midshelf seaward of Savannah and Charleston, where

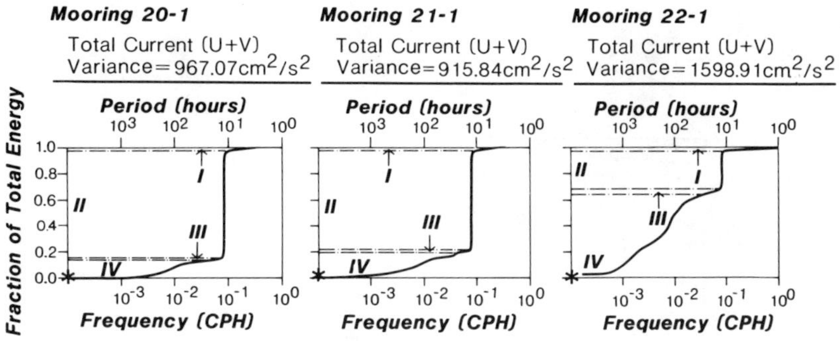

Fig. 1. Normalized energy spectra calculated from current meter data collected offshore of Savannah, showing the fraction of kinetic energy due to (I) shallow water tidal harmonics, seiches, internal waves, turbulence; (II) semidiurnal tides; (III) diurnal tides and inertial currents; and (IV) atmospheric and ocean-forced motions. Spectra from left to right are for currents 17 m below the surface at the 28-, 40-, and 75-m isobaths, respectively.

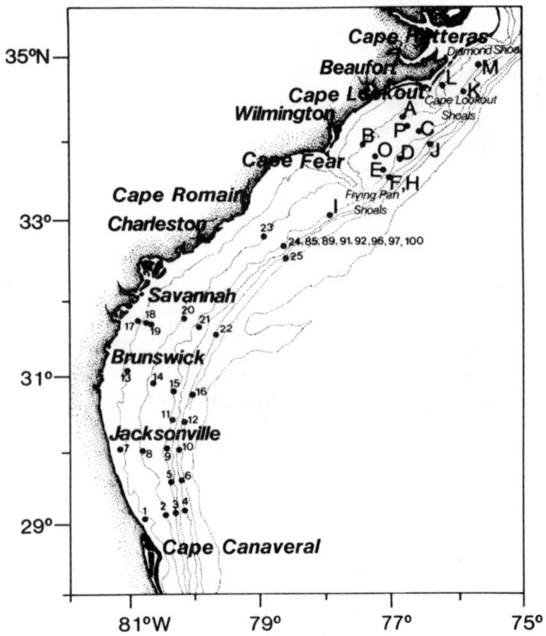

Fig. 2. The South Atlantic Bight (SAB) study area, showing current meter mooring locations for the period 1975–1981. The region from Cape Canaveral to Cape Romain is referred to as the Georgia Bight, while the region from Cape Romain to Cape Hatteras is referred to as the Carolina Capes.

the shelf is widest. In general, high-energy values occupy an alongshore midshelf region extending from Charleston southward almost to Cape Canaveral, with energy decreasing both seaward and shoreward. Minimum energy values occur at both the northern and southern boundaries of the

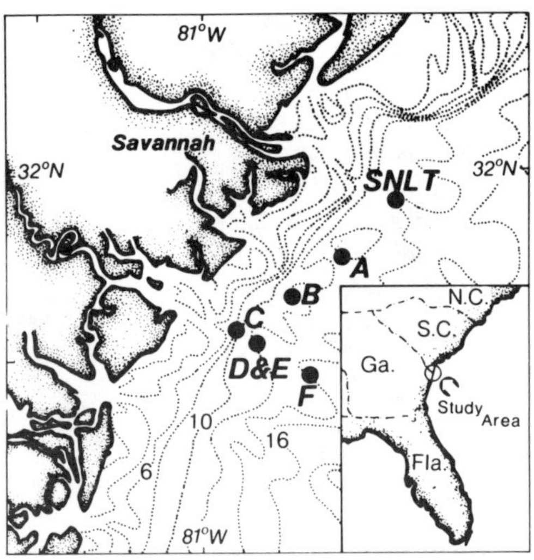

Fig. 3. Locations of coastal tidal stations. SNLT is the Savannah Navigational Light Tower.

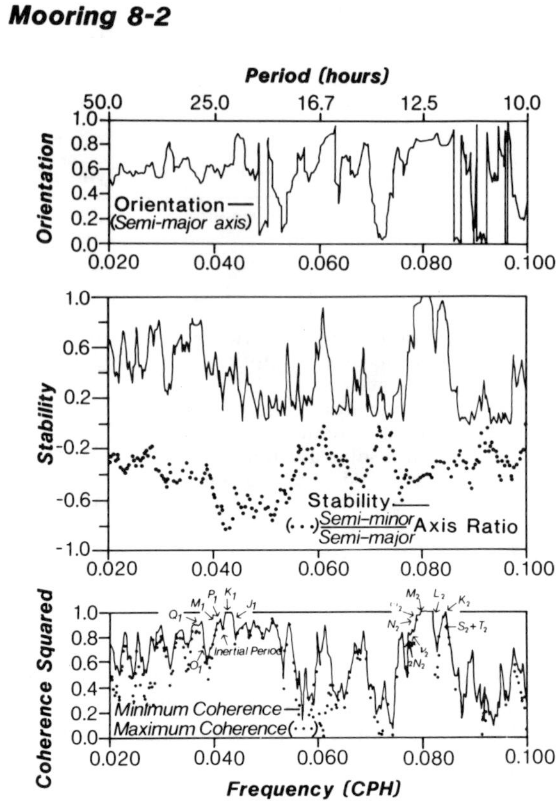

Fig. 4. Characteristics of tidal ellipses at mooring (instrument) 8 (2).

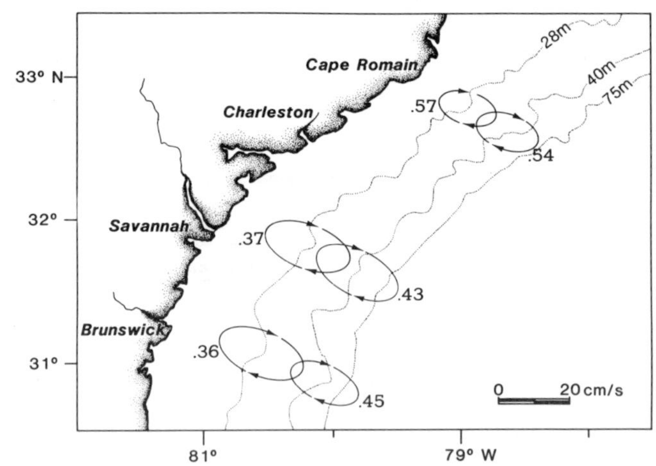

Fig. 5. Principal lunar semidiurnal tidal harmonic component (M_2) velocity ellipses along three representative transects in SAB. Minor to major ellipse ratios v/u are given. Data are taken 17 m below the surface on the 28- and 40-m isobaths. Ellipse arrows indicate direction of particle rotation.

TABLE 2. M_2 Tidal Ellipse Characteristics at the 40-m Mooring Offshore Charleston From November 1977 to October 1979

Mooring-Instrument*	M_2 Tidal Ellipse Orientation†	M_2 Tidal Ellipse Stability	M_2 Tidal Current Amplitude, cm/s	$v/u \times 100$ Ellipse Axis Ratio	Minimum (Coherence)^2	Maximum (Coherence)^2	Time Period
85-1	0.7238	0.93	6.12	34.6	0.90	0.96	Nov. 1977–Jan. 1978
85-2	0.7463	0.91	6.00	32.4	0.86	0.94	
85-3	0.7542	0.93	4.50	24.2	0.78	0.94	
89-1	0.7058	0.90	6.54	50.2	0.95	0.96	Jan. 1978–April 1978
89-2	0.7212	0.93	6.34	50.7	0.96	0.97	
89-3	0.7356	0.99	4.39	34.7	0.98	0.99	
91-1	0.7569	0.93	4.88	36.4	0.91	0.96	April 1978–Aug. 1978
91-3	0.7099	0.99	5.09	37.3	0.99	1.00	
92-1	0.7914	0.87	4.63	43.1	0.90	0.94	Aug. 1978–Nov. 1978
92-2	0.7710	0.96	5.21	40.0	0.96	0.98	
92-3	0.7421	0.92	4.47	35.7	0.90	0.95	
96-1	0.7409	0.90	5.73	42.2	0.92	0.95	Nov. 1978–March 1979
96-2	0.7343	0.91	5.61	44.5	0.93	0.96	
96-3	0.7468	0.99	3.85	27.3	0.97	0.99	
97-1	0.7356	0.91	5.49	39.2	0.91	0.95	March 1979–May 1979
97-2	0.7404	0.88	5.69	32.7	0.82	0.92	
97-3	0.7456	0.95	4.10	29.4	0.90	0.97	
100-1	0.7657	0.97	4.90	34.9	0.96	0.98	May 1979–Oct. 1979
100-2	0.7508	0.98	5.31	33.1	0.96	0.99	
100-3	0.7193	0.98	4.44	37.3	0.98	0.99	

Instrument levels 1, 2, and 3 are 10 m below the surface, 22 m below the surface, and 3 m above the bottom, respectively.
*Refer to Figure 2 for mooring location.
†Orientation measured in fractions of π: 0.0 is east, 0.5 is north, and 1.0 is west.

SAB. These energy calculations support *Redfield's* [1958] findings that the semidiurnal tidal currents along the U.S. east coast are strongest on the wide shelf off Savannah.

Figure 4 illustrates some of the characteristics of tidal ellipses at mooring (instrument) 8 (2), which is typical of the SAB. The orientation of the major axis was calculated according to *Fofonoff* [1969] in fractions of π radians, where 0.0 is east, 0.5 is north, and 1.0 is west. Ellipse stability is a measure of the isotropy of the wave field, with low-stability values indicating no predominant directionality of the wave field. Highest-stability values occur in the SAB at semidiurnal frequencies. The ratio v/u of minor to major axis is a measure of the eccentricity of the ellipse, where 0 indicates rectilinear motion and 1 indicates circular motion. Negative values indicate counterclockwise rotation. The bottom panel shows minimum and maximum rotary coherences. Simultaneously high minimum and maximum coherences denote coherent elliptical motion. High maximum and low minimum coherences correspond to coherent rectilinear motion, while simultaneously low minimum and maximum coherences indicate incoherent motion.

Tables 2, 3, and 4 list M_2 tidal ellipse characteristics for various regions of the SAB. These ellipses are relatively flat, with minor to major axis ratios of approximately 0.4, oriented with major axes perpendicular to the coast in alignment with shelf topography, as shown in Figure 5. At a particular location there is usually a turning of the ellipse orientation from surface to bottom. Table 2 indicates that at the 40-m mooring off Cape Romain ellipse orientation rotates counterclockwise during the fall, winter, and spring and clockwise during the summer. The seasonal patterns of ellipse orientation rotation seen off Cape Romain are not evident off Florida and Georgia. Instead, as shown in Table 3, there is a top to bottom clockwise rotation during the spring, opposite to that found off Charleston during the same period in previous years. The significance of these small changes in ellipse orientation with depth is not certain, although the persistence of the seasonal patterns suggests a real phenomenon.

During winter and spring the tidal ellipses become more rectilinear, with the ratio v/u decreasing with increasing depth, as shown in Table 3. Figure 6 is a schematic illustration of this seasonal variation of ellipse orientation. M_2 ellipses are more rectilinear in the winter and spring than in the summer and fall. Also, during winter and spring, near-bottom ellipses are more rectilinear than are near-surface ellipses. Summer and fall ellipses are of about equal size at both surface and bottom. Absolute values of v/u ratios listed in Tables 3 and 4 are generally less then f/ω, indicating the influence of bottom friction on these tidal ellipses [*Battisti and Clarke,* 1982]. During summer the ratios are more nearly equal, in accord with Poincaré wave theory [*Platzman,* 1971; *Battisti and Clarke,* 1982], suggesting that bottom friction is not strongly felt. SAB inner shelf waters are highly stratified in the summer [*Atkinson et al.,* 1983; *Stefansson et al.,* 1971] so that bottom frictional effects are effectively confined to near bottom. Thus during summer the M_2 tide behaves like a pure Poincaré wave throughout the water column, while during the winter it behaves like a frictionally modified Poincaré wave.

TABLE 3. M_2 Tidal Ellipse Characteristics for April 1980

Mooring* (Instrument)	M_2 Tidal Ellipse Orientation	M_2 Tidal Ellipse Stability	M_2 Tidal Current Amplitude, cm/s	$v/u \times 100$ Ellipse Axis Ratio	Ratio of f/ω	Minimum (Coherence)^2	Maximum (Coherence)^2
1 (2)	0.8402	0.98	3.06	12.1	0.502	0.79	0.98
2 (1)	0.9501	0.97	7.24	40.9	0.504	0.97	0.99
2 (2)	0.9094	0.98	5.05	35.9	0.504	0.97	0.99
2T (2)	0.9104	0.98	4.21	39.4	0.504	0.98	0.99
3 (1)	0.9177	0.99	5.45	43.0	0.505	0.99	0.99
3 (2)†	0.8997	0.98	2.64	29.2	0.505	0.94	0.98
4 (1)	0.5559	0.72	7.49	62.8	0.505	0.94	0.95
4 (2)	0.8503	0.71	3.02	44.7	0.505	0.81	0.87
4 (3)	0.8491	0.93	1.26	11.4	0.505	0.44	0.94
5 (2)	0.8857	0.99	5.54	41.4	0.511	0.99	0.99
6 (1)	0.9470	0.25	6.13	44.3	0.512	0.62	0.66
6 (2)	0.9078	0.97	3.20	27.5	0.512	0.91	0.97
7 (1)	0.7953	0.99	7.38	27.8	0.518	0.98	0.99
8 (2)	0.8401	1.00	9.48	40.0	0.518	0.99	1.00
8 (4)	0.8281	1.00	4.66	34.8	0.518	1.00	1.00
9 (1)	0.8910	0.98	9.89	36.0	0.518	0.97	0.99
9 (2)	0.9328	0.99	9.93	45.7	0.518	0.99	1.00
9 (3)	0.8856	0.99	10.43	44.1	0.518	0.99	1.00
9 (4)	0.8636	0.99	7.69	36.6	0.518	0.99	1.00
10 (1)	0.5155	0.64	7.69	76.7	0.518	0.97	0.98
10 (3)	0.1956	0.49	6.87	75.9	0.518	0.96	0.96
10 (4)	0.1193	0.95	6.40	56.2	0.518	0.98	0.99
10 (5)	0.9585	0.89	6.34	45.1	0.518	0.92	0.95
10 (6)	0.8843	0.98	5.57	30.9	0.518	0.96	0.99
11 (1)†	0.9511	0.99	11.56	49.8	0.524	1.00	1.00
11 (2)†	0.9092	1.00	8.35	39.0	0.524	0.99	1.00
12 (2)	0.9226	0.98	6.47	37.3	0.524	0.97	0.99
14 (2)	0.8257	0.99	9.59	32.5	0.532	0.98	0.99
15 (2)	0.9081	0.99	7.16	34.3	0.531	0.98	0.99
16 (2)	0.9945	0.87	5.47	11.5	0.529	0.30	0.88
20 (1)	0.8295	0.99	13.34	36.7	0.545	0.99	1.00
20 (3)	0.8150	0.99	9.16	33.4	0.545	0.98	0.99
21 (1)	0.9618	1.00	11.92	42.6	0.544	1.00	1.00
21 (2)	0.9374	1.00	9.12	32.7	0.544	1.00	1.00
22 (1)	0.9262	0.92	9.31	61.7	0.542	0.98	0.98
22 (2)	0.9487	0.97	7.93	51.7	0.542	0.99	0.99
22 (3)	0.9361	0.99	4.43	29.8	0.542	0.98	0.99
23 (1)	0.7021	0.99	8.13	57.0	0.561	1.00	1.00
23 (2)	0.7065	0.98	6.13	50.8	0.561	0.99	0.99
24 (1)	0.7451	1.00	7.93	54.3	0.558	1.00	1.00
24 (2)	0.7006	0.97	5.42	42.0	0.558	0.97	0.99
25 (1)	0.7891	0.95	3.77	28.9	0.557	0.88	0.96
25 (2)	0.8202	0.93	2.62	24.0	0.557	0.78	0.94
25 (3)	0.6346	0.92	2.70	0.9	0.557	0.00	0.92

*Refer to Figure 2 for mooring locations.
†Data are of poor quality with some data drop.

Phase propagation of the M_2 tide is summarized in Table 5. Tidal currents on the outer shelf lead those on the inner shelf by 6–7 hours. Offshore of St. Augustine there is an onshore tidal current propagation of about 9 cm/s in the midshelf region and about 13 cm/s in the inner shelf. This shoreward tidal propagation supports the theory that the SAB shelf tide is a cooscillation of the North Atlantic Ocean tide. Bottom frictional forces diminish with distance above the bottom, so near-bottom tidal currents will peak sooner than near-surface currents. The apparent upward phase speed propagation was approximately 0.8 cm/s at mooring 8 and 0.5 cm/s at mooring 10. This results in a lag at mooring 8 of about 17 min between the tidal current at 25 m below the surface and that at 17 m below the surface.

Characteristics of nearshore tidal currents at the stations indicated in Figure 3 are summarized in Table 6. The baroclinic nature of the inner shelf off Georgia [*Blanton*, 1981] complicates the probable effects of friction on the structure of the horizontal tidal currents [*Magnell et al.*, 1980]. Quantitative differences exist over relatively short distances on the inner shelf probably because of irregular bottom topography and the proximity of tidal inlets [*Blanton*, 1980]. Tidal ellipses become narrower and smaller as the bottom is approached, and there is a phase lead of near-

TABLE 4. M_2 Tidal Ellipse Characteristics in Onslow Bay for July–September 1976

Mooring (Instrument)	M_2 Tidal Ellipse Orientation	$v/u \times 100$ Ellipse Axis Ratio	f/ω Ratio
C(1)	0.7667	55.0	0.5804
C(2)	0.7389	43.0	0.5804
E(1)	0.7333	55.0	0.5735
F(2)	0.6000	26.0	0.5724
A(1)	0.7333	45.0	0.5833
B(1)	0.8611	44.0	0.5783

bottom over near-surface currents. The principal axis amplitudes usually increased with distance above the bottom, except at site C, which showed a slight decrease in amplitudes at distances greater than 6 m above bottom. Maximum flood currents occurred at near-bottom locations 1 to 2 hours before near-surface locations, in agreement with findings on the midshelf and outer shelf and at other locations, such as the North Siberian Shelf [*Sverdrup,* 1926; *Fjeldstad,* 1929, 1936], the English Channel [*Proudman,* 1953] Chesapeake Bay [*Cannon,* 1969], and Narragansett Bay [*Weisberg and Sturges,* 1976]. The magnitude of the phase lag is somewhat greater nearshore than that found on the midshelf and outer shelf, suggesting that the influence of bottom friction in shallow water extends upward throughout the water column [*Kundu et al.,* 1981].

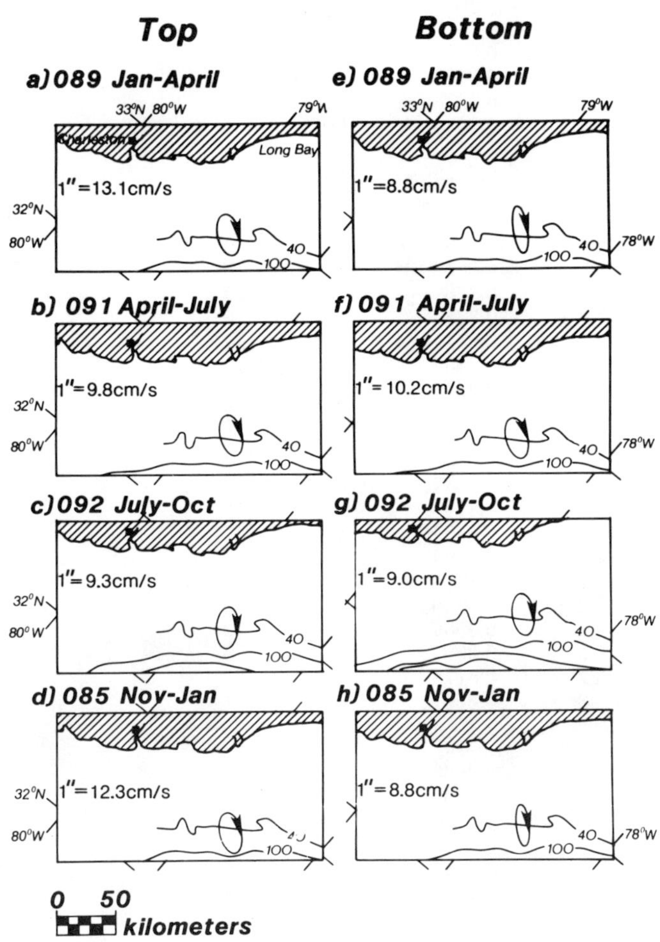

Fig. 6. Schematic representation of the seasonal characteristics of M_2 tidal ellipses in the near surface (Figures 6*a*–6*d*), 10 m below the surface, and near bottom (Figures 6*e*–6*h*), 3 m above bottom, at a mooring site off Cape Romain, South Carolina.

The shape of the tidal ellipse varied greatly from site to site. Tidal currents tended to rotate clockwise ($v/u > 0$), although counterclockwise rotation ($v/u < 0$) occurred near bottom at sites D and E. Lower values of v/u near bottom represent a trend to almost rectilinear tidal currents. Figure 7 summarizes differences in ellipse configuration on the inner shelf for near-surface and near-bottom tidal currents. Topographic features are thought to be primarily responsible for variations in ellipse orientation that occur over relatively short distances [*Blanton,* 1980].

Tidal currents adjacent to the coast apparently flood toward different locations. Near-bottom flood currents at SNLT and at site A are toward the Savannah River inlet, while those at site B are toward either Ossabaw or Wassaw Inlet. This suggests that the sounds and tidal creeks are momentum sinks with influence extending at least 15 km offshore. A particle advected during the ebb flow in a constant direction with maximum amplitude V travels a distance $L = VT/\pi$. With a tidal period $T = 12.42$ hours and $V = 0.35$ m/s the tidal excursion L is 5 km. Thus the zone of influence extends outward at least 3 times the tidal excursion for a particle advected from the inlet over one-

TABLE 5. Phase Propagation of M_2 Tidal Currents Between Current Meters on the Florida/Georgia Shelf

First Mooring No.* (Instrument Depth) Versus Second Mooring No. (Instrument Depth)	Tidal Current Vector Component	Phase Difference,† hours	Phase Propagation Speed, cm/s
7 (5 m) versus 8 (17 m)	u	−2.70	13.5
7 (5 m) versus 8 (17 m)	v	−2.75	13.3
8 (17 m) versus 9 (17 m)	u	−5.16	8.0
8 (17 m) versus 9 (17 m)	v	−4.64	8.9
8 (17 m) versus 8 (25 m)	u	−0.35	0.79
8 (17 m) versus 8 (25 m)	v	−0.34	0.82
10 (7 m) versus 10 (72 m)	u	−4.03	0.46
10 (7 m) versus 10 (72 m)	v	−3.39	0.55
20 (17 m) versus 21 (17 m)	u	−2.30	9.0

*Refer to Figure 2 for mooring locations.

†Negative hourly values imply first mooring instrument lags second mooring instrument.

TABLE 6. Characteristics of Nearshore M_2 Tidal Currents Based on Harmonic Analysis of Smoothed Velocity Profiles

Site	Depth Above Bottom	CMA, cm/s	CMI, cm/s	$v/u \times 100$	Phase, hours	Tidal Ellipse Orientation	RMAJ	RMIN
				May 1978				
A	0.5	16.8	1.2	7	2.0	0.5444	77	8
	1.5	23.6	2.6	11	2.0	0.5556	89	22
	2.5	22.4	6.3	28	2.3	0.5611	72	33
	3.5	29.1	3.2	11	2.5	0.5667	84	36
	4.5	31.4	4.4	14	2.6	0.5722	86	37
	5.5	31.6	6.7	21	2.6	0.5556	85	27
	6.5	31.4	7.9	25	2.6	0.5556	84	28
	7.5	32.8	8.9	27	2.8	0.5722	83	29
	8.5	35.2	9.1	28	3.2	0.5778	83	32
	9.5	38.2	7.8	20	2.5	0.5722	84	32
	10.5	38.7	8.7	23	3.7	0.5500	84	24
	11.5	38.2	11.6	30	3.8	0.5333	81	21
B	0.5	10.3	0.3	3	1.5	0.8333	55	44
	1.5	16.8	0.2	−1	2.1	0.8222	60	44
	2.5	20.2	0.6	3	2.0	0.8389	76	52
	3.5	22.1	0.1	1	1.9	0.8333	66	67
	4.5	25.2	0.8	−3	2.2	0.8389	80	71
	5.5	30.7	1.7	6	2.5	0.8611	89	61
	6.5	33.7	6.5	19	2.9	0.8667	88	58
	7.5	35.6	10.0	28	3.3	0.8667	86	59
	8.5	36.5	12.8	35	2.6	0.8833	87	63
	9.5	36.6	14.7	40	3.1	0.9278	87	60
	10.5	34.1	15.5	46	3.3	0.9333	80	54
SNLT*	2	18.8	4.4	24	2.3	0.6667	86	93
	10	37.9	13.7	36	3.3	0.9222	83	69
				November 1979				
C	1	17.7	0.6	3	2.9	0.8399	75	0
	2	21.6	2.7	12	3.3	0.8556	78	6
	3	23.3	3.8	16	3.4	0.8667	80	10
	4	24.9	4.8	19	3.4	0.8833	80	15
	5	26.5	4.7	18	3.5	0.9111	80	17
	6	27.0	4.1	15	3.5	0.9333	79	13
	7	26.7	3.5	13	3.6	0.9444	78	9
	8	25.3	3.3	13	3.6	0.9500	75	7
	9	22.3	3.2	15	3.7	0.9556	71	5
	10	21.7	2.9	13	3.7	0.9667	70	4
D	1	17.6	−3.1	−18	2.6	0.8667	81	17
	2	20.1	−1.6	−8	2.9	0.8778	80	4
	3	22.6	0.5	2	3.0	0.8667	85	1
	4	24.5	3.0	12	3.2	0.8667	86	25
	5	26.3	5.4	20	3.2	0.8611	88	50
	6	27.7	7.8	28	3.3	0.8556	89	63
	7	29.4	10.8	37	3.3	0.8556	90	72
	8	31.4	12.9	41	3.3	0.8667	91	76
	9	31.8	13.9	44	3.2	0.8778	89	74
	10	30.8	13.9	45	3.3	0.8667	86	71
				April 1981				
E	1	19.4	−2.2	−11.0	2.6	0.8500	89	20
	2	19.5	−2.1	−11.0	2.6	0.8500	89	19
	3	20.4	−1.5	−7.3	2.6	0.8500	89	13
	4	21.9	−0.6	−2.7	2.7	0.8500	89	8
	5	23.1	0.6	2.5	2.7	0.8444	88	8
	6	24.3	2.0	8.4	2.9	0.8444	88	24
	7	25.9	3.0	12.0	3.1	0.8389	88	23
	8	28.5	5.4	19.0	3.3	0.8500	91	37
	9	30.9	7.6	25.0	3.4	0.8667	92	40
	10	32.6	10.3	32.0	3.6	0.8889	91	52
	11	33.8	13.6	40.0	3.7	0.8944	89	64
	12	33.7	13.7	41.0	3.8	0.8944	87	61

TABLE 6. (continued)

Site	Depth Above Bottom	CMA, cm/s	CMI, cm/s	$v/u \times 100$	Phase, hours	Tidal Ellipse Orientation	RMAJ	RMIN
				April 1981 (continued)				
F	1	25.5	2.9	11.0	2.6	0.8333	91	18
	2	26.1	2.9	11.0	2.6	0.8333	91	19
	3	27.7	2.3	8.4	2.7	0.8222	90	16
	4	29.5	1.8	6.3	2.7	0.8222	91	10
	5	30.4	2.8	9.1	2.7	0.8222	91	14
	6	31.8	3.9	12.0	2.8	0.8222	91	21
	7	33.2	5.7	17.0	2.8	0.8222	91	43
	8	35.1	7.2	21.0	2.9	0.8222	91	50
	9	36.8	9.4	25.0	3.0	0.8278	92	58
	10	38.4	11.4	30.0	3.1	0.8278	92	66
	11	40.6	12.8	32.0	3.3	0.8389	92	74
	12	41.3	14.5	35.0	3.5	0.8611	93	79
	13	40.9	15.6	38.0	3.8	0.8611	88	69
	14	42.0	16.2	39.0	3.9	0.8611	86	62
	15	44.2	16.2	37.0	3.8	0.8778	89	63

CMI and CMA are amplitudes of the minor and major axes and v/u = CMI/CMA. Phases are measured relative to time of low or high water and designate the maximum tidal currents for the major axis components. The orientation angle of the ellipses is measured as a fraction of π (see Table 2). The factors RMAJ and RMIN denote the percent of total variance recovered by the harmonic analyses for the east and north components, respectively. Depths are measured relative from bottom. See Figure 3 for locations.

*SNLT (Savannah Navigational Light Tower) data were obtained at time periods different from that at sites A and B. Data 2 m above bottom represent the average of 12 spring tides in 1977. Data 10 m above bottom represent an 11-day record in May 1978.

half tidal cycle. The zones of influence of each inlet are probably related to the magnitude of the tidal volumes (tidal prisms) passing through each inlet.

Tidal Flow Model

A vertically integrated two-dimensional model was used to predict winter tidal currents on the SAB shelf. Model application was limited to winter nonstratified conditions so that the vertically integrated equations of motion could be used. Formulation of the model has been described in detail by *Wang and Conner* [1975] and *Wang* [1978]. The simplified continuity and horizontal momentum equations are

$$\frac{\partial \eta}{\partial t} + \frac{\partial (uH)}{\partial x} + \frac{\partial (vH)}{\partial y} = 0 \tag{8}$$

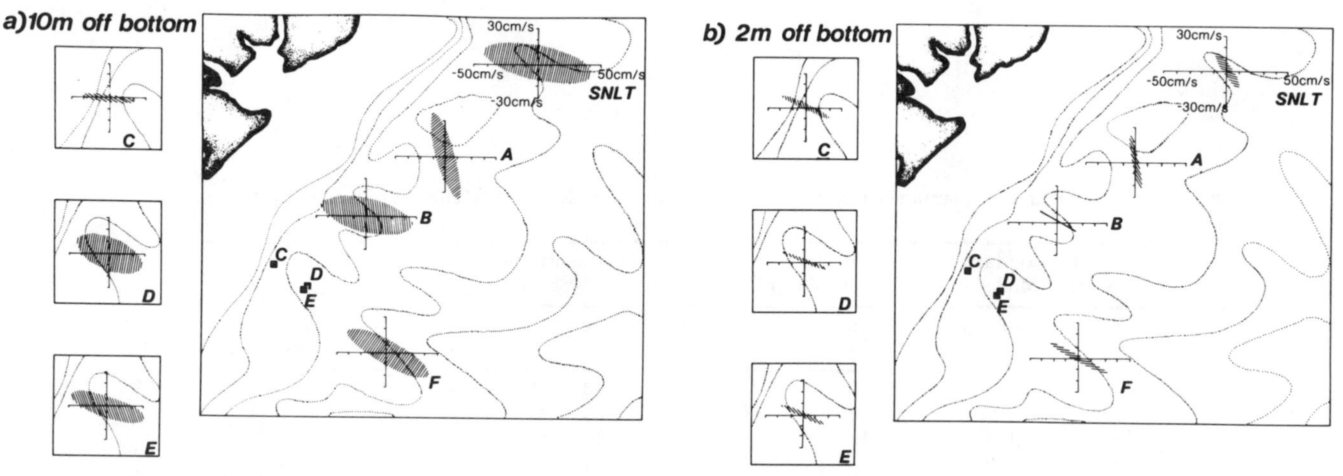

Fig. 7. Tidal ellipse configuration on the inner shelf for near-surface (top panel) and near-bottom (lower panel) tidal currents. See Figure 3 for station locations.

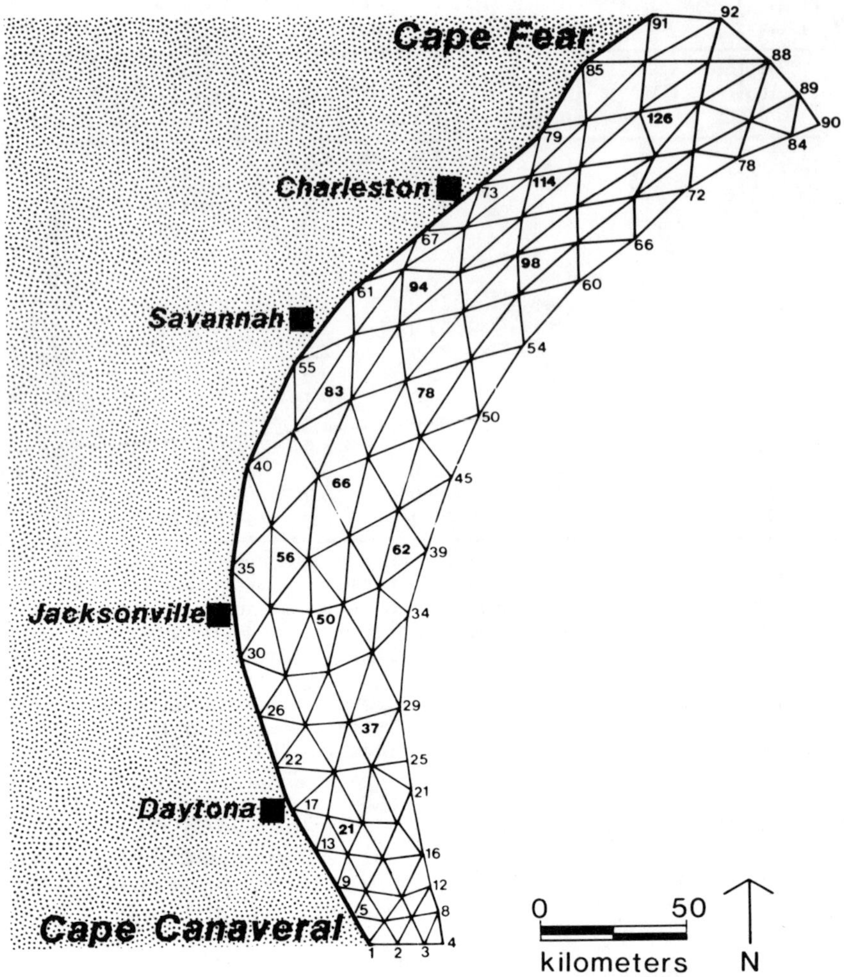

Fig. 8. Finite element grid of model domain. The SAB is divided into 144 linear triangular elements defined by 93 node points.

TABLE 7. Comparison of Predicted Tidal Amplitudes With Mean Amplitudes From Tide Tables

Ocean Boundary Global Model		Coastal Sea Level Global Model		Coastal Sea Level Shelf Model		
Node No.	Amplitude, m	Node No.	Amplitude, m	Node No.	Amplitude, m	Coastal Mean Tidal Amplitude (Tide Tables)
16	0.55	13	0.58	13	0.57	0.62
29	0.60	26	0.70	26	0.71	0.79
45	0.64	40	0.88	40	0.90	1.10
54	0.64	67	0.80	67	0.81	0.88
78	0.56	91	0.72	91	0.73	0.79
90	0.52	92	0.66	92	0.65	0.68

$$\frac{\partial u}{\partial t} + \frac{\partial u^2}{\partial x} + \frac{\partial uv}{\partial y} = -g\frac{\partial \eta}{\partial x} - \frac{1}{\rho H}\tau_{xb} + fv \quad (9)$$

$$\frac{\partial v}{\partial t} + \frac{\partial uv}{\partial x} + \frac{\partial v^2}{\partial y} = -g\frac{\partial \eta}{\partial y} - \frac{1}{\rho H}\tau_{yb} - fu \quad (10)$$

where x and y are horizontal coordinates, t is time, $H = h + \eta$ is total water depth at mean sea level (msl), η is the deviation of the water surface from msl, u,v are the vertically averaged velocity components in the x and y directions, respectively, g is gravitational acceleration, τ_{xb} and τ_{yb} are bottom stress in the x and y directions, respectively, and f is the Coriolis parameter. Details of model adaptation to the SAB are given by *Kourafalou et al.* [1983] and *Wang et al.* [1984]. A numerical finite element method in space was coupled with a simple finite difference scheme in time to integrate the equations. The model is predictive in that, apart from boundary conditions, only bottom friction coefficients need be prescribed. In the application of the model to tidal flow conditions, results of a 1° × 1° global model [*Schwiderski*, 1980] were used to prescribe the boundary conditions for surface elevations.

Tides vary considerably in the SAB because of the varying shelf width. The maximum coastal tidal range of 2.2 m occurs at Savannah, where the shelf is widest, and decreases to 1.3 m at Cape Fear and 1.1 m at Cape Canaveral. The M_2 component dominates sea level and current tidal fluctuations throughout the region, accounting for roughly 80% of the tidal variation. To model the tidal currents, the SAB was divided into 144 triangular elements as shown in Figure 8. The 75-m isobath, which defines the shelf break, was taken as the open boundary of the model. Typical node separation in the grid was approximately 40 km, with 93 nodes. Since natural depth variations are relatively smooth, this grid provided sufficient resolution to accurately represent both the shoreline and the bathymetry. Mean sea level along the open boundaries was assumed to be level.

The model was spun up with surface elevations and velocities initialized to zero everywhere and allowed to reach a periodic state, which was achieved in less than one tidal period. The run was then continued to another tidal period so that results would not be influenced by the arbitrary initial conditions. A comparison of predicted coastal tidal amplitudes with the data analyzed by *Schwiderski* [1980], given in Table 7, shows good agreement with differences of only a few centimeters. With careful adjustment of ocean boundary conditions a better fit to the data could be obtained. To verify model currents, the tide on an arbitrary day of the current observations (March 23, 1980) was simulated. The chosen tide resulted in a tidal range of 2 m at Savannah, and the model boundary conditions were scaled to reflect this range. The scaling factor was the ratio between the predicted total tide (2 m) and the M_2 tide (1.76 m) at Savannah. This factor is essentially constant along the coast of the SAB. Plots of modeled velocity fields at hourly intervals of the tidal period are shown in Figures 9*a*, 9*b*, and 9*c*.

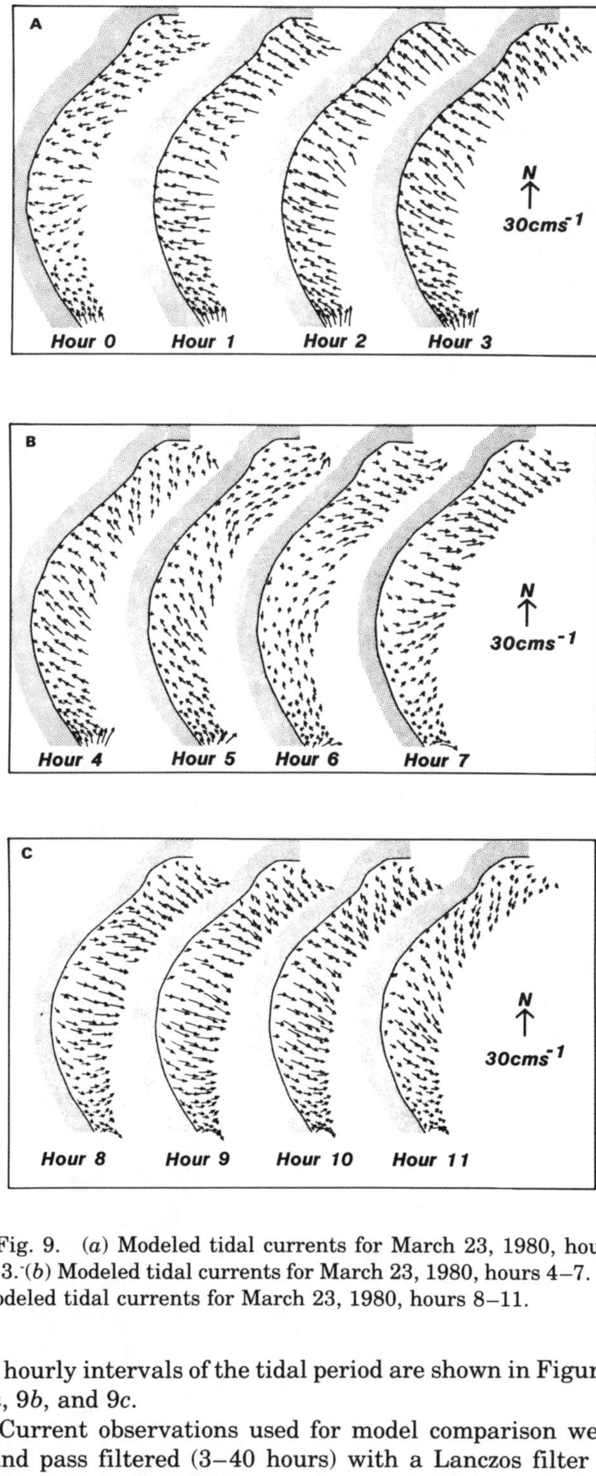

Fig. 9. (*a*) Modeled tidal currents for March 23, 1980, hours 0–3. (*b*) Modeled tidal currents for March 23, 1980, hours 4–7. (*c*) Modeled tidal currents for March 23, 1980, hours 8–11.

Current observations used for model comparison were band pass filtered (3–40 hours) with a Lanczos filter to isolate tidal effects and were then vertically averaged. A detailed hour by hour comparison of model versus measured data is presented in Figures 10*a*, 10*b*, and 10*c*. Excellent agreement was found in both magnitude and direction at the 28- and 40-m isobaths, with magnitudes within

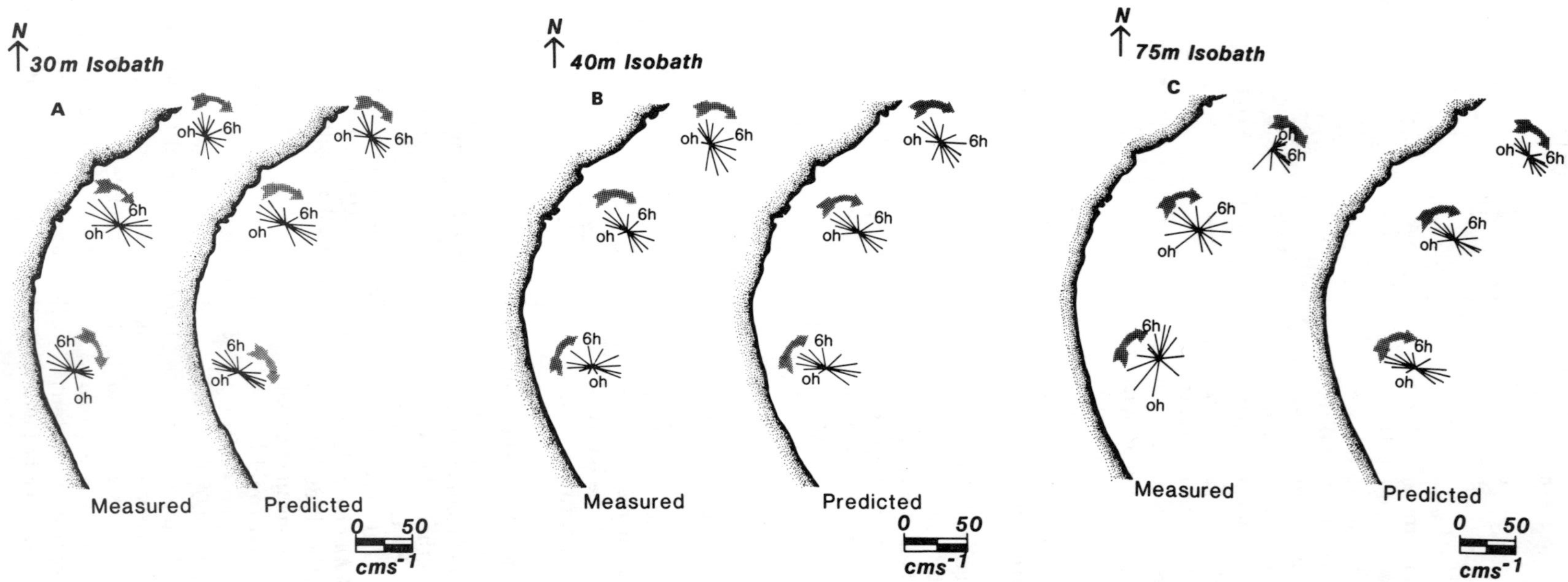

Fig. 10. (*a*) Comparison of measured and predicted tidal ellipses at the 30-m isobath for current meter moorings 8, 20, and 23. (*b*) Comparison of measured and predicted tidal ellipses at the 40-m isobath for current meter moorings 9, 21, and 24. (*c*) Comparison of measured and predicted tidal ellipses at the 75-m isobath for current meter moorings 10, 22, and 25.

±10% and directions within ±20° of measured values [*Wang et al.,* 1984]. However, agreement was poor at the 75-m isobath, probably because of Gulf Stream effects in the data, and perhaps because of internal tides caused by stratification at the shelf break. Such strong stratification is not generally found on the inner shelf and midshelf.

Summary

Tidal currents contribute a major portion of the total kinetic energy of ocean currents on the SAB shelf, particularly near the coast. Investigations of the SAB tidal regime show that tidal currents are most coherent at the M_2 frequency. M_2 tides dominate the cross-shelf variance on the inner shelf and midshelf but decrease in magnitude on the outer shelf. In the alongshelf direction, M_2 tides account for about 50% of the midshelf variance and about 10–20% of the outer shelf variance. Friction clearly influences the character of the tidal ellipses.

While there appear to be some seasonal differences in the character of tidal currents, these differences are subtle. Stratification, which follows a seasonal pattern in many areas of the SAB, may modify the frictional environment and account, to some extent, for the observed seasonal differences.

A relatively simple vertically integrated two-dimensional tidal model has been used to reproduce important characteristics of tidal ellipses in the midshelf region during well-mixed winter conditions. The model was less successful in predicting outer shelf characteristics.

Acknowledgments. This work was carried out under Department of Energy contract AS09-F6EY00902; the University of North Carolina Sea Grant, NOAA, Department of Commerce contract 04-6-158-44-54; the State of North Carolina, Department of Administration; and Minerals Management Service (formerly Bureau of Land Management) contract 209-04.

References

Airy, G. B., *Tides and Waves,* Encyclopedia Metropol, London, 1842.

Atkinson, L. P., T. N. Lee, J. O. Blanton, and W. S. Chandler, Climatology of the southeastern United States continental shelf waters, *J. Geophys. Res., 88,* 4705–4718, 1983.

Battisti, D. S., and A. J. Clarke, A simple method for estimating barotropic tidal currents on continental margins with specific application to the M_2 tide off the Atlantic and Pacific coasts of the United States, *J. Phys. Oceanogr., 12,* 8–16, 1982.

Blanton, J. O., Transport of freshwater off a multi-inlet coastline, in *Modeling of Estuaries and Wetland Processes,* edited by P. Hamilton and K. B. MacDonald, pp. 49–64, Plenum, New York, 1980.

Blanton, J. O., Ocean currents along a nearshore frontal zone on the continental shelf of the southeastern United States, *J. Phys. Oceanogr., 11,* 1627–1637, 1981.

Cannon, G. A., Observations of motion at intermediate plain estuary, *Tech. Rep. 52,* 114 pp., Chesapeake Bay Institute, Shadyside, Md., 1969.

Clarke, A. J., and D. S. Battisti, The effect of continental shelves on tides, *Deep Sea Res., 28,* 665–682, 1981.

Fjeldstad, J. E., Dynamics of free progressive tidal waves, *Sci. Results Norw. North Polar Exped. Maud 1918–1925, 4*(3), 1929.

Fjeldstad, J. E., Results of tidal observations, *Sci. Results Norw. North Polar Exped. Maud 1918–1925, 4*(4), 1936.

Fleming, R. H., Tides and tidal currents in the Gulf of Panama, *J. Mar. Res., 1,* 192–206, 1938.

Fofonoff, N. P., Spectral characteristics of internal waves in the ocean, *Deep Sea Res., 16,* suppl., 59–71, 1969.

Haight, F. J., Coastal currents along the Atlantic coast of the United States, *Spec. Publ. 230,* 73 pp., U. S. Dep. of Commer., Coast and Geod. Surv., Rockville, Md., 1942.

Kourafalou, V., J. D. Wang, T. N. Lee, and L. J. Pietrafesa, Modeling of winter circulation on the southeast U.S. continental shelf, *Tech. Rep. 83009,* 97 pp., Univ. of Miami, Miami, Fla., 1983.

Kundu, P. K., J. O. Blanton, and M. M. Janopaul, Analysis of current observations on the Georgia shelf, *J. Phys. Oceanogr., 11,* 1139–1149, 1981.

Laplace, P. S., Recherches sur plusieurs points du système du monde, *Mem. Acad. R. Sci. Paris, 75,* 1775.

Lee, T. N., and D. Brooks, Initial observations of current, temperature, and coastal sea level response to atmospheric and Gulf Stream forcing, *Geophys. Res. Lett., 6,* 321–324, 1979.

Magnell, B. A., S. L. Spiegel, R. I. Scarlet, and J. B. Andrews, The relationship of tidal and low-frequency currents on the north slope of Georges Bank, *J. Phys. Oceanogr., 10,* 1200–1212, 1980.

Mofjeld, H. O., Tidal currents, in *Marine Sediment Transport and Environmental Management,* edited by D. J. Stanley and D. J. P. Swift, pp. 53–64, John Wiley, New York, 1976.

Newton, I., *Philosophiae Naturalis Principia Mathematica,* lib, III, prop. 24, 36, 37, London, 1687.

Pietrafesa, L. J., Shelfbreak circulation, fronts and physical oceanography: East and west coast perspectives, in *The Shelfbreak: Critical Interface on Continental Margins, Spec. Publ. 33,* edited by D. J. Stanley and G. J. Moore, pp. 233–250, Society of Economic Paleontologists and Mineralogists, Tulsa, Okla., 1983.

Platzman, G. W., Ocean tides and related waves, in *Mathematical Problems in the Geophysical Sciences, 2, Inverse Problems, Dynamo Theory and Tides, Lect. in Appl. Math.,* vol. 14, pp. 239–291, American Mathematical Society, Providence, R. I., 1971.

Proudman, J., *Dynamical Oceanography,* 409 pp., Methuen, London, 1953.

Redfield, A. C., The influence of the continental shelf on

the tides of the Atlantic coast of the United States., *J. Mar. Res., 17,* 432–448, 1958.
Redfield, A. C., *Introduction to Tides: The Tides of the Waters of New England and New York,* 108 pp., Marine Science International, Woods Hole, Mass., 1980.
Schwiderski, E. W., Ocean tides, 2, A hydrodynamical interpolation model, *Mar. Geod., 3,* 219–255, 1980.
Stefansson, V., L. P. Atkinson, and D. F. Bumpus, Hydrographic properties and circulation of the North Carolina shelf and slope waters, *Deep Sea Res., 18,* 383–420, 1971.
Sterneck, R., Hydrodynamische Theorie der halbtagigen Gezeiten des Mittelmeeres, *Sitzungsber. Kais. Akad. Wiss. Wien Math. Naturwiss. Kl., 124,* 905 pp., 1915.
Sverdrup, H. U., Dynamics of tides on the North Siberian Shelf, *Geophys. Publ. 4,* 77 pp., Norske Videns. Akad., Oslo, 1926.
Tebeau, P. A., and T. N. Lee, Wind induced circulation on the Georgia shelf, *Tech. Rep. 79003,* 177 pp., Univ. of Miami, Miami, Fla., 1979.
Wang, J. D., Real-time flow in unstratified shallow water, *J. Waterway Port Coastal Ocean Div. Am. Soc. Civ. Eng., WW1,* 53–68, 1978.
Wang, J. D., and J. J. Connor, Mathematical modeling of near coastal circulation, *Tech. Rep. 200,* 272 pp., R. M. Parsons Lab., Dep. of Civ. Eng. Mass. Inst. of Technol., Cambridge, Mass., 1975.
Wang, J. D., V. Kourafalou, and T. N. Lee, Circulation on the continental shelf of the southeastern United States, 2, Model development and application to tidal flow, *J. Phys. Oceanogr., 14,* 1013–1021, 1984.
Weisberg, R. H., and W. Sturges, Velocity observations in the west passage of Narragansett Bay: A partially mixed estuary, *J. Phys. Oceanogr., 6,* 345–354, 1976.

Hydrography and Nutrients of the Southeastern U.S. Continental Shelf

L. P. Atkinson

Skidaway Institute of Oceanography, Savannah, Georgia

The distribution of properties such as temperature and salinity, of nutrients such as nitrate, phosphate, and silicate, and of derived properties such as buoyancy in the South Atlantic Bight depends on complex interactions of shelf waters with rivers, coastal estuaries and sounds, atmospheric forcing, and, most importantly, the Gulf Stream. Seasonal variations in temperature depend on the seasonal heating cycle, air-sea exchange processes, and Gulf Stream interaction. The result is extreme variations in temperature both seasonally and spatially. Salinity variations are mainly related to seasonal river flow, which peaks in the spring, and subsequent shelf circulation processes. Variations in temperature and salinity occur over shorter time scales because of Gulf Stream meanders and frontal eddies that cause intrusions of surface or subsurface water into the shelf waters. Intrusion frequency and cross-shelf excursion vary seasonally and latitudinally. The supply of nutrients mainly depends on Gulf Stream intrusions. Gulf Stream water colder than about 20°C contains significant amounts of nutrients. Such waters invade the shelf more frequently during the summer in the southern bight. Chemical tracers such as phenolic aldehydes, humic compound fluorescence, and tritium have been used successfully to trace coastal waters.

Introduction

The distribution of conservative and nonconservative variables such as temperature, salinity, and of properties such as buoyancy in the South Atlantic Bight (SAB) depends principally on river runoff, interaction of the Gulf Stream with shelf water, and atmospheric forcing. The characteristic periods over which significant changes can occur vary from hours to years. A general description of the distributions of temperature, salinity, stability, and nutrients is presented here. The major point discussed is that nutrient inputs to, and the distribution of other variables in, the SAB are controlled primarily by interactions between the Gulf Stream and shelf waters.

Seasonal Variations

The SAB extends over more than 6° of latitude (29°–36°N). The southern region is near tropical, while the northern is more temperate in climate. Air temperatures have large annual variations (Figure 1). Average surface water temperatures (Figure 2) follow air temperatures quite well in inner shelf waters, but those on the outer shelf are moderated by the Gulf Stream. Minimum temperatures in the inner shelf occur in February. Water in the central SAB at the same time is nearly as cold as inner shelf water farther north because the central SAB shelf is wider and Gulf Stream—induced moderation is less pronounced. Also, atmospheric cooling has more effect on water temperatures over the wide, shallow central SAB water. Cross-shelf thermal gradients are high in winter and low in summer. Summer surface temperatures are rather uniform near 28°C, excluding the north Florida outer shelf where subsurface intrusions occur more frequently. Another common feature is outer shelf minimum temperatures at 32° to 33°N in February through May. This is related to circulation around a topographic irregularity, the Charleston Bump, which causes the Gulf Stream to veer offshore [*Pashinski and Maul,* 1973; *Legeckis,* 1979; *Brooks and Bane,* 1978].

Average monthly surface salinities (Figure 3) are influenced by variations in runoff and the effect of shelf circulation. High runoff in spring causes lowered inner shelf salinities and higher cross-shelf salinity gradients. In summer runoff is low, salinities increase, and cross-shelf gradients are small. However, during fall, when runoff is low, inner shelf salinities are often reduced, especially off north Florida and Georgia. We hypothesize that the cause is high southward wind stress that restricts cross-shelf flow and advects coastal water southward [*Blanton and Atkinson,* 1983].

Variations in subsurface continental shelf waters are affected by interactions with the deeper ocean more than by runoff or atmospheric processes. Monthly mean near-bottom temperatures (Figure 4) reflect atmospheric interaction with inner shelf water, with minimum temperatures over the inner shelf of the central and upper SAB. Temperatures in middle and outer shelf waters reflect the effect of the Gulf Stream, which causes outer shelf water to be

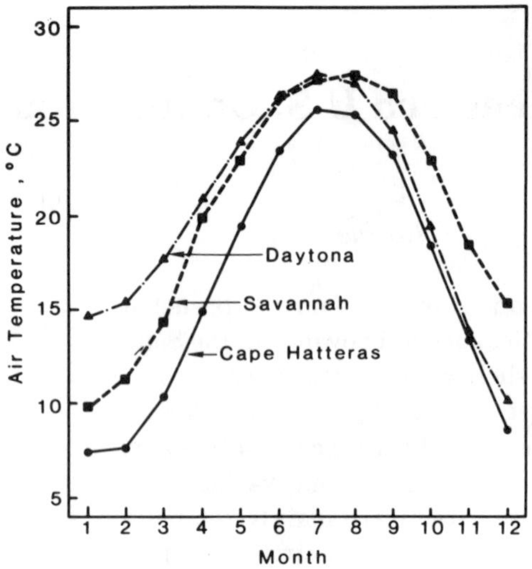

Fig. 1. Mean monthly air temperature at various coastal stations.

warmer in winter and colder in summer relative to the middle and inner shelf. This effect is especially dramatic off north Florida where cold water upwells in summer, reducing outer shelf bottom temperatures to <18°C. Bottom temperature distributions are anomalous between 32° and 33°N, as is surface salinity, again showing the effect of Gulf Stream circulation in the area of the Charleston Bump.

Vertical water column stability varies horizontally (Figure 5). In February strong winds create a well-mixed water column. By May inner shelf waters are stratified by surface heating, runoff, and decreased wind mixing. In August stratification is high in all shelf waters. Off north Florida, where summer upwelling occurs, very high stability is caused by warm surface waters overlaying cold upwelled water. By November strong southward winds and reduced surface heating reduce water column stability to typical winter conditions.

Water Column Characteristics

Gulf Stream Water Over the Slope

Waters in the SAB range from well-defined Gulf Stream water to highly variable and modified shelf waters. Temperature and salinity are normally used to identify water masses, but Gulf Stream water is particularly well defined by oxygen and nutrients (summarized by *Wennekens* [1959], *Richards and Redfield* [1955], and *Morrison and Nowlin,* [1982]). Figure 6 shows a typical section across the Gulf Stream at 29°N. Density profiles indicate that the Gulf Stream proper was located between stations 107 and 115, which includes the shelf break, slope, and western Blake Plateau. Within this region, isotherms rise steeply to the west. The 18°C isotherm, for example, rises from 400 m over the Blake Plateau to 100 m where it intersects the shelf break. Vertical temperature gradients increase from deep water to the slope and shelf break; thus vertical stability is higher over the shelf break than farther offshore. The salinity structure is dominated by Subtropical Underwater that appears as a salinity maximum at about 200 m over the Blake Plateau. Over the upper slope the isohalines rise in the Gulf Stream Front. Various salinity maxima represent disassociated masses of Subtropical Underwater. Oxygen generally decreases with decreasing temperature, reaching a mid-depth minimum of 3.0 ml l^{-1} in Antarctic Intermediate Water, seen near bottom at station 111. Nutrient concentrations follow the distribution of apparent oxygen utilization (AOU) and upward sloping isopycnals. Over the Blake Plateau, concentrations of biologically important nutrients are high below depths of 200 m, but over the slope and outer shelf significant concentrations can be found above 100 m and in the euphotic zone (~40 m).

Plots of several key variables in Gulf Stream waters are useful aids to analyze characteristics of both slope and shelf waters (Figure 7). The temperature/salinity (T/S) relationship shown is classic, showing deeper Western North Atlantic Water, Subtropical Underwater, Surface Water, and the altered Continental Edge Water. The core of Subtropical Underwater is at the subsurface salinity maximum, coinciding with approximately 22°C or about 25.2 σ_t units. Maximum salinities in the Gulf Stream between Florida and North Carolina are typically less than 36.7‰, although 37.0‰ has occasionally been observed. Because the T/S relationship of Gulf Stream water is predictable, especially at temperatures of <22°C, the identification of Gulf Stream water in the shelf environment is straightforward. Since the temperature of Antarctic Intermediate Water is <12°C, it seldom affects adjacent shelf waters and is not discussed further. Surface Gulf Stream temperatures vary from 21° to 24°C in the winter to 27° to 29°C in the summer [*Fuglister,* 1947; *Atkinson et al.,* 1983]. These variations determine, in part, the modes by which intruding Gulf Stream waters interact with adjacent shelf waters [*Atkinson,* 1977; *Atkinson et al.,* 1982].

Nitrate, phosphate, and silicate are all closely related to temperature within given temperature ranges. Below the 20°C isotherm, nutrient concentrations increase proportionally with decreasing temperature and, along with increasing AOU, are indicative of biochemical oxidative processes. Several points are made concerning these distributions. The first is that at depths below about the 20°C isotherm, nutrient concentrations begin to increase. Considerable scatter is related mainly to the presence or absence of Western North Atlantic Water of Caribbean origin [*Richards and Redfield,* 1955] because this water has lower oxygen and higher nutrient concentrations than Western North Atlantic Water of the same density. Second, there are some differences between inflection points of the three nutrients at temperatures near 20°C. For nitrate the inflection point is at about 20°C, and for phosphate and sil-

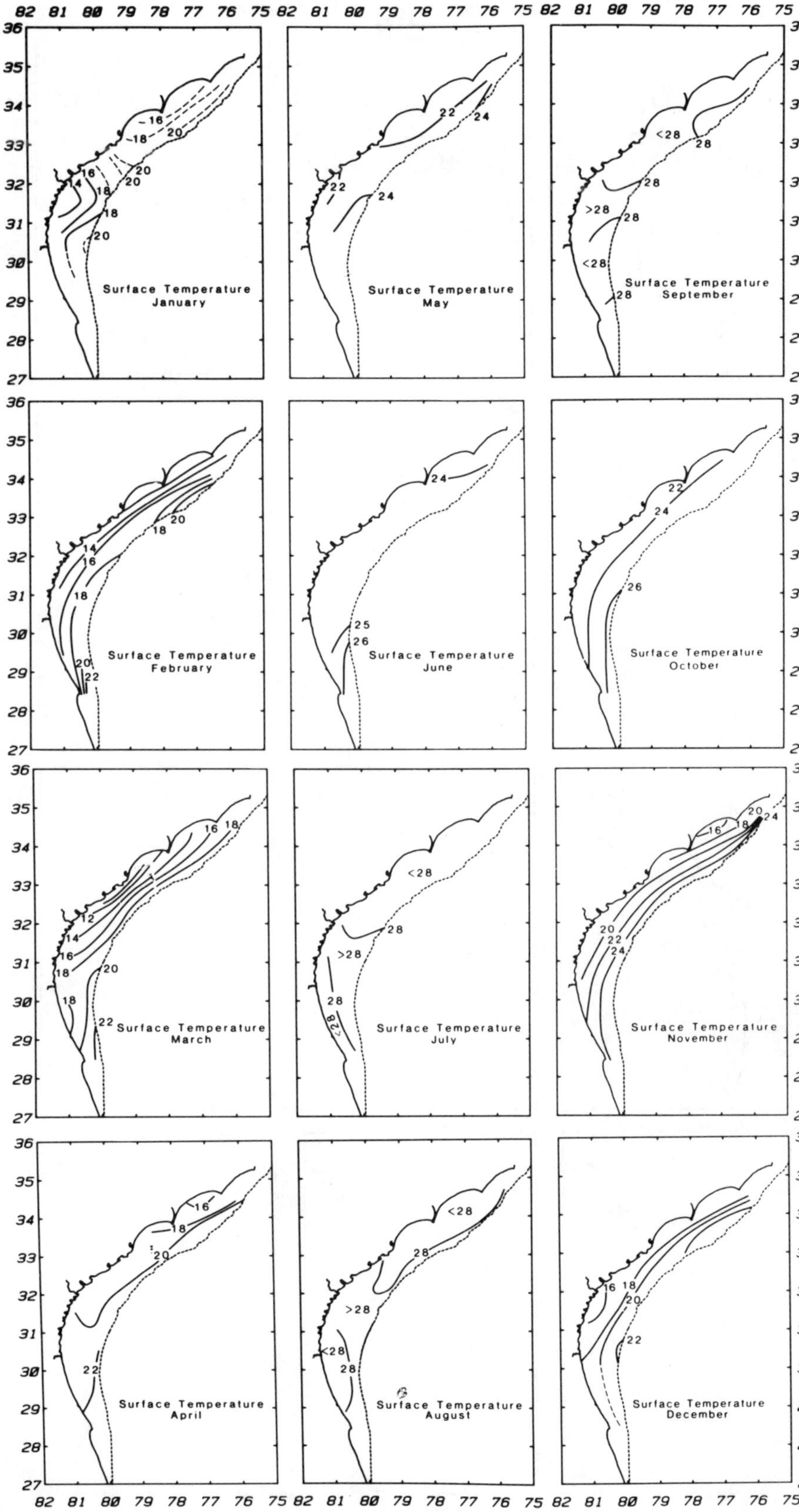

Fig. 2. Average monthly surface temperature from *Atkinson et al.* [1983].

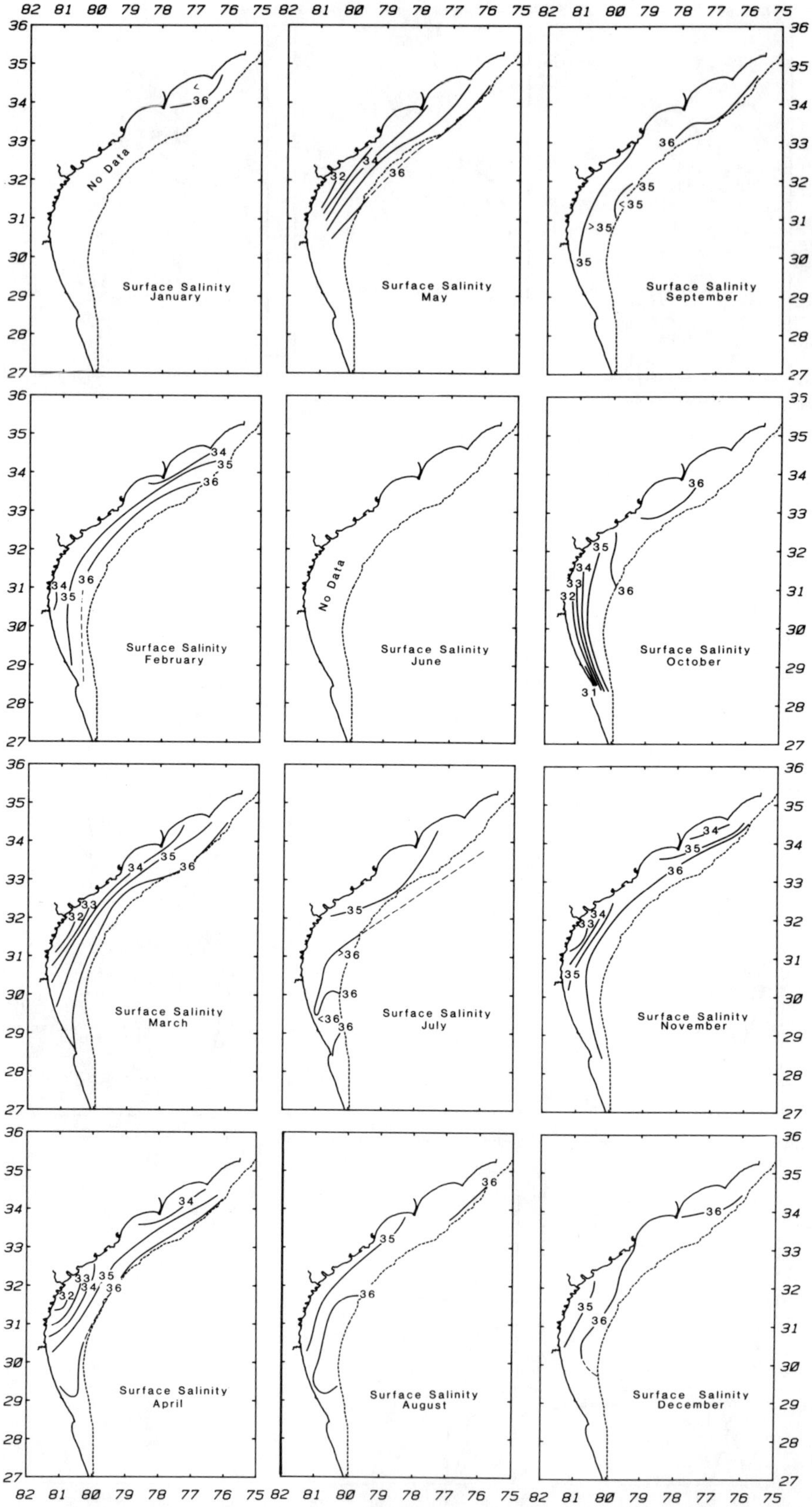

Fig. 3. Average monthly surface salinity from *Atkinson et al.* [1983].

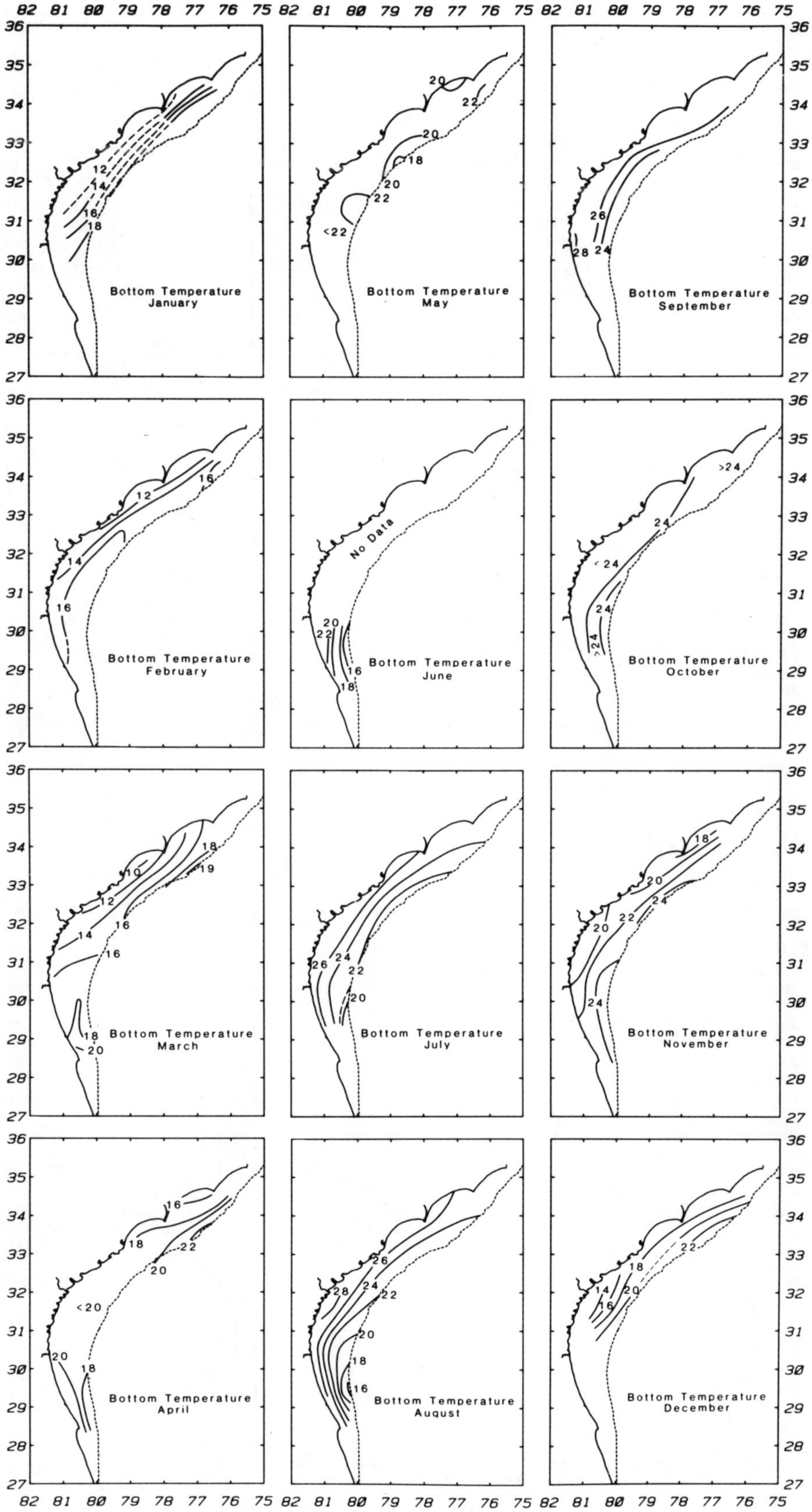

Fig. 4. Average monthly bottom temperature from *Atkinson et al.* [1983].

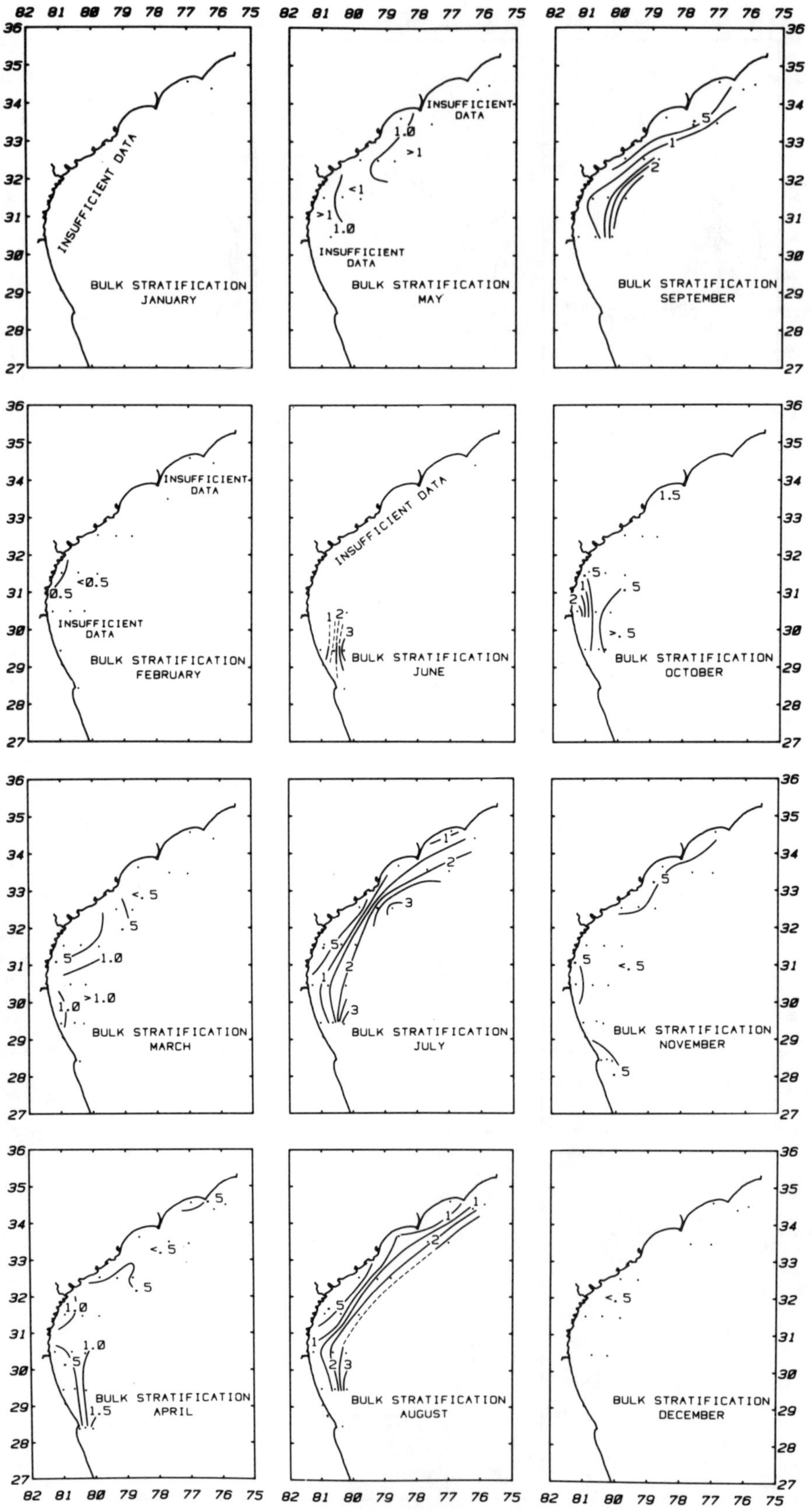

Fig. 5. Average bulk stratification from *Atkinson et al.* [1983]. Bulk stratification is the simple difference between near-surface and near-bottom density in sigma-*t* units.

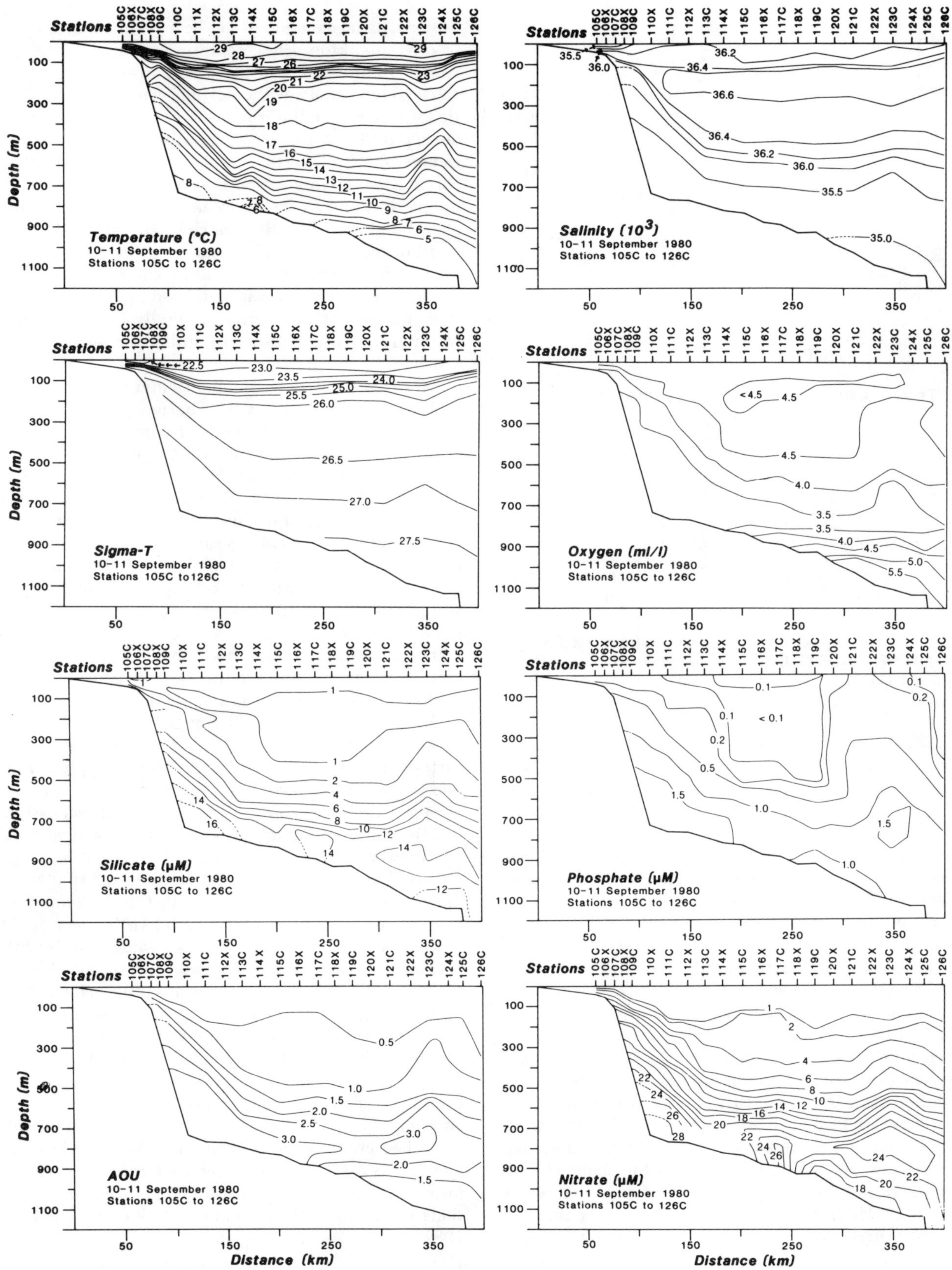

Fig. 6. Section across outer shelf, Gulf Stream, and Blake Plateau at 29°N. Data from September 1980.

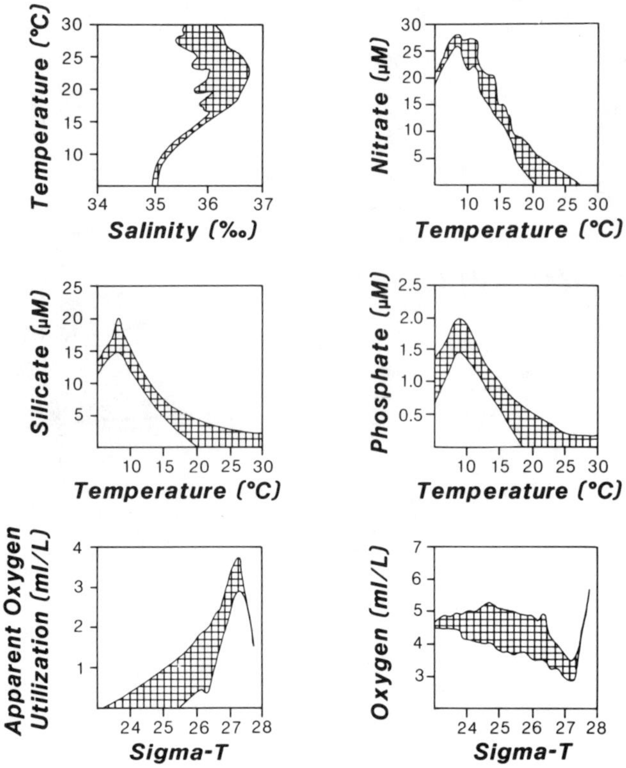

Fig. 7. Relationship between various parameters in Gulf Stream water. Data from shelf, Gulf Stream, and Blake Plateau cruise in September 1980.

icate it is at about 18°C. Thus shelf and slope water masses can be characterized by temperature, salinity, and, if unaffected by biological activity, by nutrients.

Shelf Waters

Having defined characteristic water masses over the slope, we now examine water column characteristics in adjacent shelf waters. Figures 8 and 9 show typical cross-shelf sections in the southern and northern SAB during summer and winter. Figure 8 shows a typical cross-shelf section during the winter off north Florida and North Carolina. Vertical profiles of temperature, salinity, and density off Florida indicate well-mixed shelf waters, particularly in the north Florida middle shelf. The position of the 24°C isotherm, the downward sloping isotherms, the presence of a subsurface salinity maximum over the upper slope, and high-nitrate water (5–10 µM) over the middle and outer shelf all indicate the presence of Gulf Stream water. Off north Florida, where waters are less dense, upwelled Gulf Stream water has intruded across the shelf. The intruded water was colder (18°C) than resident shelf water (18° to 20°C). High nitrate concentrations imply that the intrusion occurred, at most, a few days prior to the observations, as levels had not yet been reduced by the activity of phytoplankton. Shelf waters were more stratified off North Carolina than in the previous example, mainly because river runoff in the spring formed a lens of low-salinity water-over the inner shelf (Figure 8). Middle shelf densities were >26.5 σ_T, or 0.5 σ_t units higher than on the middle shelf off north Florida. The position of the Gulf Stream was farther offshore than in the north Florida example, but the water between the Gulf Stream and the shelf break had Gulf Stream characteristics. High nitrate and low oxygen concentrations over the upper slope indicate active upwelling, but unlike the north Florida situation, upwelled water was restricted to the outer shelf. The higher shelf densities gravitationally block the upwelled water from moving onshore, unlike the situation farther south.

By summer (Figure 9), shelf waters in both the north and south warmed to >28°C. Thermal structures off north Florida indicate that the Gulf Stream was over the upper slope, and domed isotherms over the shelf break imply the presence of a frontal eddy [*Lee and Atkinson,* 1983; *Lee et al.,* 1981]. The presence of the Gulf Stream is confirmed by the appearance of a subsurface salinity maximum. The outer shelf was vertically stratified, but stratification was minimal over the inner and middle shelf. Domed isotherms and stratified shelf waters led to upwelling of nutrient rich water at the shelf break which, in this case, elevated nitrate concentrations halfway across the shelf. Because nitrate concentrations were high, it can be assumed that the intrusion was relatively "new." Low oxygen concentrations confirm the intrusion origin as the Gulf Stream.

Distribution patterns off North Carolina differed from those off north Florida because here the Gulf Stream is further offshore. Isotherms did not slope steeply downward, and a dome structure was not present, nor was there a strong salinity maximum. The outer shelf was stratified, but the inner and middle shelves were less stratified. Nutrients were abundant at the shelf break but had not been advected shoreward. Other studies showed that nutrients are less often advected into North Carolina shelf waters than in southern parts of the SAB [*Atkinson et al.,* 1982].

Nutrient concentrations in river water flowing into the SAB vary depending on the river and season [*Windom et al.,* 1975]. Maximum concentrations are discharged during periods of high runoff in spring, and minimum amounts are discharged in July through October. The total load of nutrients discharged is not large, supplying only enough phosphorus and less nitrogen than is required to support the observed marsh grass growth [*Windom et al.,* 1975]. Nutrient concentrations in estuaries of the Savannah and Ogeechee rivers show that mixing alone reduces concentrations to very low levels, and, therefore, even though the nutrient flux from rivers may be significant, concentrations in coastal waters remain low [*Willey and Atkinson,* 1982].

WINTER

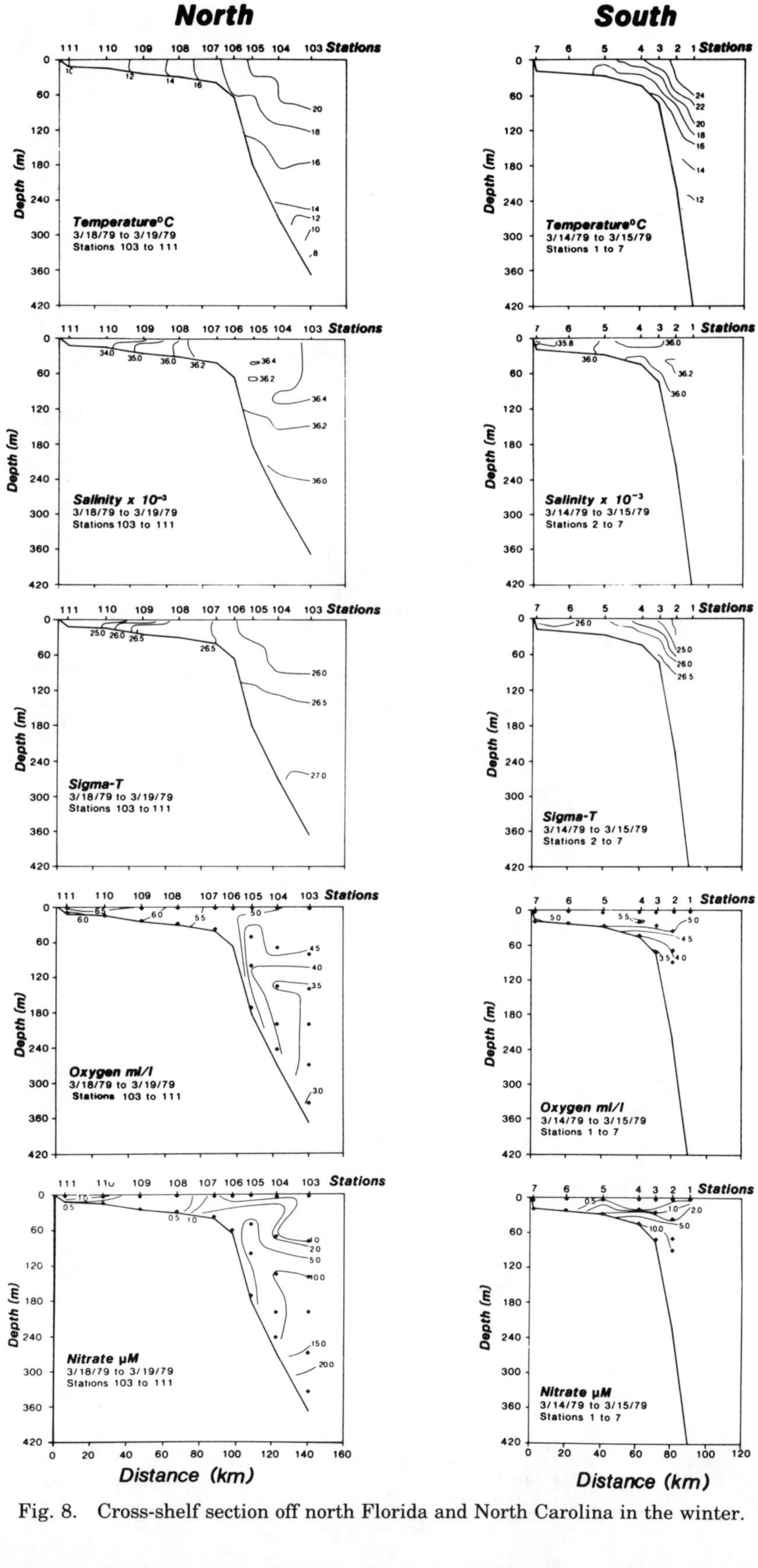

Fig. 8. Cross-shelf section off north Florida and North Carolina in the winter.

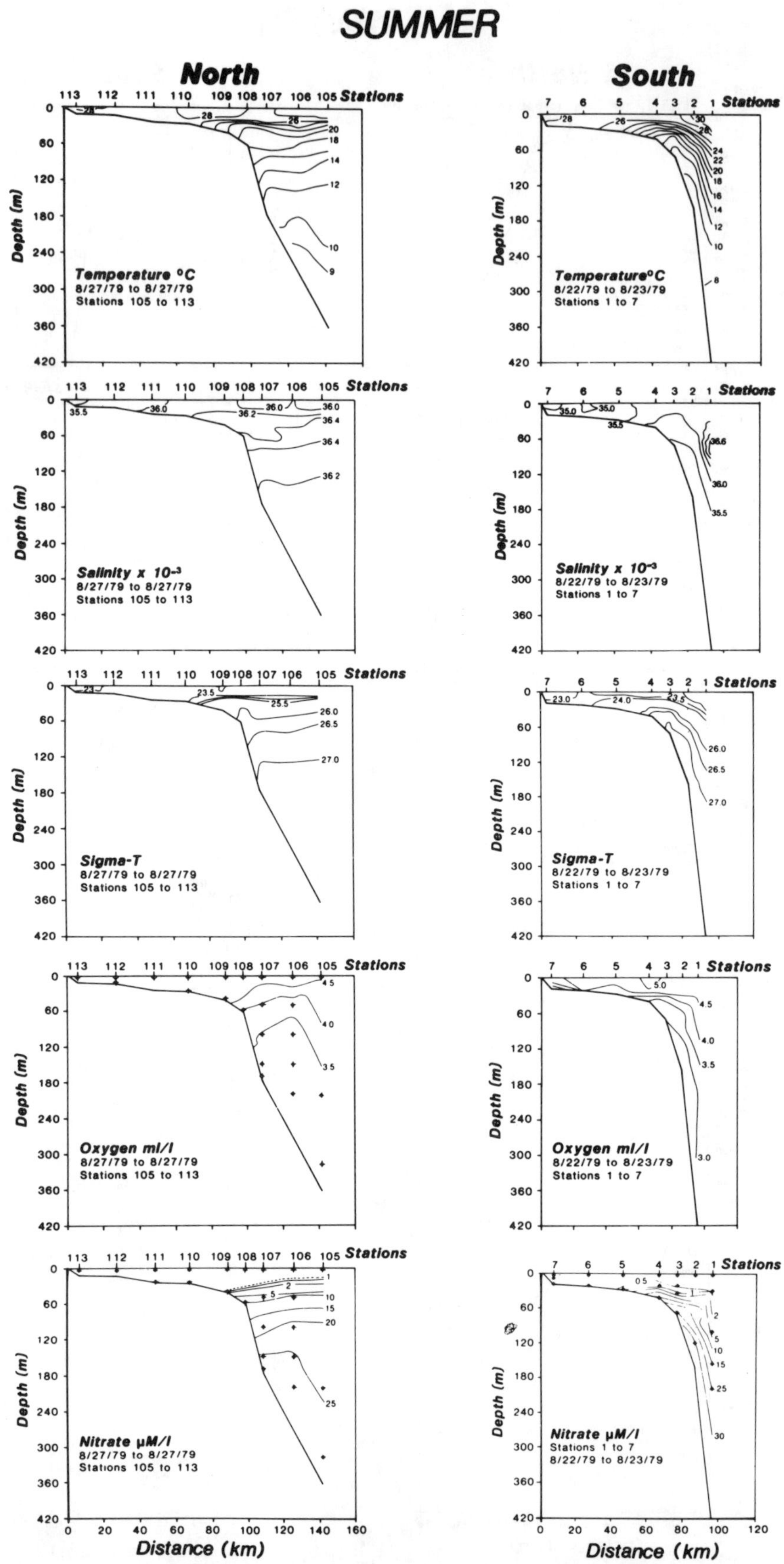

Fig. 9. Cross-shelf section off north Florida and North Carolina in the summer.

Variability

Seasonal variations have been discussed above and in other papers. Here we discuss variations in shelf water properties on shorter time scales, such as those associated with the passage of frontal eddies, stranded intrusions, or variations in coastal waters.

Short-Term Variations During Mixed Conditions. During colder months, relatively dense shelf water tends to restrict surface and upwelled Gulf Stream waters to the outer shelf [*Atkinson*, 1977]. Thus periodic northward propagating frontal events of the Gulf Stream change outer shelf temperature, salinity, and nutrient fields. In spring 1981 we organized a large-scale study (Georgia Bight Experiment (GABEX) I) to examine the effect of these events. Selected results are presented here. Changes in properties with time at the shelf break at 30°N (Figure 10) indicate that we observed four upwelling events. The first event, characterized by water with temperatures of <17°C, salinities between 36.0 and 36.1‰, and nitrate concentrations of >10 μM, was initiated before our sampling began. A second event on April 17–18 was characterized by temperatures of <18°C, salinities of >36.1‰, and nitrate concentrations of >10 μM. On April 22–24 a third event was characterized by temperatures of <18°C, salinities between 36.0 and 36.1‰, and nitrate concentrations above 15 μM. On April 24 the passage of a frontal eddy caused strong upwelling [*Lee and Atkinson*, 1983]. Near-bottom temperatures decreased to <16°C, and nitrate concentrations were between 10 and 15 μM. Surface water was well mixed with temperatures between 20° and 22°C, except on April 23–26 when the warm filament associated with the frontal eddy caused surface temperatures to rise above 24°C.

The effect of short-term events on adjacent shelf waters at 30°N is shown in Figure 11. On April 24 the Gulf Stream front, indicated by the 23° to 25°C isotherms, was over the outer shelf at 30°N. There was a stranded pool of 19°C water at midshelf, and vertical mixing apparently lowered surface temperatures. Shelf salinities were <36‰, and nitrate concentrations were <5 μM. Less than a day later (Figure 12) the Gulf Stream meandered offshore, a characteristic cold core/warm tongue feature appeared over the outer shelf, lower layer temperatures decreased, and salinity and nitrate concentrations increased. The position of the 18°C isotherm and 5-μM nitrate isopleth dramatically indicate the vertical motion that occurred. Vertical velocities reached 10^{-2} cm s^{-1} [*Lee and Atkinson*, 1983].

Short-Term Variations During Stratified Conditions. In the preceding section we discussed shelf break upwelling during seasons when upwelled nutrient-rich water is restricted to the outer shelf. Here we discuss the other situation, when upwelled water penetrates across the shelf during stratified conditions (summer) and periods of up-

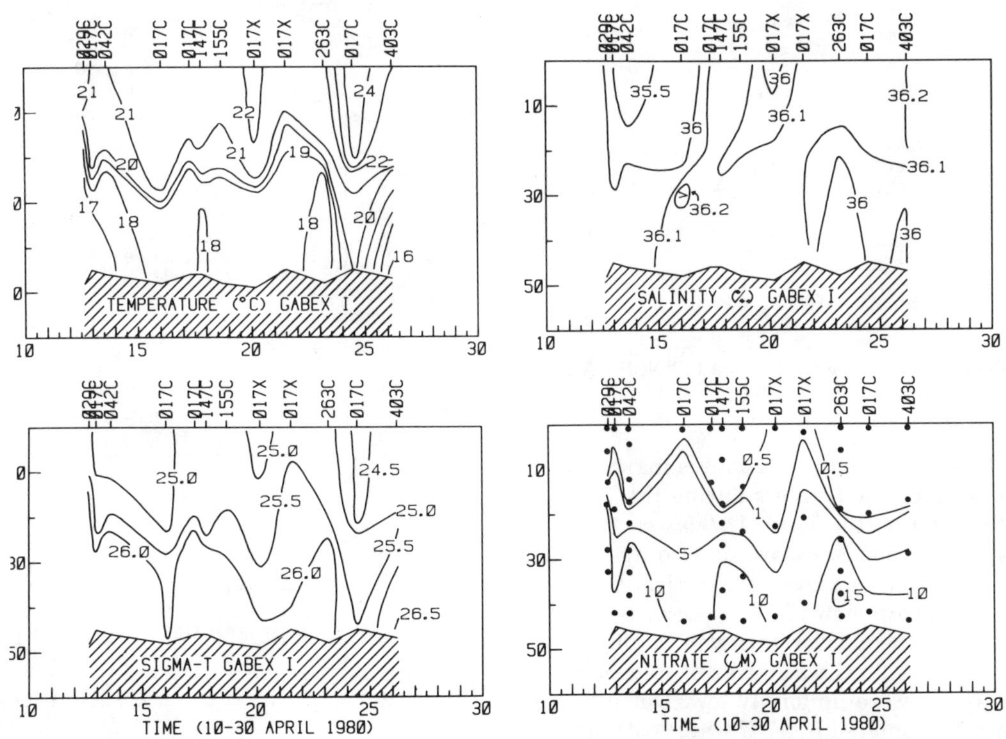

Fig. 10. Time/depth changes in temperature, salinity, density, and nitrate at shelf break off Florida. Data from GABEX I.

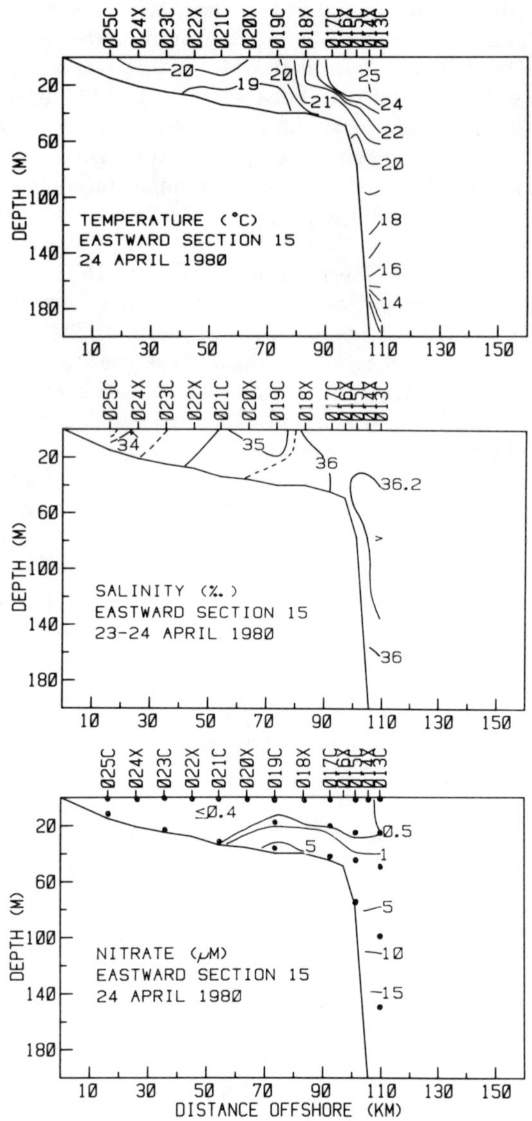

Fig. 11. Cross-shelf section off north Florida (30°N, St. Augustine) on April 24, 1980.

welling-favorable winds. The process causes marked cooling of middle and inner shelf waters during the summer that is most intense off north Florida [*Atkinson*, 1977].

In 1981 off North Florida, a succession of nutrient rich intrusions were observed. In Figure 13, vertical temperature and nitrate profiles are shown for four successive weeks. Unlike during spring, the shelf waters were highly stratified. During week four, 23°C water was at midshelf and nitrate concentrations were uniformly low. During week five, 23°C water was displaced by 17°C water with nitrate concentrations above 9 μM. During week six, the cold water was still present and nitrate concentrations remained high, indicating continuing upwelling and cross-shelf transport. If the water had been on the shelf and disconnected from its upwelling source for more than a few days, the nitrate would have been depleted. During week seven, 17° to 18°C water was advected farther onshore and nitrate concentrations were 7 μM or less. A new water mass with a temperature of 17°C and nitrate concentration of 14 μM appeared at the shelf break. Similar sequences have been observed many times during our studies in the SAB.

Nutrient Distribution and Fluxes

Outer Shelf

The goal of many investigators has been to determine the effect of physical variability on biological activities and, more specifically, on the rates at which nutrients are advected into SAB shelf waters. Earlier studies indicated that primary productivity is high only near the coast

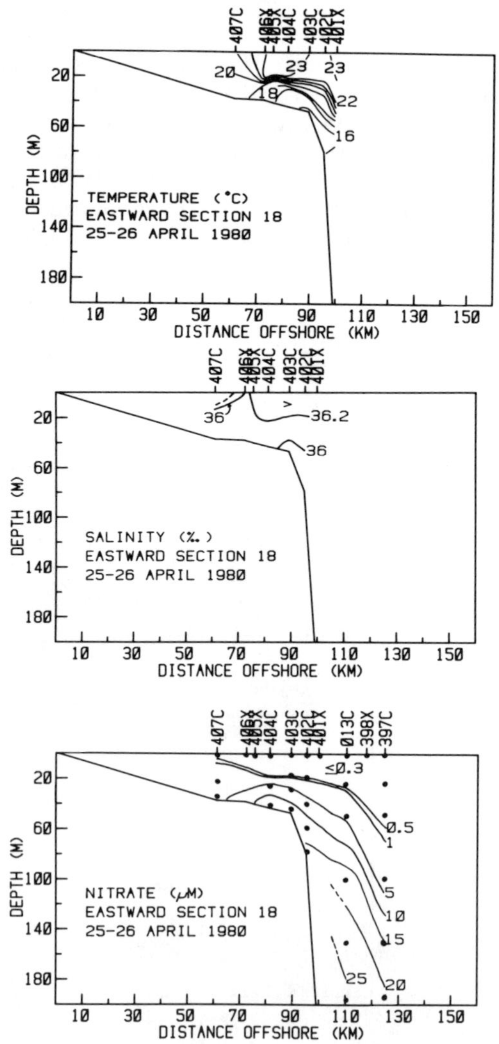

Fig. 12. Cross-shelf section off north Florida (30°N. St. Augustine) on April 25–26, 1980.

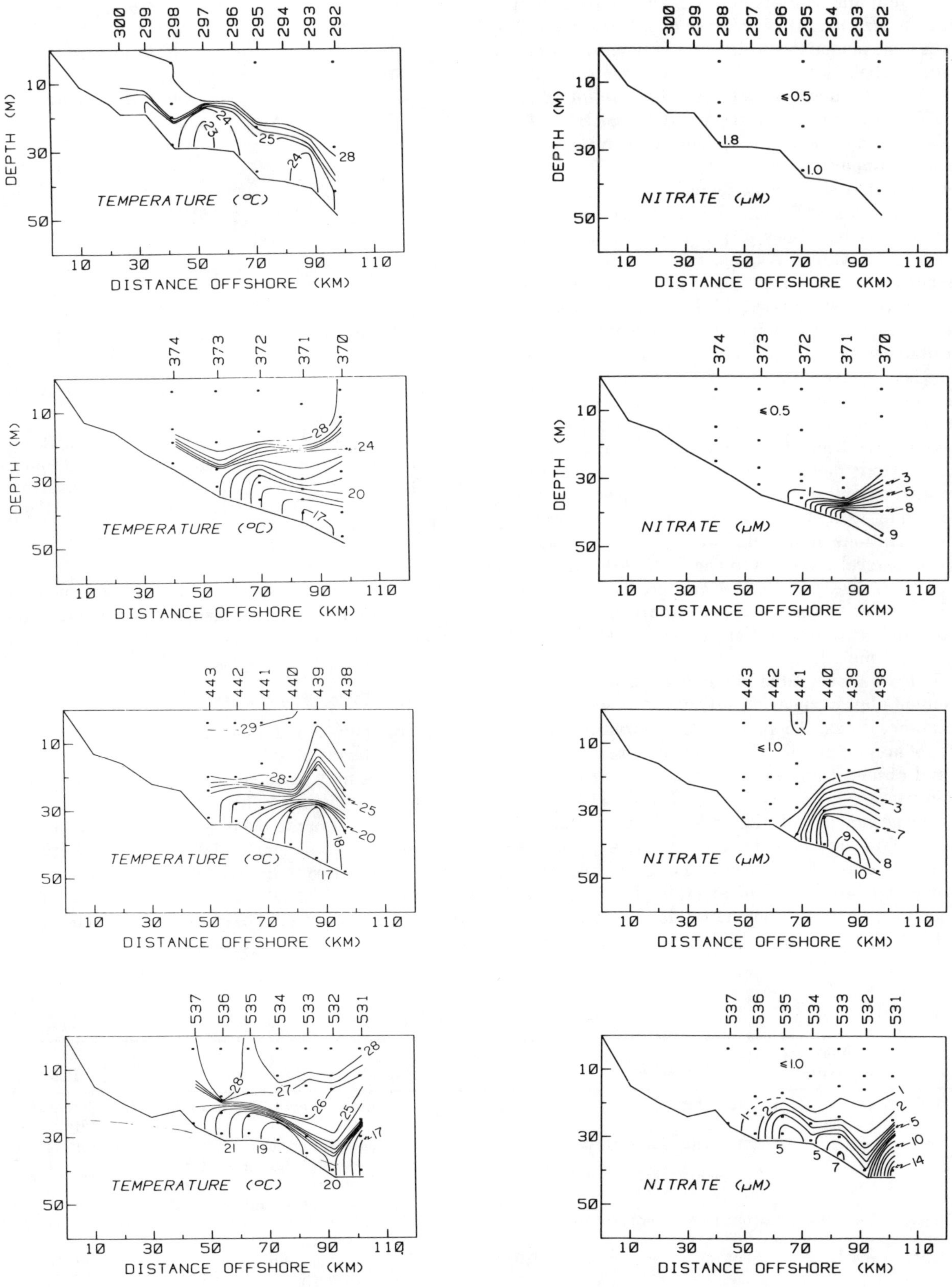

Fig. 13. Cross-shelf section off north Florida (30°N, St. Augustine) on successive weeks (top to bottom) during the summer of 1979.

[*Thomas,* 1966] and that offshore waters are particularly unproductive [*Haines,* 1974; *Haines and Dunstan,* 1975]. Later it became evident that high production is not limited to the inner shelf, forcing us to change our concepts of nutrient fluxes [*Yoder,* this volume]. *Dunstan and Atkinson*'s [1976] data on nitrogen sources to the SAB were the first indication that "new" nitrogen upwelled from the Gulf Stream is an important source of nutrients. When it became obvious that significant amounts of nutrients may be advected into the SAB from offshore, we devised cruises to determine the frequency of the events that pulse nutrients into shelf waters. A quasi-synoptic view of the region permitted quantitative determination of water mass volumes and nutrient masses, while current meters indicated the frequency at which advective events occur.

The nutrient/temperature relationship of Gulf Stream water has been a key element in our flux analyses (Figure 7). The relationship is consistent for Gulf Stream water that has not been in the euphotic zone (<50 m) for more than several days. Thus, for example, temperature sensors placed on current meters at the shelf break can reliably indicate nitrate concentrations in newly upwelled water. This technique, using temperature to calculate initial nitrate concentrations in upwelled water, has been used successfully at several locations in the SAB [*Atkinson et al.,* 1982; *Lee et al.,* 1981; *Lee and Atkinson,* 1983]. Knowing the onshore velocity component (u) and using eddy flux techniques and depth integration of discrete values, nitrate flux can be estimated.

Nitrate flux values (Table 1) include not only eddy flux measurements but also earlier estimates based on hydrographic observations. An immediate observation is that estimates of annual nutrient fluxes have increased as techniques and observations were refined. *Lee and Atkinson* [1983] discuss some reasons for this. The important point is that a large amount of nitrate is transported into the euphotic shelf waters where it is used by primary producers [*Yoder,* this volume]. The fate of this onshore nitrate flux was resolved by *Lee and Atkinson* [1983, Figure 17], who show a net onshore flow off north Florida and a net offshore flow north of 31°N (discussed by *Lee et al.* [this volume]).

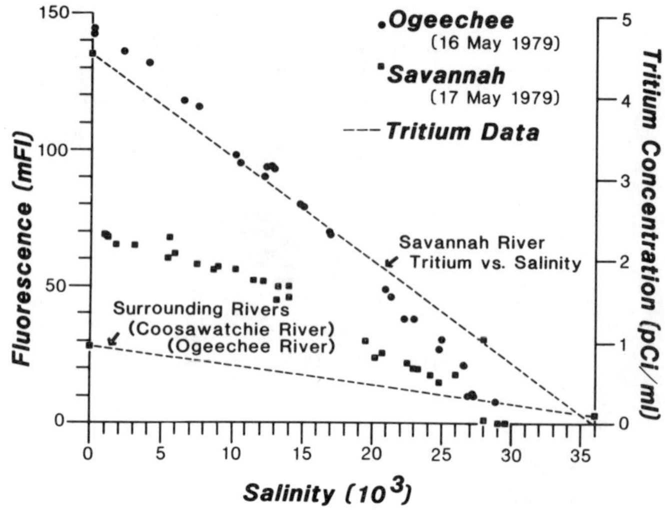

Fig. 14. Natural fluorescence in two adjacent Georgia rivers (Ogeechee and Savannah) versus salinity. See *Willey and Atkinson* [1982] for details. (Tritium concentrations in Savannah River, adjacent rivers, and offshore waters versus salinity. Data supplied by David W. Hayes.)

From the discussion above and unpublished data, it is obvious that there are alongshore variations in the onshore advective flux of nutrients. All evidence suggests decreasing flux rates from south to north.

Chemical Tracers in Coastal Waters

The distribution of freshwater in the SAB was discussed earlier in terms of salinity distributions, but other chemical tracers may in some cases provide more information, especially in the inner shelf region. The distribution of clay in coastal sediments clearly shows the offshore extent of riverine inorganic particulate matter [*Bingham,* 1973]. Phenolic aldehydes in sediments show a similar pattern [*Gardner and Menzel,* 1974], implying that much of the matter carried into the SAB by rivers is deposited in inner shelf sediments. Other tracers besides clays and phenolic

TABLE 1. Nitrite Flux Estimates

Area	Time Interval	Technique	Result	Comment	Reference
Georgia Bight	yearly	hydrographic	13.3 $\mu M\ L^{-1}\ yr^{-1}$	not convertible to linear flux	*Dunstan and Atkinson* [1976]
Georgia Bight	yearly	current meter/temperature	51,000 $tN\ yr^{-1}$	calculation at 55-μM isobath, based on flux observed during 1 eddy; assume 1 eddy/2 weeks	*Lee et al.* [1981]
Onslow Bay	1 event	current meter/temperature	170 tN	calculation at 40-m isobath	*Atkinson et al.* [1982]
Georgia Bight	yearly	current meter/temperature	220,000 tN	calculated for 75-m isobath	*Lee and Atkinson* [1983]

aldehydes can be used to indicate the mixing of individual rivers into the SAB.

Tritium is at elevated concentrations in the Savannah River because of releases from the Department of Energy Savannah River Plant. Concentration levels 3 times above typical river values were common in the mid-1970's and levels are still elevated but decreasing, so tritium is a unique tracer of Savannah River water [*Hayes,* 1980]. A plot of tritium versus salinity for the Savannah River and some adjacent rivers shows that tritium may be a useful tracer in coastal waters (Figure 14).

The natural fluorescence of humic compounds in river water also varies from river to river [*Willey and Atkinson,* 1982]. Coastal plain rivers drain freshwater marshes and have high humic concentrations, while piedmont rivers drain uplands and have low humic concentrations. The result is that adjacent rivers may have different fluorescence/salinity relationships (Figure 14).

Other tracers such as ^{18}O and D have been tested in the SAB but show little promise as good tracers [*Torgerson,* 1979]. Dissolved silica is of limited use because of its nonconservative behavior.

Conclusion

Distributions of variables such as temperature, salinity, and nutrients are affected by such normal processes as air-sea interaction and riverine influx, but in the South Atlantic Bight the effect of the Gulf Stream is significant and unique. Temperature varies basically because of air-sea interaction, but the variations caused by interaction with the Gulf Stream can be equally important, especially in moderating temperatures. Salinity variations are related to seasonally-varying river flow. However, the shelfwide salinity distribution is dependent on shelf flushing, which depends on the Gulf Stream. Nutrient supplies to the shelf depend on the riverine input, local regeneration, and advective influx from the Gulf Stream, with the latter being the most important.

Acknowledgments. I would like to acknowledge the support of the Department of Energy (contract DE-AS09-76EV00889) and the Bureau of Land Management. The technical assistance of William Chandler, James Singer, and Patrick O'Malley is greatly appreciated.

References

Atkinson, L. P., Modes of Gulf Stream intrusion into the South Atlantic Bight shelf waters, *Geophys. Res. Lett., 4,* 538–586, 1977.

Atkinson, L. P., L. J. Pietrafesa, and E. E. Hofmann, An evaluation of nutrient sources to Onslow Bay, North Carolina, *J. Mar. Res., 40,* 679–699, 1982.

Atkinson, L. P., T. N. Lee, J. O. Blanton, and W. S. Chandler, Climatology of southeastern United States shelf waters, *J. Geophys. Res., 88,* 4705–4718, 1983.

Bingham, G. N., Zone of influence—Inner continental shelf of Georgia, *J. Sediment. Petrol., 43,* 207–214, 1973.

Blanton, J. O., and L. P. Atkinson, Transport and fate of river discharge on the continental shelf of the southeastern United States, *J. Geophys. Res., 88,* 4730–4738, 1983.

Brooks, D. A., and J. M. Bane, Gulf Stream deflection by a bottom feature off Charleston, South Carolina, *Science, 201,* 1225–1226, 1978.

Dunstan, W. M., and L. P. Atkinson, Sources of new nitrogen for the South Atlantic Bight, in *Estuarine Processes, Uses, Stresses and Adaptation to the Estuary,* vol. 1, edited by M. Wiley, pp. 69–78, Academic, Orlando, Fla., 1976.

Fuglister, F., Average monthly sea surface temperatures of the western North Atlantic Ocean, *Pap. Meteorol. Oceanogr., 10,* 1–25, 1947.

Gardner, W. S., and D. W. Menzel, Phenolic aldehydes as indicators of terrestrially derived organic matter in the sea. *Geochim. Cosmochim. Acta, 38,* 813–822, 1974.

Haines, E. B., Processes affecting production in Georgia coastal waters, Ph.D. dissertation, 118 pp., Duke Univ., Durham, N.C., 1974.

Haines, E. B., and W. M. Dunstan, The distribution and relation of particulate organic material and primary productivity in the Georgia Bight, *Estuarine Coastal Mar. Sci., 3,* 431–441, 1975.

Hayes, D. W., Tritium in the Savannah River estuary and adjacent marine waters, *Sci. Memo. 232/80,* pp. 271–281, Int. At. Energy Agency, Vienna, 1980.

Lee, T. N., and L. P. Atkinson, Low-frequency current and temperature variability from Gulf Stream frontal eddies and atmospheric forcing along the southeast U.S. outer continental shelf, *J. Geophys. Res., 88,* 4541–4567, 1983.

Lee, T. N., L. P. Atkinson, and R. Legeckis, Detailed observations of a Gulf Stream frontal eddy on the Georgia continental shelf, April 1977, *Deep Sea Res., 28,* 347–378, 1981.

Lee, T. N., V. Kourafalou, J. D. Wang, W. J. Ho, J. O. Blanton, L. P. Atkinson, and L. J. Pietrafesa, Shelf circulation from Cape Canaveral to Cape Fear during winter, in *Oceanography of the Southeastern U.S. Continental Shelf, Coastal and Estuarine Sci.,* vol. 2, edited by L. P. Atkinson, D. W. Menzel, and K. A. Bush, AGU, Washington, D. C., this volume.

Legeckis, R., Satellite observations of the influence of bottom topography on the seaward deflection of the Gulf Stream off Charleston, South Carolina, *J. Phys. Oceanogr., 9,* 483–497, 1979.

Morrison, J. M., and W. D. Nowlin, Jr., General distribution of water masses within the eastern Caribbean Sea during the winter of 1972 and fall of 1973, *J. Geophys. Res., 87,* 4207–4229, 1982.

Pashinski, D. J., and G. A. Maul, Use of ocean temperature while coasting between the Straits of Florida and Cape Hatteras, *Mariners Weather Log, 17,* Jan. 1–3, 1973.

Richards, F. A., and A. C. Redfield, Oxygen-density relationships in the western North Atlantic, *Deep Sea Res., 2,* 182–199, 1955.

Thomas, J. P., The influence of the Altamaha River on Primary productivity beyond the mouth of the river, M.S. thesis, 88 pp., Univ. of Ga., Athens, 1966.

Torgerson, T., Isotopic composition of river runoff on the U.S. east coast: Evaluation of stable isotope versus salinity plots for coastal water mass identification, *J. Geophys., Res., 84,* 3773–3775, 1979.

Wennekens, M. P., Water masses properties of the Straits of Florida and related waters, *Bull. Mar. Sci. Gulf Caribb., 9,* 1–52, 1959.

Willey, J. D., and L. P. Atkinson, Natural fluorescence as a tracer for distinguishing between piedmont and coastal plain river water in the nearshore waters of Georgia and North Carolina, *Estaurine Coastal Shelf Sci., 14,* 49–59, 1982.

Windom, H. L., W. M. Dunstan, and W. S. Gardner, River input of inorganic phosphorus and nitrogen to the southeastern salt-marsh estuarine environment, in *Mineral Cycling in Southeastern Ecosystems, ERDA Symp. Ser., CONF-7400513,* edited by F. G. Howell, J. B. Gentry, and M. H. Smith, pp. 309–313, National Technical Information Service, Springfield, Va., 1975.

Yoder, J. A., Environmental control of phytoplankton production on the southeastern U.S. continental shelf, in *Oceanography of the Southeastern U.S. Continental Shelf, Coastal and Estuarine Sci.,* vol. 2, edited by L. P. Atkinson, D. W. Menzel, and K. A. Bush, AGU, Washington, D. C., this volume.

Environmental Control of Phytoplankton Production on the Southeastern U.S. Continental Shelf

J. A. YODER

Skidaway Institute of Oceanography, Savannah, Georgia

Different processes affect the rate and dynamics of primary production across the southeastern U.S. continental shelf of the South Atlantic Bight. Within relatively low salinity waters near the coast, primary production is relatively constant throughout the year and is sustained by nutrients emanating from rivers and salt marshes that line the coast. High turbidity of coastal waters limits the amount of light available for photosynthesis. Gulf Stream–induced upwelling at the shelf break and intrusion of upwelled waters across the shelf are the dominant processes affecting the dynamics of phytoplankton production on the middle and outer shelf. During colder months of the year, upwelled waters generally do not penetrate far across the shelf, and thus phytoplankton blooms and high rates of production are restricted to the outer shelf. Subsurface intrusions of nutrient-rich water penetrate across the shelf during warmer months of the year when the shelf is stratified, particularly in the northernmost and southernmost regions of the South Atlantic Bight. Phytoplankton production and biomass are more than 10 times higher within intrusions than in surrounding resident shelf waters.

Introduction

The purpose of this paper is to relate the rate and dynamics of phytoplankton production to the physical and chemical environment of the southeastern continental shelf. Of particular interest are processes that affect the supply of plant nutrients to the euphotic zone. Resident middle and outer shelf waters are nutrient poor. Nutrients essential for phytoplankton production result from recycling in the water column and/or sediments and are introduced to shelf waters at the coastal or offshore boundary [*Atkinson*, this volume]. The dominant mechanisms which introduce "new" nutrients (in sensu *Dugdale and Goering* [1967]) to the euphotic zone at the two boundaries are quite different. For the purpose of this discussion, the continental shelf off the Georgia coast can be divided into three zones: the inner, middle, and outer shelf [*Bishop et al.*, 1980]. Production within inner shelf waters with salinities of $<35 \times 10^{-3}$ is principally affected by recycling and by nutrients originating from rivers, marshes, or estuaries. In contrast, production on the outer shelf is principally affected by upwelled nutrients. The dynamics of production of the middle shelf are principally controlled by processes that transport nutrients across the inner and outer zones. With some exceptions, which will be discussed, this zonation scheme applies to the entire southeastern shelf.

Inner Shelf

The inner shelf is roughly defined as that portion of the shelf inshore of the 20-m isobath which is about one third of the total area of the shelf [see *Atkinson and Menzel*, this volume]. Four characteristics of the inner shelf are important to phytoplankton productivity. First, with the exception of the Florida coast south of Jacksonville (approximately 30°20′N), phytoplankton production on the inner shelf is not affected by upwelled nutrients. Instead, "new" nutrients enter this zone from the numerous estuaries and salt marshes that line the Georgia/South Carolina coast. However, most of the nitrogen exported from marshes is in the form of dissolved (DON) or particulate organic nitrogen (PON) rather than inorganic forms (DIN) [*Dunstan and Atkinson*, 1976; *Haines*, 1979; *Kjerfve and McKellor*, 1980; *Bishop et al.*, 1984]. Thus a large proportion of the exported N is not available to phytoplankton until mineralized by heterotrophs. Second, much of the inner shelf is highly turbid, particularly along the Georgia and South Carolina coasts. Off Georgia the high turbidity zone extends 5–10 km offshore, or roughly to the 5- to 10-m isobath [*Oertel and Dunstan*, 1981]. The concentration of total suspended matter in the turbid zone is 9–200 mg/l. Concentrations decrease to <2 mg/l at a distance of 20–35 km offshore, which corresponds to the 13- to 16-m isobaths

[*Oertel and Dunstan*, 1981]. A third important characteristic of the inner shelf is the existence of coastal fronts which separate waters with salinities of $<35 \times 10^{-3}$ from more saline waters offshore [e.g., *Blanton*, 1981]. These fronts inhibit the cross-shelf exchange of dissolved and suspended materials [*Blanton*, 1981; *Yoder et al.*, 1981*a*], thereby affecting the residence time of nutrients and phytoplankton on the inner shelf. The fourth characteristic of the inner shelf of potential importance to phytoplankton production is the large tidal amplitude (approximately 2–3 m), particularly off Georgia [see *Pietrafesa et al.*, this volume]. The importance of these four characteristics will be explained in the following discussion of phytoplankton productivity.

Off Georgia one of the most striking features of the inner shelf is a rapid decrease in chlorophyll *a* concentration with distance from shore and with increasing salinity [*Turner et al.*, 1979; *Bishop et al.*, 1980; *Oertel and Dunstan*, 1981]. At the 20-m isobath (approximately 40 km offshore), chlorophyll *a* concentration is generally only a fraction of the levels closer to shore (Figure 1). Within lower-salinity waters close to shore, chlorophyll *a* levels and rates of primary production are relatively high throughout the year. Within 15 km of the mouth of the Altamaha River (Georgia coast, 31°20′N), the annual range in chlorophyll *a* concentration and primary production was 2–7 mg/m^3 and about 1–3 g C/m^2/d, respectively, and annual areal primary production was 600 g C/m^2 [*Thomas*, 1966]. About 50 km north of the Altamaha area, there are similar annual ranges in chlorophyll *a* concentration at distances of 7 km off the Georgia coast [*Bishop et al.*, 1980]. Annual primary production for the inner shelf (0- to 20-m isobaths, approximately 40–50 km wide) is 285 g C/m^2 [*Haines and Dunstan*, 1975]. Thus relatively fresh water of the inner shelf off Georgia is highly productive throughout the year. Centric diatoms are the dominant phytoplankton taxa in these waters [*Bishop et al.*, 1980; *Oertel and Dunstan*, 1981].

Another feature of Georgia coastal waters is that a band of high-turbidity water extends several kilometers offshore (Figure 1). Because of the high turbidity, light limits production out to at least 11–15 km offshore, and maximum productivity on an areal basis occurs just seaward of the highly turbid waters [*Thomas*, 1966]. Contrary to Thomas' findings, *Oertel and Dunstan* [1981] showed that areal production was highest in the turbid zone in spite of the very shallow compensation depth (1% light level at 1 m). High production in the turbid zone may result from strong vertical mixing. The 5-m isobath is generally about 5 km offshore, and the large tidal amplitudes (approximately 2–3 m) off Georgia mean that the turbid zone is usually well mixed. Thus light limitation of production within the turbid zone is overcome to some extent because "Individual organisms must constantly be moving in and out of the surface where light is sufficient for photosynthesis" [*Oertel and Dunstan*, 1981].

"New N" (in sensu *Dugdale and Goering* [1967]) is introduced to coastal waters in a number of different forms

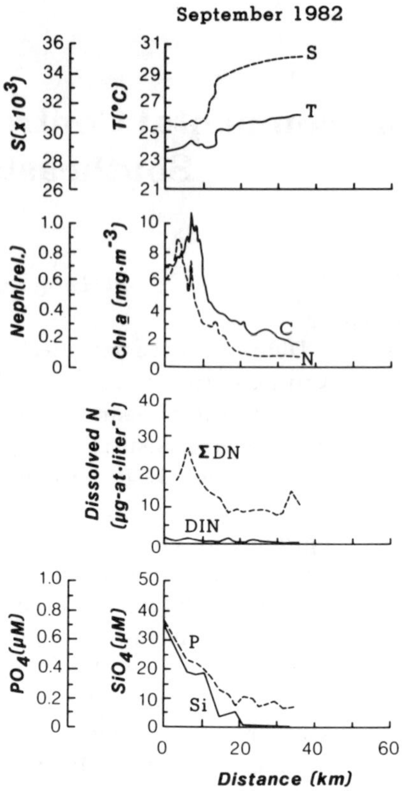

Fig. 1. Surface distributions along a transect extending offshore from the mouth of Wassaw Sound, Georgia (31°51′N, 80°55′W), to the 20-m isobath. "Neph" refers to relative turbidity as determined by nephelometry, DIN = NO_3 + NO_2 + NH_4, and DN = DIN + dissolved organic nitrogen (DON).

(i.e., NO_3, NH_3, DON, PON). Nitrogen export (mostly DON and PON) from salt marshes is highly seasonal, and net export to shelf waters occurs from about May to October [*Kjerfve and McKellor*, 1980]. Using the data of *Kjerfve and McKellor* [1980] and an estimate of salt marsh area along the South Carolina and Georgia coasts, *Turner* [1981] estimated that the total N export to the shelf from marshes is 1.2×10^5 metric tons per year. Assuming that all of this N is directly utilized by phytoplankton (highly unlikely given the high proportion of DON and PON) and using *Haines and Dunstan*'s [1975] estimate of primary productivity for the inner shelf (0- to 20-m isobath, approximately 40 km wide), the annual C:new N uptake ratio is 35:1 implying a maximum (and unrealistically high) rate of "new" production of only ~17% of the total if we assume a phytoplankton C:N ratio of 5.9 (Redfield ratio). This calculation clearly indicates the importance of nutrient recycling on the inner shelf.

Two characteristics of the inner shelf may contribute to efficient nutrient recycling. Coastal salinity fronts inhibit the rate of cross-shelf exchange of dissolved and suspended materials [*Blanton*, 1981], thereby affecting the residence time and favoring the recycling of nutrients and organic materials. Second, strong tidal mixing may release nutrients from sediments into the shallow euphotic zone of the turbid zone. The importance of sediment-derived nutrients is presently under investigation (C. S. Hopkinson, personal communication, 1982).

Most studies of inner shelf primary production have been conducted off the Georgia coast. Conclusions, however, probably apply to the entire Georgia/South Carolina coastline because extensive salt marshes are common to both. Because salt marshes are not extensive south of Jacksonville, Florida (approximately 30°20′N), the dynamics of phytoplankton production of the inner shelf there may be quite different. During warmer months, cold and nutrient-rich waters upwelled at the shelf break move across the shelf as subsurface intrusions and reach the coast south of Jacksonville. High concentrations of chlorophyll (>5 mg/m^3) occurred in upwelled waters at the coast near 30°N in April 1980 and during summer 1981 [*Yoder et al.*, 1984]. The importance of these upwellings to the inner shelf in this region has not been evaluated. However, the lack of large rivers and extensive salt marshes south of Jacksonville may mean that summer upwelling exerts a major control on plant production.

Middle and Outer Shelf

Quantitative studies of phytoplankton productivity of New England continental shelf waters were initiated almost 3 decades before similar studies in the southeast [*Riley*, 1946, 1947]. One of the principal conclusions of these early studies, and of others concerned with the continental shelf north of Cape Hatteras [e.g., *Ryther and Yentsch*, 1958; *Riley*, 1959, 1967; *Smayda*, 1973, 1976; *Walsh et al.*, 1978; *Malone et al.*, 1983], was the importance of seasonal phytoplankton blooms, particularly in the winter-spring. *Sverdrup*'s [1953] model was often used to relate seasonal changes in phytoplankton production to processes affecting nutrient supply to the euphotic zone and to the depth of the mixed layer in relation to the depth of the euphotic zone. Briefly, the model showed that primary production was low in the winter because incident irradiance was low and the mixed layer extended well below the euphotic zone. With the onset of stratification in the spring and an increase in incident irradiance, phytoplankton "bloomed" until nutrients were depleted from the mixed layer. Productivity in summer and fall was principally controlled by nutrient regeneration or by storm mixing of nutrients across the pycnocline and into the euphotic zone.

In contrast to North Atlantic and Middle Atlantic areas, the Sverdrup model does not apply to the southeastern shelf [*Bishop et al.*, 1980]. First, resident shelf waters are nutrient poor throughout the water column, and thus stratification does not "trap" nutrients. Second, the euphotic zone on the middle and much of the outer shelf extends to the bottom (1% light depth ~45–50 m). Nutrients, therefore, do not accumulate in near-bottom waters because they are continuously taken up by phytoplankton throughout the water column. In addition, we now know that the dominant process affecting the supply of "new" nutrients to the middle and outer southeastern shelf are specific events which cause upwelling at the shelf break and intrusion of upwelled waters across the shelf [*Atkinson*, this volume].

Early studies of phytoplankton productivity of the southeastern shelf [*Haines*, 1974; *Haines and Dunstan*, 1975; *Turner et al.*, 1979; *Turner*, 1981] were conducted before the importance of upwelling was recognized and before the dominant temporal and spatial scales of upwelling events were known. For lack of other hypotheses (and data), the *Sverdrup* [1953] model of phytoplankton production was assumed. Thus initial investigations looked for seasonal trends and assumed a high degree of spatial homogeneity. These assumptions dictated a sampling strategy with relatively short (days) cruises spaced months to seasons apart and stations spaced evenly and relatively far apart. This sampling strategy is poorly suited for evaluating, or even recognizing, the importance of short-period (days to weeks) and mesocale processes such as the upwelling that we now know occurs at the shelf break. As a result the conclusions of earlier studies contrast sharply with those conducted after about 1975 when sampling strategies were changed to better resolve the effects of upwelling.

At the time of *Haines*' [1974] first interpretation of phytoplankton production on the middle and outer southeastern shelf, only a relatively small amount of physical and biological data were available, and no important source of "new" nitrogen had been identified. As a result it was concluded that recycling met more than 90% of the phytoplankton demand for nitrogen. Thus the first interpretation of processes affecting phytoplankton productivity was that the middle and outer shelf were essentially similar to oligotrophic ocean waters. From the results of four quarterly cruises, *Haines and Dunstan* [1975] estimated an annual productivity of about 130 g C/m^2 for shelf waters offshore of the 20-m isobath. Primary production was highest during December (~1 g C/m^2/d) and lowest in April (~0.05 g C/m^2/d). High rates observed in December were interpreted as a seasonal phytoplankton bloom. On two of the four cruises, the average daily rate of primary production was higher in the Gulf Stream just seaward of the shelf than in shelf waters. Based on $^{13}C/^{12}C$ ratios and other considerations, *Haines and Dunstan* [1975] concluded that POC found on the shelf originated from in situ phytoplankton production, rather than from estuarine-derived detritus.

Using the productivity measurements of *Haines and Dunstan* [1975] and other data, others concluded that primary production decreased with distance from shore, that the principal phytoplankton bloom was during the summer, and that seasonal patterns in productivity were coherent across the shelf [*Turner et al.*, 1979; *Turner*, 1981]. Thus it was concluded that plankton productivity on the shelf was closely coupled with that of the estuaries. No underlying mechanism was presented to explain the apparent coupling of productivity from the estuaries to the outer shelf, but the implication was that outwelling of estuarine/marsh nutrients was responsible [*Turner et al.*, 1979].

From the investigations discussed above, one discerns two conflicting views of the processes controlling primary productivity of the middle and outer southeastern shelf: (1) the middle and outer shelf is essentially a shoreward extension of the oceanic regime that lies seaward of the 200-m isobath and (2) the middle and outer shelf is essentially a seaward extension of the estuarine/marsh regime. Neither view is consistent with current information which shows that the dominant process controlling the supply of "new" nutrients to the middle and outer shelf is upwelling at the shelf break [*Atkinson*, this volume]. Quantifying the importance of the upwelling process is a relatively recent and ongoing area of research.

Dunstan and Atkinson [1976] first hypothesized that upwelled nutrients had an important effect on the productivity of southeastern shelf waters. They used the annual shelf productivity estimate of *Haines and Dunstan* [1975], assumed 10 upwelling events each year, and showed that upwelled nitrate could provide as much as 25% of the annual shelfwide phytoplankton demand for nitrogen. Thus *Dunstan and Atkinson* [1976] concluded that the Georgia shelf was different from open ocean regimes in that a high proportion of the nitrogen required to support primary production was supplied by nitrate imported to the shelf by upwelling at the shelf break. This conclusion led to interdisciplinary studies of the middle and outer shelf which focused on the dynamics of upwelling and its effect on biological productivity.

Since 1975, two types of upwelling events (hereinafter referred to as types 1 and 2) important to understanding biological productivity of the middle and outer shelf were identified. For both types, upwelling occurs at the shelf break primarily in response to eddies and meanders in the Gulf Stream front. However, the distance that upwelled waters penetrate across the shelf depends on wind velocity, local topography, and the density of resident shelf waters [see *Pietrafesa et al.*, this volume; *Lee et al.*, this volume; *Atkinson*, this volume]. During all seasons of the year, eddies and meanders associated with the cyclonic Gulf Stream front cause upwelling at the shelf break. These events occur at characteristic periods of 2–14 days and

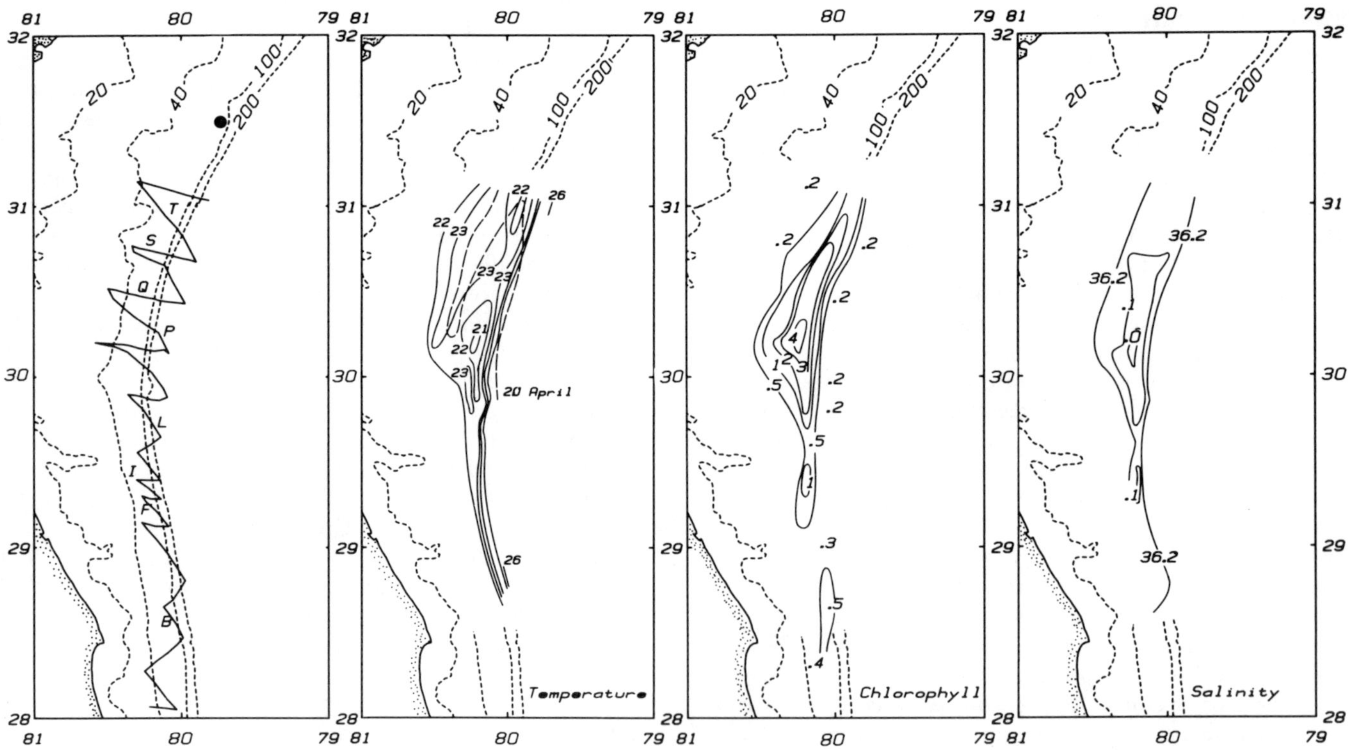

Fig. 2. Cruise track (April 20–22, 1979) and surface temperature (°C), chlorophyll *a* (mg m^{-3}), and salinity ($\times$ 10^{-3}) as determined by continuous shipboard measurements of a frontal eddy. Dashed line in temperature frame indicates Gulf Stream surface thermal front as determined by satellite on April 20. Note that chlorophyll *a* maximum (4 mg m^{-3}) coincides with coldest (21°C) and lowest-salinity (36 $\times$ 10^{-3}) water that forms the upwelled core of the frontal eddy [from *Yoder et al.*, 1981*b*].

primarily affect the outer shelf (i.e., 40- to 200-m isobaths). As a result of the frequent upwelling, nutrient-rich waters are present on the outer shelf (type 1) approximately 50% of the time [*Yoder et al.*, 1983]. The second type of upwelling event (type 2) occurs when the shelf is thermally stratified during approximately May–October. Upwelled waters then penetrate across the shelf as subsurface intrusions of cold, nutrient-rich water. Intrusions are most frequent south of Jacksonville, Florida (30°30′N), and in the cape region north of 33°N [*Blanton et al.*, 1981; *Janowitz and Pietrafesa*, 1982]. Intrusions penetrate all the way to the coast south of Jacksonville, Florida, but only to the middle shelf off Georgia and South Carolina.

Effects of Upwelling Near the Shelf Break (Type 1)

The biological effects of type 1 upwelling are summarized in terms of the following categories: (1) spatial scales, (2) daily rates of total and "new" primary production, (3) types of phytoplankton that dominate upwelled waters, and (4) the importance of upwelling to seasonal productivity.

Upwelling associated with large Gulf Stream frontal eddies leads to mesocale patches (~1000 km²) of relatively high surface chlorophyll *a* concentrations (~4 mg/m³) as illustrated in Figure 2. High chlorophyll concentrations are initially centered within colder water that forms the upwelled core of the eddy [*Yoder et al.*, 1981*b*]. As the event "ages," wind effects and circulation of the eddy cause a more complicated distribution of surface chlorophyll [*McClain et al.*, 1984]. Often, highest concentrations of chlorophyll are beneath the surface mixed layer within subsurface intrusions of nutrient-rich water (*Atkinson et al.* [1978], *Bishop et al.* [1980], *Yoder et al.* [1981*b*, 1983]; see Figure 3). The cross-shelf dimension of relatively high chlorophyll levels (>1 mg/m³) during type 1 upwelling is about 40 km [*Yoder et al.*, 1981, 1983].

Effects of type 1 upwelling on phytoplankton productivity [*Yoder et al.*, 1983] are summarized in the following paragraph. Daily rates of primary production are highly variable during upwelling and depend upon both the amount of chlorophyll *a* in the euphotic zone and upon the chlorophyll *a*–specific carbon uptake rate. The latter is sensitive to the vertical distribution of chlorophyll *a*, i.e., whether most of the euphotic zone chlorophyll is within or below the mixed layer. Figure 4 illustrates rates of primary production and other measurements during two studies of upwelling on the outer shelf. In the first (Figure 4, April 1979), upwelling was very strong, most euphotic zone chlorophyll *a* was within the mixed layer, and primary production was relatively high, ranging from 1 to 6 g C/m²/d (x = 2.7). During the second study (April 1980), most euphotic zone chlorophyll *a* was below the mixed layer, maximum primary production occurred beneath the mixed layer, and average primary production (0.7 g C/m²/d) was much lower than in 1979. The results from three spring studies (1978–1980) showed that average primary production on the outer shelf during upwelling was about 1.8 g C/m²/d with high day-to-day variability.

Figure 4 also illustrates the close relation between primary production and NO_3 uptake during April 1980 [*Yoder et al.*, 1983]. The average C:NO_3 uptake ratio was about 7.5 (by weight). Using this uptake ratio and the Redfield ratio (5.9) to represent C:total N uptake, it was estimated that new production exceeded 50% of the total during upwelling [*Yoder et al.*, 1983].

In general, centric diatoms are the most abundant phytoplankters on the middle and outer shelf [*Hulbert*, 1967; *Marshall*, 1971]. High densities of blue-green algae, *Oscillatoria sp.*, are found during summer [*Dunstan and Hosford*, 1977]. The centric diatom, *Thalassiosira subtilis*, or similar species, is often found within subsurface intrusions of upwelled waters [*Yoder et al.*, 1983]. This diatom forms

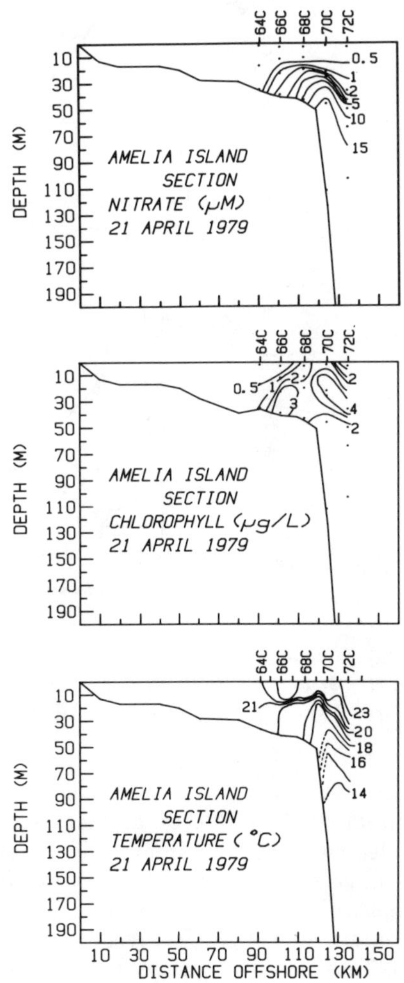

Fig. 3. Outer shelf vertical sections of nitrate (μM), chlorophyll *a* (mg m⁻³), and temperature (°C) during upwelling associated with a Gulf Stream frontal eddy (April 21, 1979) [from *Yoder et al.*, 1981*b*].

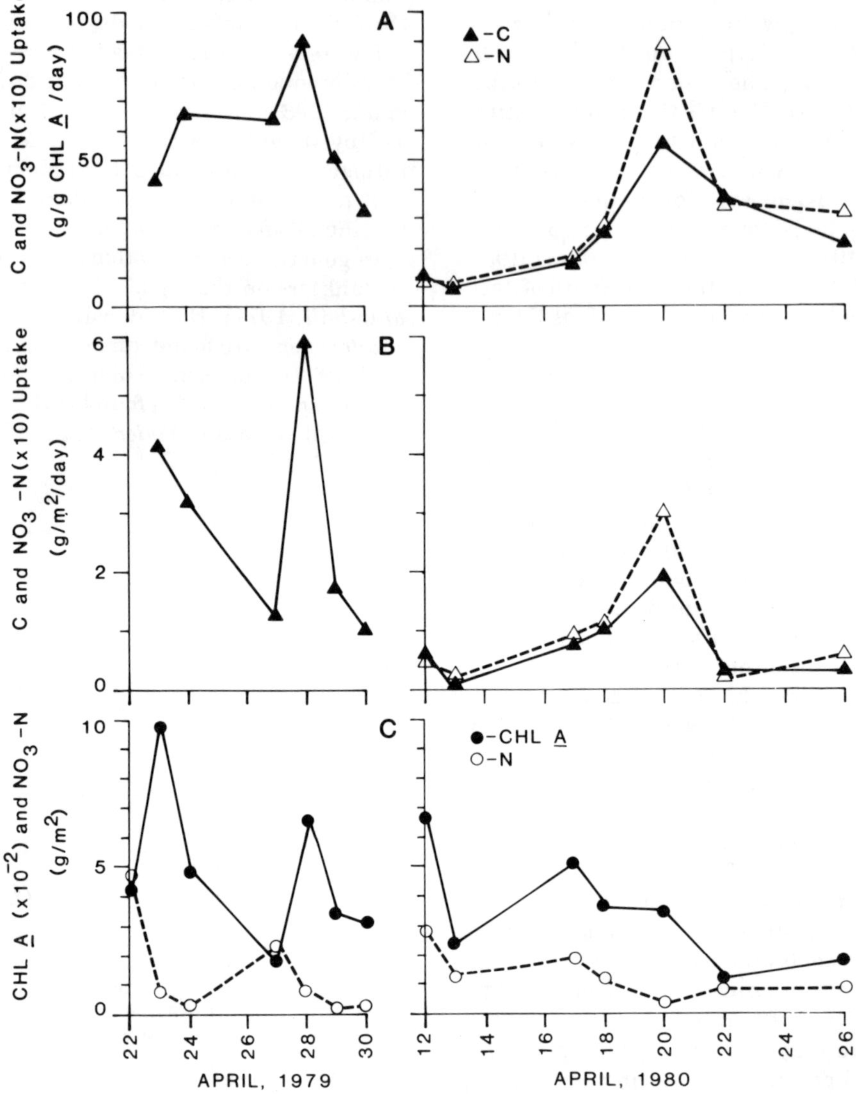

Fig. 4. Euphotic zone time series at the shelf break of (*a*) chlorophyll *a*–specific C and NO_3-N uptake, (*b*) C and NO_3-N uptake, and (*c*) chlorophyll *a* and NO_3-N during upwelling conditions similar to those illustrated in Figures 2 and 3 [from *Yoder et al.*, 1983].

large (~1-mm diameter) gelatinous colonies containing many individual cells (~10-μm diameter). Also reported within upwelled waters are chain-forming diatoms such as *Skeletonema costatum, Asterionella japonica,* and *Rhizosolenia fragilissima* [*Bishop et al.*, 1980; *Yoder et al.*, 1981*b*, 1983; T. Ishimaru and J. A. Yoder, unpublished manuscript, 1982]. Total diatom concentrations as high as 10^6 cells/l were found within upwelled waters [*Yoder et al.*, 1981*b*]. The chain-forming species listed above are generally considered coastal species, and why they dominate upwelled waters, which have an essentially oceanic origin, is unresolved.

The episodic nature of upwelling at the shelf break makes it difficult to estimate the net seasonal or annual effect on production of such phenomena. Ideally, one would like a long time series (months) of frequent (every day) measurements at a number of locations across and along the shelf to resolve their effects on productivity. For practical reasons, such a series of measurements has not been obtained. However, the frequency and duration of upwelling events can be estimated from current meter records [*Lee et al.*, this volume]. Using current meter records to estimate the frequency and duration of upwelling and using average rates of primary production during upwelling and non-upwelling conditions suggest that primary production of the outer shelf is about 180 g C/m²/6 months between November and April [*Yoder et al.*, 1983]. This estimate implies that more than 90% of the production occurs during the 50% of the time that upwelled waters are on the outer shelf. Extrapolating this to an annual rate implies an annual production of about 360 g C/m² for the outer shelf (40- to 200-m isobaths) which is about 3 times higher than earlier estimates [*Haines and Dunstan*, 1975]. The new estimate is also higher than the average annual productivity of the inner shelf (285 g C/m² for the 0- to 20-m isobaths) [*Haines and Dunstan*, 1975].

Effects of Subsurface Intrusions That Penetrate Across the Shelf (*Type 2*)

In the paper by *Atkinson* [this volume] it is shown that upwelled waters penetrate across the shelf as subsurface intrusions of colder, nutrient-rich water during warmer months of the year (May–October). Intrusions are most frequent and penetrate farthest across the shelf between Cape Canaveral and Jacksonville, Florida, and in the cape region north of 33°N [*Blanton et al.*, 1981]. South of Jacksonville, relatively high chlorophyll *a* concentrations are found within subsurface intrusions that extend from the shelf break to the coast (Figure 5). General features of subsurface intrusions that will be discussed below are as follows: (1) phytoplankton blooms rapidly deplete available nitrate and other nutrients, (2) at the peak of the bloom subsurface intrusions contain most of the euphotic zone chlorophyll *a*, (3) productivity maxima occur within the intrusion, (4) productivity is relatively high when intrusions are on the shelf, and (5) intrusions are dominated by diatoms.

When intrusions penetrate inside the 40-m isobath, nitrate and other nutrients are depleted within days to 2 weeks [*Hofmann et al.*, 1980; *O'Malley*, 1981; *Yoder et al.*, 1983]. Figure 6 illustrates a 10-day time series of NO_3 and chlorophyll *a* at a drogue station near the center of an intrusion. The intrusion moved onto the shelf a few days prior to the beginning of the study [*Yoder et al.*, 1983]. The time series shows that all of the NO_3 within the intrusion was converted into phytoplankton biomass within about 7 days. Other studies of summer intrusions confirm the very rapid (days) depletion of NO_3 [*Hofmann et al.*, 1980; *O'Malley*, 1981].

When subsurface intrusions are present on the shelf, the chlorophyll *a* concentration in overlying mixed waters remains more or less constant with time, whereas chlorophyll *a* increases dramatically within the intrusion (Figure 6). As a result, at the peak of the phytoplankton bloom more than 90% of euphotic zone chlorophyll *a* is within the intrusion (see also *Yoder et al.* [1983]). Other investigators reported that the concentration of particulate matter within subsurface intrusions in Onslow Bay, North Carolina, was at least twice that of the overlying surface mixed layer, and the particle size spectra showed maxima between 30 and 100 μm ESP [*Paffenhöfer et al.*, 1980].

Productivity in intrusions is generally many times greater than that in the overlying mixed layer, particularly at or near the peak of the phytoplankton bloom. For example, at the peak of a bloom in summer 1981, about 80% of the primary production and >95% of the NO_3 uptake occurred within the intrusion [*Yoder et al.*, 1984]. High subsurface productivity is possible because of the clarity of the surface mixed layer overlying the intrusion. A typical irradiance attenuation coefficent (400–700 nm) of the mixed layer is 0.1 m^{-1} or less, and thus the 10% isolume is at a depth of >20 m [*Yoder et al.*, 1983].

During the development of the phytoplankton bloom, primary productivity changed roughly in proportion to the temporal change in mean chlorophyll *a* concentration [*Yoder et al.*, 1983]. Primary productivity averaged over the bloom was high in comparison with average summer productivity reported by *Haines and Dunstan* [1975]. For example, during 10 days (Figure 6; *Yoder et al.*, [1983]), production in the intrusion (~15–30 m thick) averaged 1.9 g C/m²/d. Productivity of the surface mixed layer (0–15 m) at the same time was about 0.4 g C/m²/d. A longer time series acquired near the 30-m isobath within an intrusion during summer 1981 yielded an average productivity of ~2 g C/m²/d over 35 days. Productivity decreased when NO_3 was

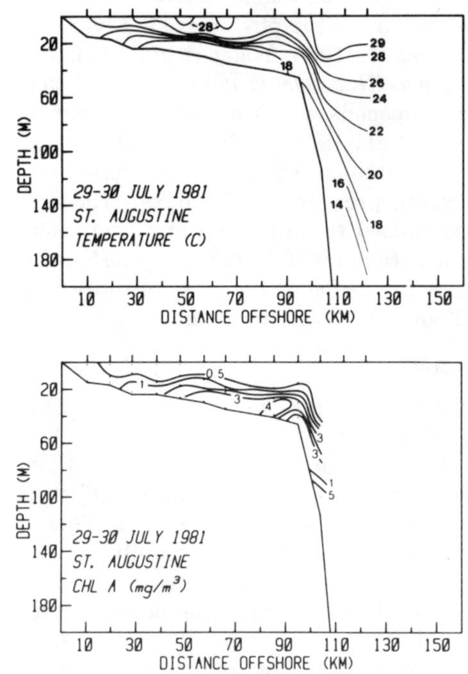

Fig. 5. Temperature and chlorophyll *a* sections across the southeastern shelf at 30°N in summer 1981 following upwelling and intrusion of relatively cold (less than 22°C) and nutrient-rich waters [*Yoder et al.*, 1984].

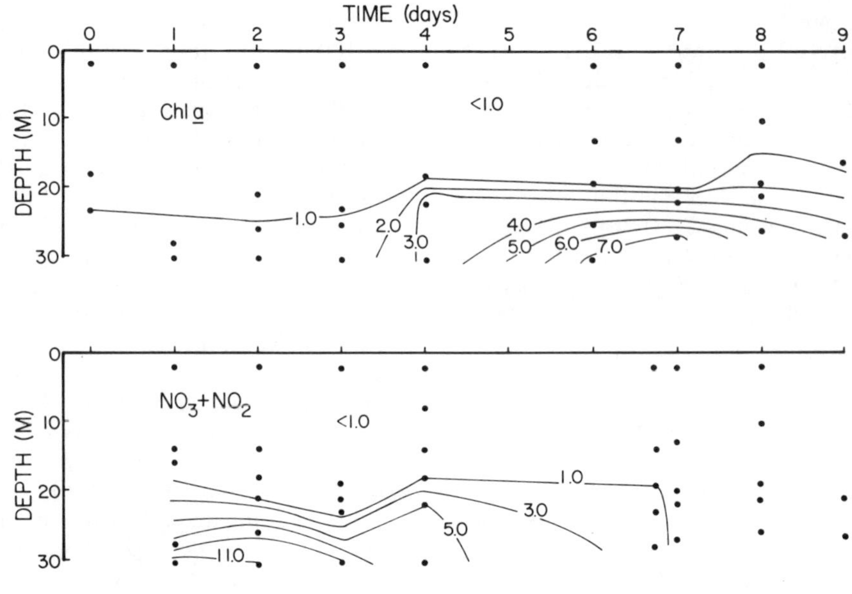

Fig. 6. Chlorophyll *a* and NO_3 times series at a drogue station located near the center of an intrusion studied in August 1978 which was similar to (although smaller than) that illustrated in Figure 5 [from *Yoder et al.*, 1981].

exhausted after about 16 days but remained >1 g C/m²/d for an additional 19 days, suggesting that nutrient recycling was rapid enough to sustain relative high rates of productivity after NO_3 was depleted.

Diatoms dominate the phytoplankton within intrusions [*Paffenhöfer et al.*, 1980; *Yoder et al.*, 1983, 1984]. Large diatoms, such as *Rhizosolenia alata*, are not efficiently grazed by juvenile copepods which initially dominate intrusions [*Paffenhöfer et al.*, 1980; *Paffenhöfer*, this volume]. Thus large diatoms become intrusion dominants because of selective grazing pressure. *Yoder et al.* [1983] found that dominance shifted to large cells and large colonies (e.g., the gelatinous colonies of *Thalassiosira subtilis*) during the 10-day study illustrated in Figure 6. Initially, the intrusion contained both small and large diatoms.

Conclusions

Recent studies support the contention of *Dunstan and Atkinson* [1976] that upwelled nitrate is the most important source of "new" nitrogen on the outer shelf, implying that upwelling is the dominant process affecting the rate and dynamics of production in that area. Thus the important time scale associated with variability of primary production and other autotrophic processes on the outer shelf is not seasonal because upwelling occurs throughout the year. Instead, the "event" time scale (i.e., 2–14 days) dominates. Regardless of season, primary production averages about 2 g C/m²/d during upwelling events with euphotic zone chlorophyll *a* concentrations of about 40–100 mg/m². In the absence of upwelling, primary production and chlorophyll *a* are less than 10% of the above values [*Bishop et al.*, 1980; *Yoder et al.*, 1981*a*, 1983]. At any given location on the outer shelf during any season of the year, conditions change from one extreme to the other within days to weeks.

During warmer months, subsurface intrusions penetrate various distances across the shelf, causing phytoplankton blooms on the middle and outer shelf. Penetration may reach the coast south of Jacksonville, Florida. Given relatively high productivity within intrusions, compared with that of shelf waters, one would expect that alongshelf variation in middle shelf productivity will coincide with alongshelf variability in the frequency and cross-shelf penetration distance of subsurface intrusions. Thus the productivity of the middle shelf south of Jacksonville, Florida (approximately 30°33′N), and in the cape region north of 33°N should be considerably higher than it is off Georgia. This has not been confirmed.

In the New York Bight, about 90% of the primary production that occurs during the winter-spring phytoplankton bloom is exported from the shelf [*Malone et al.*, 1983]. A similar budget has yet to be developed for the southeastern shelf. However, circumstantial evidence suggests that a significant fraction of primary production on the outer shelf may also be advected off the shelf before it is utilized by zooplankton or other animals. We mentioned earlier that when the water column on the outer shelf is vertically mixed (November–April), upwelled waters generally do not penetrate to the middle shelf. Rather, they remain on the outer shelf where alongshore currents are strong (~50 cm s^{-1} or greater) owing to the proximity of the Gulf Stream. Figure 7 shows a section across the outer shelf following an upwelling event. The subsurface chlo-

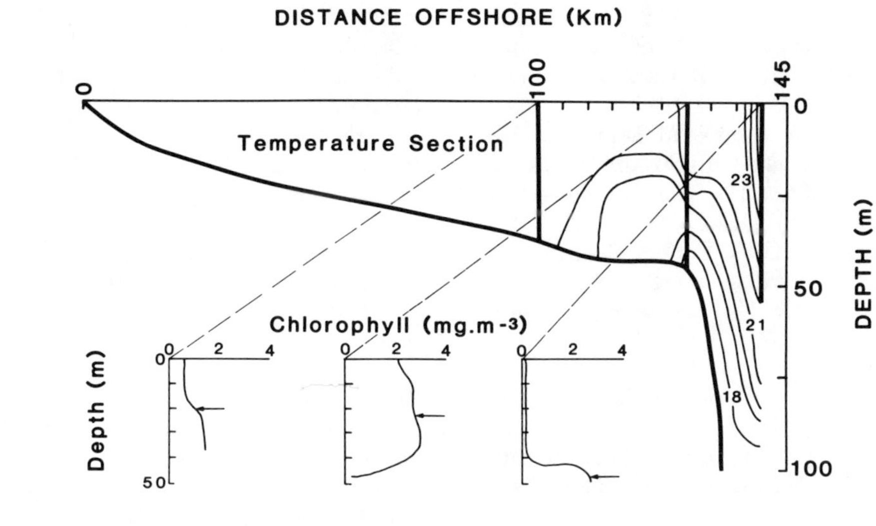

Species Composition (%)

S. Costatum	15	52	60
Chaetoceros sp.	3	14	21
Nitzschia sp.	8	8	8
R. fragilissima	0	6	3

Fig. 7. Outer shelf temperature section (°C), vertical chlorophyll *a* distribution at three stations, and diatom species composition at depths indicated by the arrows following an upwelling event in April 1979.

rophyll *a* maximum in the cyclonic front of the Gulf Stream was dominated at that time by the same species of diatom that comprised the surface bloom in upwelled waters on the shelf. Such distributions suggest a continuity of the surface bloom on the shelf with subsurface chlorophyll *a* within the Gulf Stream. In general, coastal diatoms do not characterize oceanic subsurface chlorophyll *a* maxima. Thus one possible interpretation of the results is that the bloom was "flushed" from the shelf and downwelled into the Gulf Stream. Insufficient data are available to determine the validity of this interpretation, or how frequently such a process may occur. However, similar distributions of chlorophyll *a* and diatom species composition across the outer shelf and into the Gulf Stream have been observed at other times (T. Ishimaru and J. A. Yoder, unpublished manuscript, 1982).

During seasons when the shelf is not stratified (November–April), "new production" on the outer shelf is ~100 g C m^{-2} [*Yoder et al.*, 1983]. The total area of the outer shelf (40- to 200-m isobaths) between Cape Canaveral, Florida, and Cape Hatteras, North Carolina, is $\sim 2.8 \times 10^4$ km^2. Thus 2.8×10^{12} g of carbon is available for deposition in shelf sediments, net transfer to higher trophic levels, or export from the shelf during the 6-month period. If all this material is exported from the shelf, the total mass of carbon is approximately 60% of the amount exported from the New York Bight shelf during the winter-spring diatom bloom [*Malone et al.*, 1983]. If, as in the New York Bight, 90% of outer shelf winter-spring "new production" is exported, the potential for total export from the southeastern shelf is about 50% that of the New York Bight. However, on a unit area basis, outer southeastern shelf new production between November and April is equivalent to the calculated annual carbon export from continental shelf ecosystems in general [*Walsh et al.*, 1981]. Processes affecting the removal of new production from the southeastern shelf should be addressed in the future, because if the carbon is exported from the shelf and eventually deposited in slope sediments, it represents a potentially significant sink of the global CO_2 cycle [*Walsh et al.*, 1981].

Acknowledgments. Financial support for the preparation of this manuscript was provided by U.S. Department of Energy, contract DE-AS09-76EV00936.

References

Atkinson, L. P., Hydrography and nutrients of the southeastern U.S. continental shelf, in *Oceanography of the Southeastern U.S. Continental Shelf, Coastal and Estuarine Sci.*, vol. 2, edited by L. P. Atkinson, D. W. Menzel, and K. A. Bush, AGU, Washington, D. C., this volume.

Atkinson, L. P., and D. W. Menzel, Introduction: Oceanography of the southeast United States continental shelf, in *Oceanography of the Southeastern U.S. Continental Shelf, Coastal and Estuarine Sci.*, vol. 2, edited by L. P. Atkinson, D. W. Menzel, and K. A. Bush, AGU, Washington, D. C., this volume.

Atkinson, L. P., G.-A. Paffenhöfer, and W. M. Dunstan, The chemical and biological effect of a Gulf Stream intrusion off St. Augustine, Florida, *Bull. Mar. Sci., 28*, 667–679, 1978.

Bishop, S. S., J. A. Yoder, and G.-A. Paffenhöfer, Phytoplankton and nutrient variability along a cross-shelf transect off Savannah, Georgia, USA, *Estuarine Coastal Shelf Sci.*, *11*, 359–368, 1980.

Bishop, S. S., K. A. Emmanuele, and J. A. Yoder, Nutrient limitation of phytoplankton growth in Georgia nearshore waters, *Estuaries*, in press, 1984.

Blanton, J. O., Ocean currents along a nearshore frontal zone on the continental shelf of the southeastern United States, *J. Phys. Oceanogr.*, *11*, 1627–1637, 1981.

Blanton, J. O., L. P. Atkinson, L. J. Pietrafesa, and T. N. Lee, The intrusion of Gulf Stream water across the continental shelf due to topographically-induced upwelling, *Deep Sea Res. Part A*, *28*, 393–405, 1981.

Dugdale, R. C., and J. J. Goering, Uptake of new and regenerated forms of nitrogen in primary productivity, *Limnol. Oceanogr.*, *12*, 196–206, 1967.

Dunstan, W. M., and L. P. Atkinson, Sources of new nitrogen for the South Atlantic Bight, in *Estuarine Processes,* vol. 1, edited by M. Wiley, pp. 69–78, Academic, Orlando, Fla., 1976.

Dunstan, W. M., and J. Hosford, The distribution of planktonic bluegreen algae related to the hydrography of the Georgia Bight, *Bull. Mar. Sci.*, *27*, 824–829, 1977.

Haines, E. B., Processes affecting production in Georgia coastal waters, Ph.D. dissertation, 117 pp., Duke Univ., Durham, N. C., 1974.

Haines, E. B., Nitrogen pools in Georgia coastal waters, *Estuaries*, *2*, 33–39, 1979.

Haines, E. B., and W. M. Dunstan, The distribution and relation of particulate organic material and primary productivity in the Georgia Bight 1973–74, *Estuarine Coastal Mar. Res.*, *3*, 431–441, 1975.

Hofmann, E. E., L. J. Pietrafesa, J. M. Klinck, and L. P. Atkinson, A time-dependent model of nutrient distribution in continental shelf waters, *Ecol. Modell.*, *10*, 193–214, 1980.

Hulbert, E. M., Some notes on the phytoplankton of the southeastern coast of the United States, *Bull. Mar. Sci.*, *17*, 330–337, 1967.

Janowitz, G. S., and L. J. Pietrafesa, The effects of alongshore variation in bottom topography on a boundary current—topography induced upwelling, *Cont. Shelf Res.*, *1*, 123–141, 1982.

Kjerfve, B., and H. N. McKellor, Jr., Times series measurements of estuarine fluxes, in *Estuarine Perspectives*, edited by V. S. Kennedy, pp. 341–358, Academic, Orlando, Fla., 1980.

Lee, T. N., V. Kourafalou, J. D. Wang, W. J. Ho, J. O. Blanton, L. P. Atkinson, and L. J. Pietrafesa, Shelf circulation from Cape Canaveral to Cape Fear, in *Oceanography of the Southeastern U.S. Continental Shelf, Coastal and Estuarine Sci.*, vol. 2, edited by L. P. Atkinson, D. W. Menzel, and K. A. Bush, AGU, Washington, D. C., this volume.

Malone, T. C., T. S. Hopkins, P. G. Falkowski, and T. E. Whitledge, Production and transport of phytoplankton biomass over the continental shelf of the New York Bight, *Cont. Shelf Res.*, *1*, 305–338, 1983.

Marshall, H. G., Composition of phytoplankton off the southeastern coast of the United States, *Bull. Mar. Sci.*, *21*, 806–825, 1971.

McClain, C. R., L. J. Pietrafesa, and J. A. Yoder, Observations of Gulf Stream–induced and wind-driven upwelling in the Georgia Bight using ocean color and infrared imagery, *J. Geophys. Res.*, *89*, 3705–3723, 1984.

Oertel, G. F., and W. M. Dunstan, Suspended sediment distribution and certain aspects of phytoplankton production off Georgia, USA, *Mar. Geol.*, *40*, 171–197, 1981.

O'Malley, P. G., The effect of summertime shelf break upwelling on nutrient flux in southeastern United States continental shelf waters, M.S. thesis, 83 pp., Ga. Inst. of Technol., Atlanta, 1981.

Paffenhöfer, G.-A., The abundance and distribution of zooplankton of the southeastern shelf of the U.S., in *Oceanography of the Southeastern U.S. Continental Shelf, Coastal and Estuarine Sci.*, vol. 2, edited by L. P. Atkinson, D. W. Menzel, and K. A. Bush, AGU, Washington, D. C., this volume.

Paffenhöfer, G.-A., D. Deibel, L. P. Atkinson, and W. M. Dunstan, The relation of concentration and size distribution of suspended particulate matter to hydrography in Onslow Bay, North Carolina, *Deep Sea Res.*, *27*, 435–448, 1980.

Pietrafesa, L. J., J. O. Blanton, J. D. Wang, V. Kourafalou, T. N. Lee, and K. A. Bush, The tidal regime in the South Atlantic Bight, in *Oceanography of the Southeastern U.S. Continental Shelf, Coastal and Estuarine Sci.*, vol. 2, edited by L. P. Atkinson, D. W. Menzel, and K. A. Bush, AGU, Washington, D. C., this volume.

Riley, G. A., Factors controlling phytoplankton populations on Georges Bank, *J. Mar. Res.*, *6*, 54–73, 1946.

Riley, G. A., Seasonal fluctuations of the phytoplankton population in New England coastal waters, *J. Mar. Res.*, *6*, 114–125, 1947.

Riley, G. A., Environmental control of autumn and winter diatom flowerings in Long Island Sound, in *International Oceanographic Congress*, pp. 850–851, American Association for the Advancement of Science, Washington, D. C., 1959.

Riley, G. A., The plankton of estuaries, in *Estuaries, Publ. 83*, edited by G. Lauff, pp. 316–326, American Association for the Advancement of Science, Washington, D. C., 1967.

Ryther, J. H., and C. S. Yentsch, Primary production of continental shelf waters off New York, *Limnol. Oceanogr.*, *3*, 327–335, 1958.

Smayda, T. J., A survey of phytoplankton dynamics in the coastal waters from Cape Hatteras to Nantucket, *Mar. Sci. Publ. Ser. 2*, pp. 3-1 to 3-100, Univ. of R. I., Kingston, 1973.

Smayda, T. J., Plankton processes in mid-Atlantic nearshore and shelf waters and energy-related activities, in *Effects of Energy-Related activities on the Atlantic Continental Shelf*, edited by B. Manowitz, pp. 70–95, National Technical Information Service, Springfield, Va., 1976.

Sverdrup, H. V., On conditions for the vernal blooming of plankton, *J. Cons.*, *18*, 287–295, 1953.

Thomas, J. P., Influence of the Altamaha River on primary production beyond the mouth of the river, M. S. thesis, 88 pp., Univ. of Ga., Athens, 1966.

Turner, R. E., Plankton productivity and the distribution of fishes on the southeastern U.S. continental shelf, *Science*, *214*, 351–354, 1981.

Turner, R. E., S. W. Woo, and H. R. Jitts, Estuarine influences on a continental shelf plankton community, *Science*, *206*, 218–220, 1979.

Walsh, J. J., T. E. Whitledge, F. W. Barvenik, C. D. Wirick, S. O. Howe, W. E. Esaias, and J. T. Scott, Wind events and food chain dynamics within the New York Bight, *Limnol. Oceanogr.*, *23*, 659–683, 1978.

Walsh, J. J., G. T. Rowe, R. L. Iverson, and C. P. McRoy, Biological export of shelf carbon is a sink of the global CO_2 cycle, *Nature*, *291*, 196–201, 1981.

Yoder, J. A., L. P. Atkinson, J. O. Blanton, D. R. Deibel, D. W. Menzel, and G.-A. Paffenhöfer, Plankton productivity and the distribution of fishes on the southeastern U.S. continental shelf, *Science*, *214*, 351–354, 1981*a*.

Yoder, J. A., L. P. Atkinson, T. N. Lee, H. H. Kim, and C. R. McClain, Role of Gulf Stream frontal eddies in forming phytoplankton patches on the outer southeastern shelf, *Limnol. Oceanogr.*, *26*, 1103–1110, 1981*b*.

Yoder, J. A., L. P. Atkinson, S. S. Bishop, E. E. Hofmann, and T. N. Lee, Effect of upwelling on phytoplankton productivity on the outer southeastern U.S. continental shelf, *Cont. Shelf Res.*, *1*, 385–404, 1983.

Yoder, J. A., L. P. Atkinson, S. S. Bishop, J. O. Blanton, and T. N. Lee, Phytoplankton dynamics within Gulf Stream intrusions during summer, 1981, *Cont. Shelf Res.*, in press, 1984.

The Abundance and Distribution of Zooplankton on the Southeastern Shelf of the United States

G.-A. Paffenhöfer

Skidaway Institute of Oceanography, Savannah, Georgia

Episodic upwelling of nutrient-rich subsurface water from the Gulf Stream leads, through phytoplankton production, to high zooplankton abundances on the middle and outer southeastern continental shelf (South Atlantic Bight). The most abundant zooplankton taxa in relatively young intrusions are those which reproduce asexually (e.g., doliolids and cladocerans) and small copepods with high rates of egg production. Both groups respond rapidly to increased food supply. Large copepods, with longer generation times, become abundant only after stranded intrusions "age." Zooplankton concentrations in upwelled waters can approach those in quasi-continuously upwelling environments such as off Peru, the Ivory Coast, and Oregon. High concentrations of small zooplankton temporarily support fish larvae, which occur at high average densities (5 to 10 larvae m^{-3}). It is suggested that (1) the episodic nature of upwellings results in discontinuous supplies of food for zooplankton and (2) water displacement, due to winds and motions of the Gulf Stream on the outer shelf, and the predatory behavior of many adult copepods and fish cause high losses of fish larvae. Large stocks of adult fish are not encountered on the southeastern shelf.

Introduction

Bottom topography and onshore-offshore motions of the Gulf Stream cause periodic upwellings of cold, nutrient-rich water onto the continental shelf of the South Atlantic Bight (SAB) [*Blanton et al.*, 1981]. In waters upwelled into the euphotic zone on the shelf high radiation and nutrient enrichment during summer lead to high photosynthetic rates and large standing stocks of phytoplankton [*Yoder et al.*, 1983]. Elevated levels of plant biomass are temporary, however, as nutrients are rapidly depleted, and sinking, offshore advection, and/or grazing processes remove plants from the water column.

To understand the impact of the zooplankton on available food, and vice versa, zooplankton abundances and their taxonomic composition must be known [*Jorgensen*, 1966; *Paffenhöfer and Knowles*, 1978]. Because environmental variables influence the abundance and distribution of zooplankton [*Le Borgne*, 1977; *Judkins*, 1980; *Smith et al.*, 1981], key variables such as temperature and food concentration should accompany zooplankton samplings.

Our goals were to determine abundant zooplankton taxa in and near upwellings, to relate their occurrence to circulation, hydrographic variables, and food abundance, and to correlate the abundance of fish larvae with the abundances and types of zooplankton on which they prey.

Methods

Data from July and August of 1976, 1978, and 1979 are presented. Samples in 1976 were taken in Onslow Bay, North Carolina, using an opening-closing net (Tucker-Trawl) with a mouth opening of 40 × 40 cm and 100-μm mesh [*Paffenhöfer*, 1980]. In 1978 and 1979, samples were collected on the northeastern Florida shelf using a diaphragm pump, which drew water at 140 to 200 l min^{-1} and which was passed through 103-μm and then 30-μm mesh netting. Almost all multicellular organisms were retained. Zooplankton sampling was preceded by a CTD or expendable bathythermograph (XBT) cast to describe vertical temperature and salinity structures. This information was used to establish the depth ranges at which zooplankton were sampled. In 1976, only the upper mixed layer (UML) and near bottom layer (intrusion of cold Gulf Stream water) were sampled. In 1978 and 1979 the UML, thermocline, and intrusions were sampled separately using a depth and temperature sensor at the pump intake to determine sampling depths. All samples were preserved in 2% formaldehyde and transferred after 4 weeks to a mixture of propylene glycol and propylene phenoxetol [*Steedman*, 1974]. Entire samples to 1/128 of a sample were counted. Usually at least 30 specimens of each taxon were counted resulting in confidence limits of ±10 [*Lund et al.*, 1958]. Samples collected on the 30-μm mesh were counted with an inverted microscope.

Suspended particulate matter, obtained from bottle casts, was measured as particulate organic carbon or particle volume using, respectively, a Perkin Elmer CHN Analyzer or a Coulter Counter TAII fitted with 100-μm and 400-μm diameter orifices.

Results

Information is first presented on the association of zooplankton abundances with Gulf Stream upwellings. Second, the vertical distribution of zooplankton is related to temperature and potential food. Third, the vertical and horizontal abundances of zooplankton before and during an upwelling are compared. Fourth, annual variations in zooplankton abundances will be compared. Finally, the abundance of fish larvae is addressed.

Association of Zooplankton Abundance With Gulf Stream Upwelling

On July 23, 1976, both a stranded and a shelf break intrusion were observed during an onshore-offshore transect in Onslow Bay, North Carolina. The bottom layer was characterized by high, and the surface layer by low, salinities in nearshore and midshelf areas (Figure 1*b*). The stranded intrusion (stations 51 to 54, Figure 1*a*) contained large diatoms of the genus *Rhizosolenia*, while the incoming shelf break intrusion had large diatoms other than *Rhizosolenia*. The nearshore stations (49 and 50) had higher particle loads in the surface than in the bottom layer. These higher loads presumably contributed to larger concentrations of appendicularia and calanoids near the surface at station 50 than in the bottom layer and offshore surface layers. Bottom layer maxima in numbers of calanoids, cyclopoids, and chaetognaths were within the stranded intrusion (Figure 1*b*). In the shelf break intrusion, concentrations of calanoids, cyclopoids, appendicularians, and chaetognaths decreased from a maximum near the leading edge (station 56) toward offshore (Figure 1*b*), suggesting various stages of population development (shown earlier for *Paracalanus* [*Paffenhöfer*, 1980]). Food concentrations in outer shelf waters (stations 57 and 158) should not have severely limited calanoid and appendicularian growth (Figure 1*a*). In surface waters, maximum abundances of calanoids, cyclopoids, and appendicularia were at station 55. Uniquely over the transect, virtually all calanoids were *Paracalanus* copepodid stages I to IV. Young copepods and appendicularians often occur in greater concentrations in the upper mixed layer than in intruded waters [*Paffenhöfer*, 1983], and the young copepods generally originate

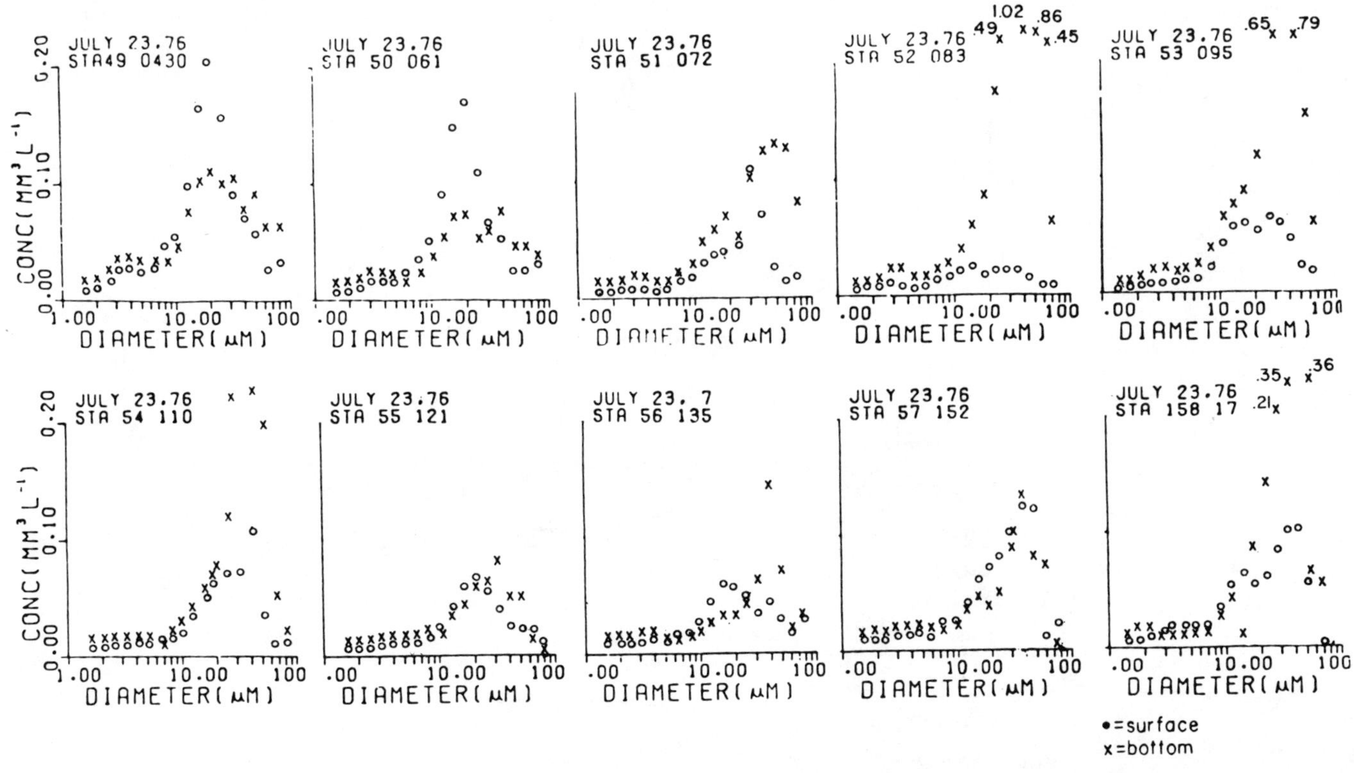

Fig. 1*a*. Near-surface (3-m depth) and near-bottom (2 m above seafloor) particle concentrations in Onslow Bay, North Carolina, July 23, 1976 (modified from *Paffenhöfer et al.* [1980]).

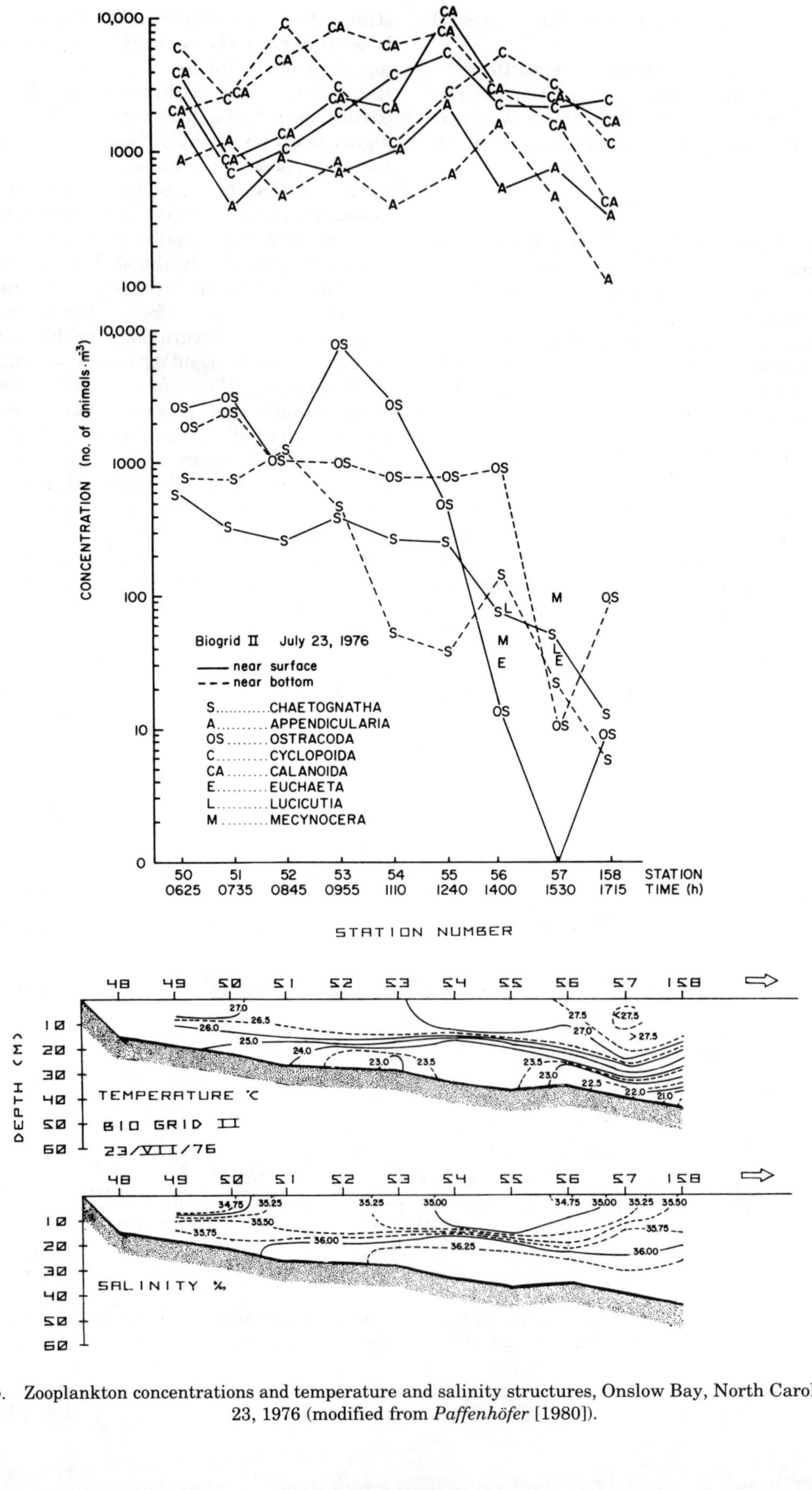

Fig. 1*b*. Zooplankton concentrations and temperature and salinity structures, Onslow Bay, North Carolina, July 23, 1976 (modified from *Paffenhöfer* [1980]).

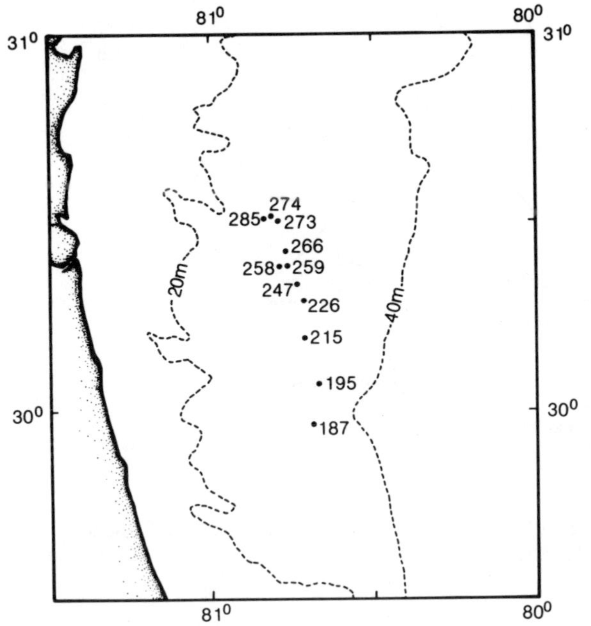

Fig. 2. Station positions, August 5–14, 1978, northeastern Florida shelf (modified from *Paffenhöfer* [1983]).

from adults which reproduced in intruded waters. Most calanoids (mostly *Paracalanus*) and cyclopoids were in the high-salinity, offshore bottom water. A strong influence of the nearshore environment on the zooplankton composition was observed only near the surface at station 50.

The Vertical Distribution of Zooplankton in Relation to Temperature and Food

Bottom intrusions of Gulf Stream water are characterized by reduced temperatures and relatively high concentrations of particulate matter. Some zooplankton, for example, *Oncaea* spp., *Temora turbinata*, and *Penilia avirostris*, are found in higher abundances in these bottom intrusions than in surface waters [*Paffenhöfer*, 1980]. Data from 1978 on the shelf off northeastern Florida (Figure 2) show a pronounced upper mixed layer, thermocline, and intrusion (Figure 3) which were sampled separately. Between August 5 and 8 the water layer from the bottom of the UML to near the 23°C isotherm was sampled as thermocline and part or most of the water below as the intrusion. After August 9 we sampled the entire UML, thermocline, and intrusion. There were periods of low (August 5 to 7) and high (from August 8 on) concentrations of particulate organic carbon in the bottom layer (Figure 4). Young *Oikopleura* spp. (Appendicularia) and *Oncaea* spp. (*Cyclopoida*) were abundant in the UML and the thermocline (Figure 5), while later stages of *Oncaea* spp. were most abundant in the bottom layer. *Penilia avirostris* (*Cladocera*), *Temora turbinata* (Calanoida), *Eucalanus* spp. (*Calanoida*), and *Branchiostoma* sp. (Cephalochordata) were located in the thermocline and bottom layer (Figure 6). The calanoid *Centropages furcatus* occurred mostly in the UML and the thermocline.

Kendall's coefficient of concordance [*Tate and Clelland*, 1957] was used to test whether zooplankton are more abundant (1) in colder (intrusion) than in warmer (UML) water and (2) near high or low particle (food) concentrations. To test hypothesis 1, each zooplankton taxon was separately ordered in the sequence of increasing temperature, and for hypothesis 2, in sequence of increasing particle concentration. Each taxon was then ranked separately in relation to its abundance. Following this the coefficient of concordance W was calculated. When W was significant, the sequence of ranks agreed between stations ($p < 0.05$). Significant concordance between stations was found for hypothesis 1 for several of the abundant taxa and groups of data (Table 1): the abundance of *Oncaea* spp., *Eucalanus* spp., *T. turbinata*, and *P. avirostris* increased with decreasing temperature. Significant concordance for hypothesis 2 (increasing abundance with increasing particle concentration) was found for the same four groups, excepting *Oikopleura* spp., *Oithona* spp., and *Centropages furcatus*. *Oikopleura* spp. was most abundant in surface waters where small particles including microflagellates dominate. *Oikopleura dioica* ingests mainly particles less than 8 μm in size (G.-A. Paffenhöfer, unpublished data, 1980). Older copepodid stages of *Oithona* spp. and *C. furcatus* are omnivores. The thermocline, with fewer large diatoms compared with the intrusion, appeared to be the favored environment of these copepods (Table 2).

The tau rank–correlation test [*Tate and Clelland*, 1957] showed that the abundance of *T. turbinata*, older copepodids, and adult *Oncaea* spp. was positively correlated with the abundance of particulate matter ($p < 0.025$). *Eucalanus* abundance was positively related to particle concentration when food concentrations were low ($p < 0.01$) but not when they were high ($p > 0.10$). At food levels above 0.6 mm^3

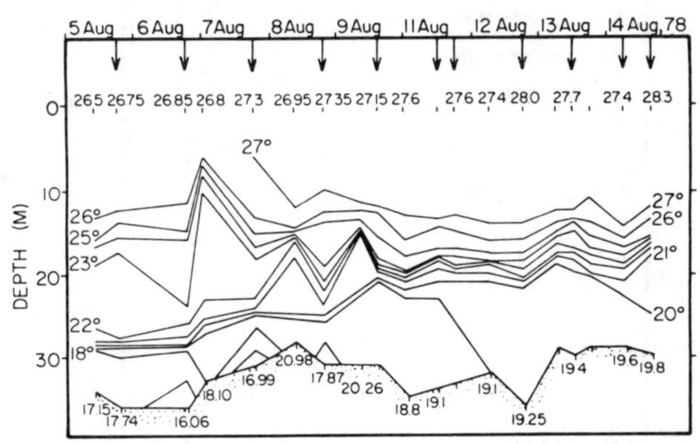

Fig. 3. Vertical and horizontal temperature structures, August 5–14, 1978, northeastern Florida shelf (modified from *Paffenhöfer* [1983]).

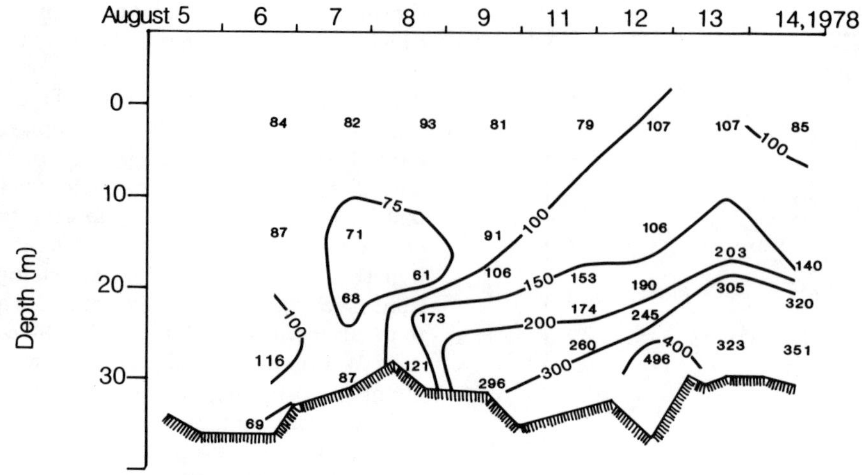

Fig. 4. Vertical concentration of particulate organic carbon, August 5–14, 1978, northeastern Florida shelf (modified from *Paffenhöfer* [1983]).

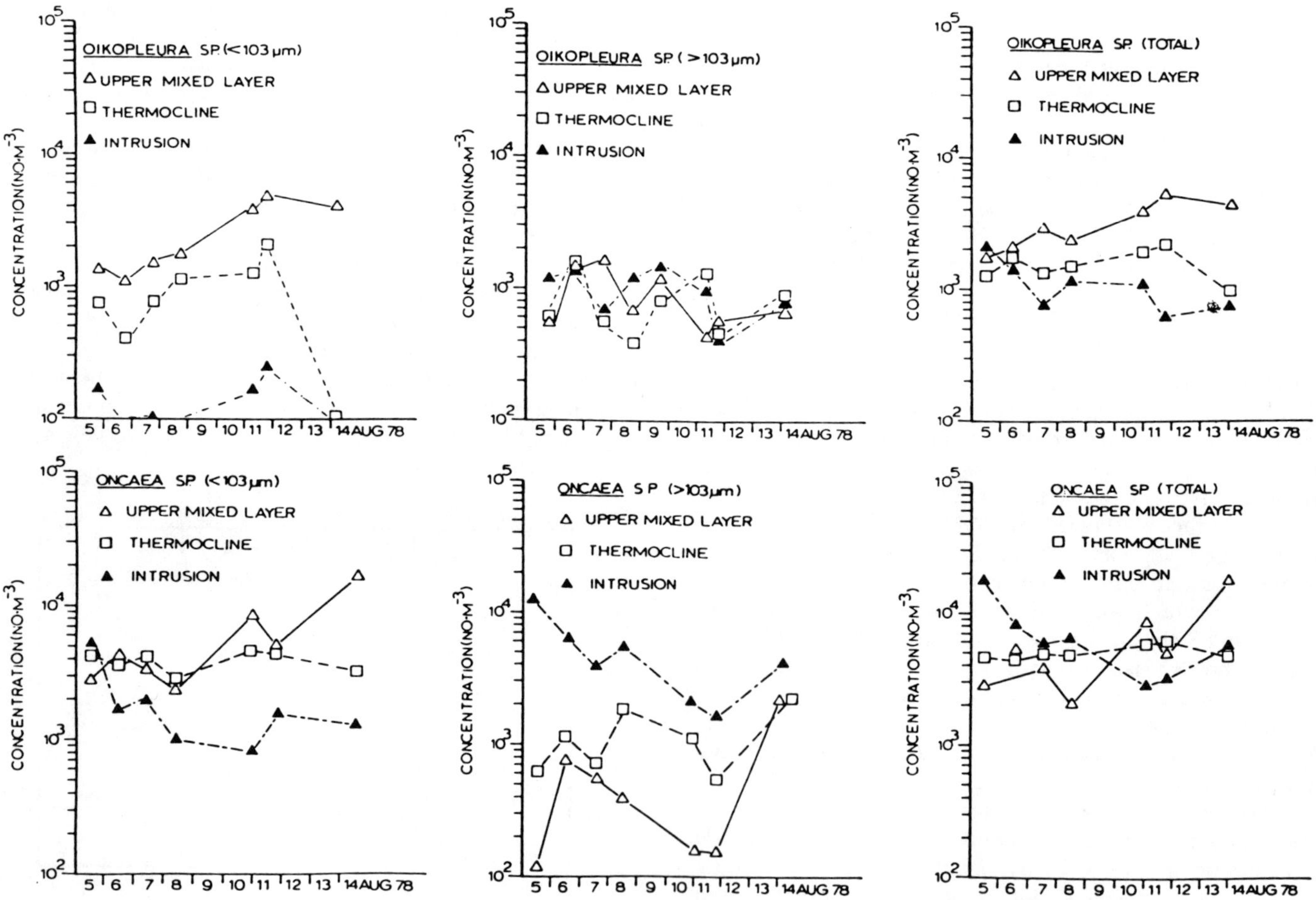

Fig. 5. Vertical distribution of abundant zooplankton taxa, August 5–14, 1978, northeastern Florida shelf (modified from *Paffenhöfer* [1983]).

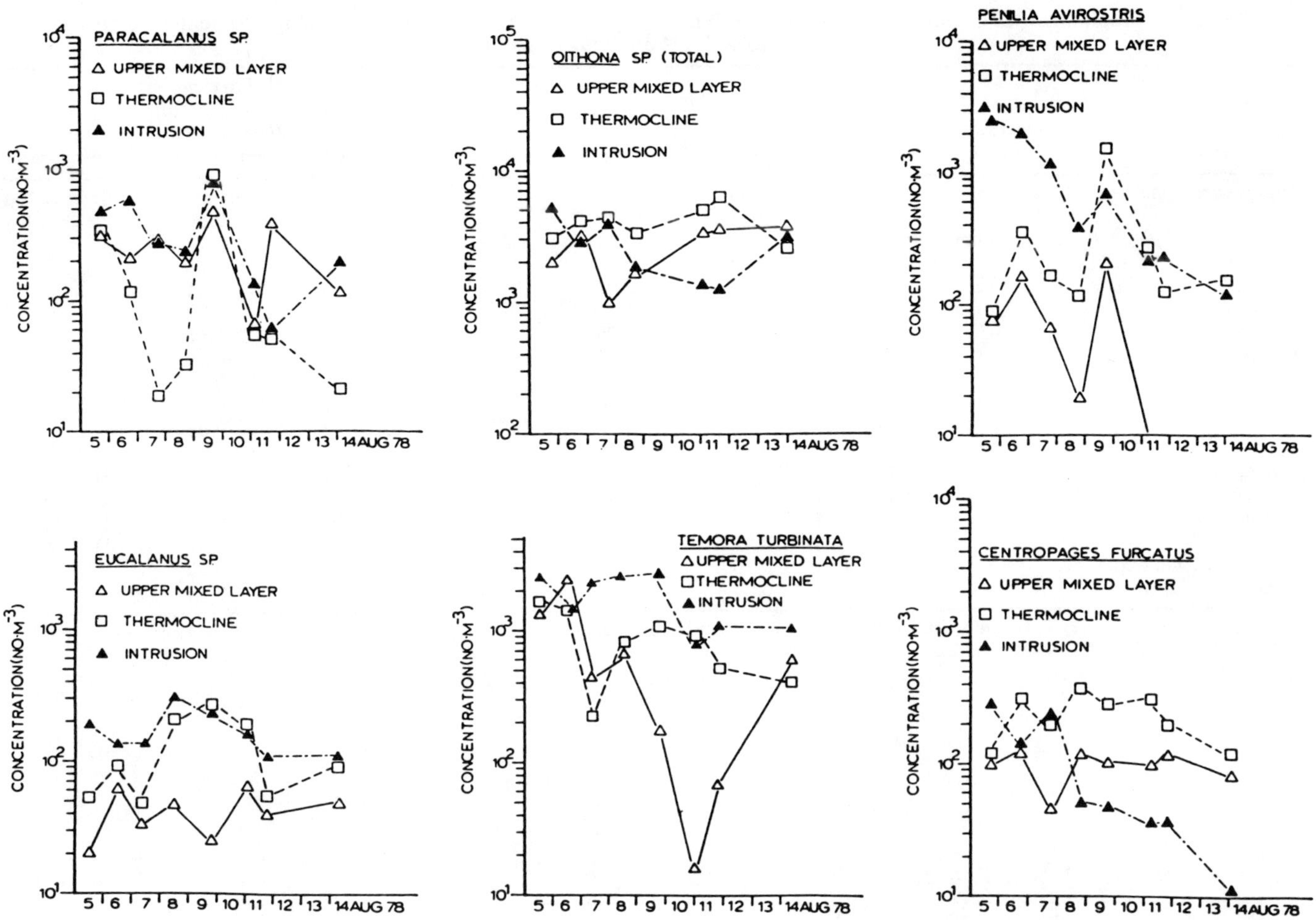

Fig. 6. Same as Figure 5.

l^{-1} the abundance of *Eucalanus* did not appear to increase with increasing food.

The Vertical and Horizontal Distribution of Zooplankton in Relation to Upwelling Events

Clear evidence on which forms of zooplankton develop in intrusions cannot be obtained unless abundances on the shelf are known before, during, or in various stages of intrusions. Data from summer 1979 [*Paffenhöfer et al.*, 1984] provide this information.

Off Jacksonville on July 17–18, 1981, shelf waters were characterized by relatively high temperatures and low concentrations of particulate matter indicating weak upwelling (Figure 7). The stronger and colder shelf break and stranded upwellings observed 2 and 3 weeks later contained up to 5 mm^3 l^{-1} of particulate matter (Figure 7), sufficient to stimulate the growth and reproduction of zooplankters (Figure 8). Comparing zooplankton distribution and abundance from three upwellings leads to the following generalizations: some zooplanktonic taxa were absent or in very low concentrations (<10 m^{-3}) in the warm, low-particle upwelling (July 17–18) but were abundant in water masses with high-particle loads. These were *Dolioletta gegenbauri* (*Tunicata*), *Temora turbinata*, and *Penilia avirostris* (Figures 9–11). The first two species require approximately 1 mm^3 l^{-1} or more particulate matter for reproduction. Doliolids and *P. avirostris* require temperatures between 16° and 23°C for optimum growth and reproduction [*Binet*, 1976]. All three species increased in numbers in response to intensive phytoplankton production in the upwellings. Different stages of development were observed at station 370 near the shelf break and station 372 further inshore. At both stations, temperatures at the depths sampled for zooplankton were about the same; yet doliolids in the intrusion and thermocline at station 372 had higher concentrations than at station 370, suggesting an advanced response in water upwelled earlier (Figure 10). A similar observation applies to *T. turbinata*

TABLE 1. Zooplankton Abundance Related to Depth/Temperature and Particle Concentration Using Kendall's Coefficient of Concordance (W) [from *Paffenhöfer,* 1983]

Taxon	Size Fraction, μm	Groups of Stations*	W	Depth/Temperature, Probability of Concordance	Sequence of Abundance	W	Particle Concentration, Probability of Concordance	Sequence of Abundance
Oikopleura spp.	<103	187–274	1.00	<0.01	U > T > I	0.75	<0.01	L > M > H
	>103		0.02	>0.20		0.11	>0.20	
	all specimens		0.88	<0.01	U > T > I	0.57	<0.05	
Oncaea spp.	<103	187–274	0.36	>0.10		0.39	>0.05	
	>103		1.00	<0.01	I > T > U	0.76	<0.01	H > M > L
	all specimens		0.04	>0.20		0.08	>0.20	
	<103	226–274	0.75	>0.05		0.81	=0.05	L > M > H
Oithona spp.	<103	187–274	0.43	>0.05		0.26	>0.20	
	>103		0.20	>0.20		0.11	>0.20	
	all specimens		0.24	>0.20		0.14	>0.20	
	<103	226–274	0.81	=0.05	T > U > I	0.75	>0.05	
	>103		0.45	<0.05	T > U > I	0.12	>0.20	
Eucalanus spp.	···	187–274	0.57	<0.01	I > T > U	0.45	<0.01	H > M > L
T. turbinata	···	187–274	0.33	<0.01	I > T > I	0.38	<0.01	H > M > L
C. furcatus	···	187–274	0.40	<0.01	T > U > I	0.26	<0.05	L > M > H
P. avirostris	···	187–274	0.79	<0.01	I > T > U	0.43	<0.01	H > M > L
Branchiostoma spp.	···	187–274	0.75	<0.01	I = T > U	0.33	>0.05	
Copepod Nauplii	···	187–274	0.25	>0.20		0.38	<0.10	

U, upper mixed layer; T, thermocline; I, intrusion. L, M, and H indicate low, medium, and high food concentration, respectively.
*Numbers 187–274 = 7–13 and 226–274 = 4–9.

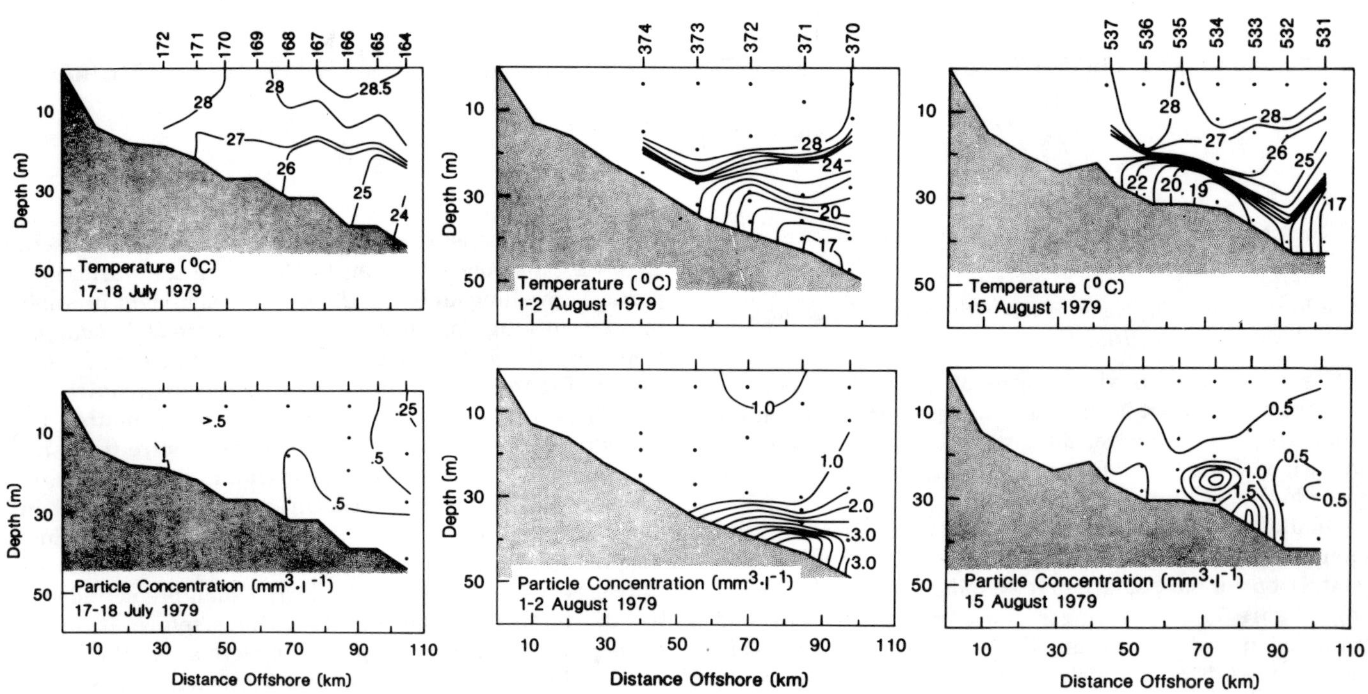

Fig. 7. Vertical structure of temperature and particulate matter on three transects off St. Augustine, Florida, during summer 1979 (modified from *Paffenhöfer et al.* [1984]).

TABLE 2. Average Concentrations of Abundant Zooplankton Taxa (Specimens per Cubic Meter) in Upwelling Areas off the Southeastern Shelf of the United States and off Peru, Northwest Africa, the Ivory Coast and Oregon

	Onslow Bay [*Paffenhöfer*, 1980]		East Florida Shelf						Other Upwelling Areas	
			Summer 1978			Summer 1979				
	UML	Intrusion	UML	Thermocline	Intrusion	UML	Thermocline	Intrusion		
Doliolida	71	26	1.5	20	27	33	384	818	up to 350	Ivory Coast [*Binet*, 1976]
*Oikopleura**	832	947	867	788	971	342	1,030	1,037	427	Peru [*Dagg et al.*, 1980]
			2,584	878	99	405	599	197		
Corycaeus	218	673	118	124	338	195	222	73		
*Oithona**	1,443	803	986	1,374	1,414	661	957	1,262	460	Peru [*Dagg et al.*, 1980]
			1,543	2,811	1,417	414	453	151		
*Oncaea**	279	1,985	576	1,356	5,310	27	410	414	3,195	Peru [*Dagg et al.*, 1980]
			6,289	4,084	1,974	102	575	172		
Paracalanus	1,429	3,033	265	191	350	973	1,079	917	2,088	Peru [*Dagg et al.*, 1980]
T. turbinata	115	166	737	904	1,871	98	835	1,745	100 to 2,000	Ivory Coast [*Binet*, 1977]
C. furcatus	287	84	99	242	105	2	21	10	up to 250	Ivory Coast [*Binet*, 1977]
Eucalanus	51	32	42	126	170	1	18	30	up to 220	Ivory Coast [*Binet*, 1977]
P. avirostris	4	22	68	355	920	28	701	916	~250 to 3,000	Ivory Coast [*Binet*, 1975]
*Branchiostoma**	39	11	105	2,461	1,553	23	338	355	917	North West Africa [*Gosselck and Kuehner*, 1973]
			70	2,342	1,327					
Chaetognatha	190	420	37	110	129	14	84	118		
Copepod Nauplii			59,800	41,800	44,500	13,400	18,300	14,800	up to 40 l^{-1} of *Calanus marshallae*	Oregon [*Peterson et al.*, 1979]

Binet [1975, 1976, 1977]: 200-μm mesh. *Dagg et al.* [1980]: 102-μm mesh. *Gosselck and Kuehner* [1973]: 200-μm mesh. *Peterson et al.* [1979]: 120-μm mesh.

*Where two lines of data are given, upper line gives concentration retained by 103- to 110-μm mesh and lower line gives concentration retained by 30-μm mesh after passing 103-μm mesh.

nauplii, copepodids and adults (Figure 10). The stranded intrusion of August 14 to 16, 1979 (stations 530, 540, and 541 near intrusion center), had been on the shelf longer than its predecessor of August 1 and 2. This was obvious by the presence of greater concentrations of opportunists like doliolids and *T. turbinata*, which had had more time to respond, by reproduction, to increased food supply (Figures 10 and 11).

The vertical distribution patterns of the various zooplanktonic taxa can be divided into two groups: those clearly preferring the colder bottom and thermocline waters and those which are evenly distributed throughout the water column. The latter include *Oithona* spp., *Paracalanus* spp., and copepod nauplii (excluding nauplii of *T. turbinata*). Nauplii and copepodids of *Paracalanus* spp. feed efficiently, grow, and reproduce on particles 10 μm and smaller at concentrations near 0.3 mm^3 l^{-1} (G.-A. Paffenhöfer, unpublished data, 1982). This partly explains their even vertical distribution. Because information on feeding and growth of juvenile *Oithona* spp. is scarce [*Petipa et al.*, 1970], a similar correlation cannot be applied to this species. Asexually reproducing species, like *P. avirostris* and *D. gegenbauri* (together with the small copepod *T. turbinata*), are the zooplankton that respond the fastest to sudden increases in phytoplankton. They and *Oncaea* spp., which reaches high concentrations in "older" intrusions, are the abundant zooplankton taxa characteristic of intrusions on the southeastern shelf.

Interannual Comparisons of Zooplankton Abundances

During 3 years of studies of summer upwellings on the southeastern shelf, only *Oikopleura* spp. and *Oithona* spp. were uniformly abundant in intruded waters (Table 2). All other taxa were in high abundance in some summers but

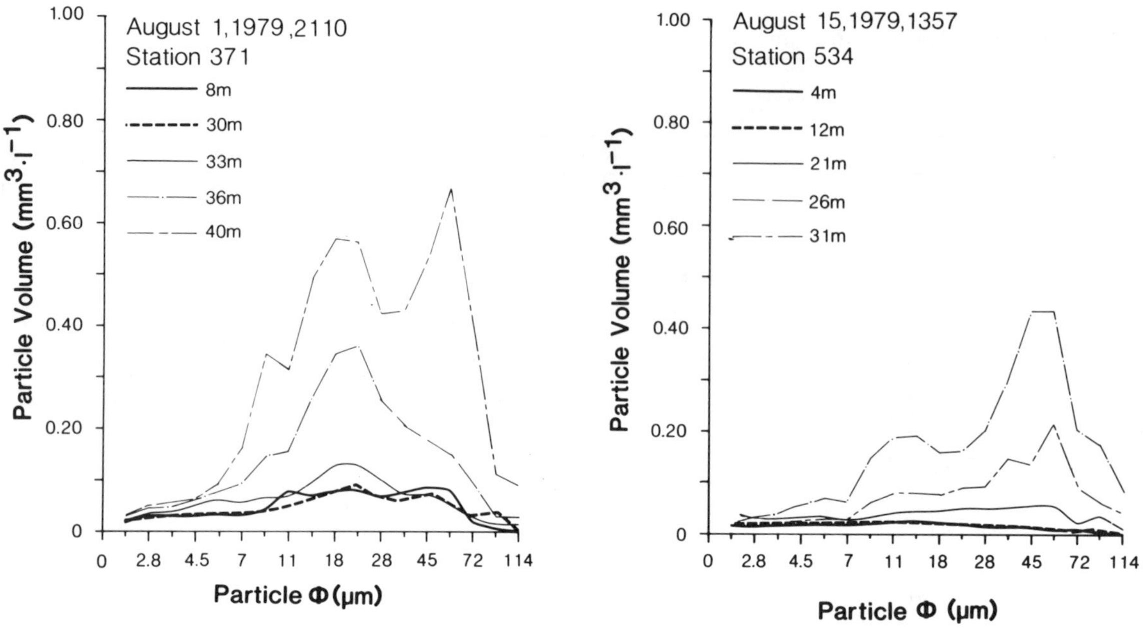

Fig. 8. Particle size distribution on two selected stations, northeastern Florida shelf [*Paffenhöfer et al.*, 1984].

not all. The following taxa consistently occurred in highest abundance in intrusions: *Oncaea* spp., *T. turbinata*, *P. avirosstris* and chaetognaths. The concentration of zooplankton in upwelled water depends on initial concentrations, subsequent reproduction, or seeding from shallower depths through immigration or displacement. In 1976 and 1978, but not in 1979, *Oncaea* spp. was very abundant, normally in "older" intrusions after the nitrate had been exhausted. Upwelled waters with temperatures of <22°C off Florida in 1979 were considered "young" intrusions, as all had detectable levels of nitrate when sampled [*O'Malley*, 1981]. One of the six major zooplankton components, which are not exclusively abundant in bottom layers (the other five are *Oikopleura* spp., *Oithona* spp., *Paracalanus* spp., *C. furcatus*, and copepod nauplii), is the juvenile of *Branchiostoma* sp. This meroplanktonic organism is scarce off North Carolina but is abundant off northeast Florida where considerable concentrations of gravid *Branchiostoma* sp. were found near the 20-m isobath in coarse sand between April and August [*Frankenberg*, 1971]. It appears that in the stratified water column, observed off northeastern Florida from April to September, planktonic *Branchiostoma* sp. larvae were displaced offshore with surface waters and migrated into particle-rich subsurface waters further offshore. Because their swimming movements attracted predators, the majority of these larvae, with almost no mortality in the absence of predators (G.-A. Paffenhöfer, unpublished data, 1982), are eaten before returning to the seafloor to metamorphose. Off northwestern Africa, larvae of *Branchiostoma* sp. have accounted for up to 65% of the stomach contents of juvenile mackerel [*Wiktor*, 1970].

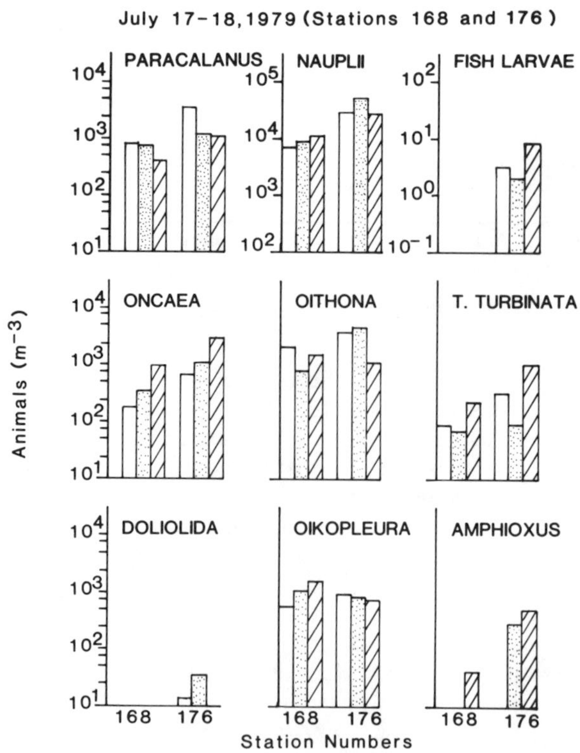

Fig. 9. Zooplankton distribution off northeastern Florida during summer 1979 [*Paffenhöfer et al.*, 1984]; open bars, UML; stippled bars, thermocline; dashed bars, intrusion.

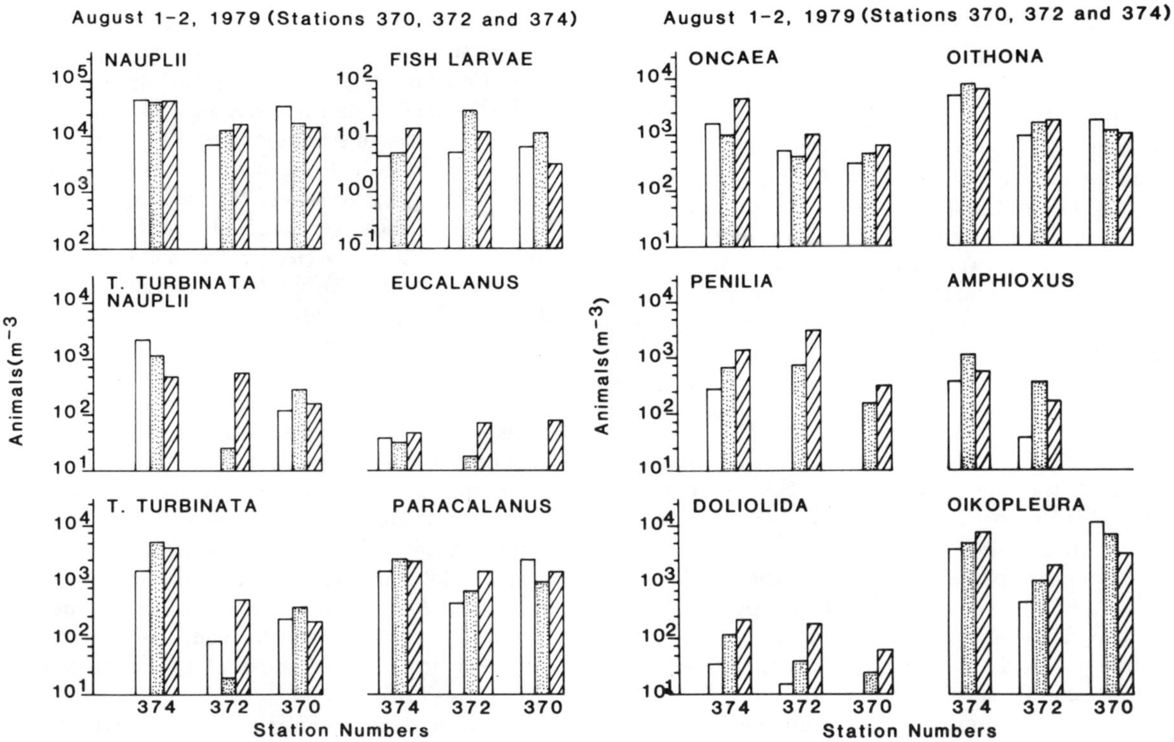

Fig. 10. Same as Figure 9.

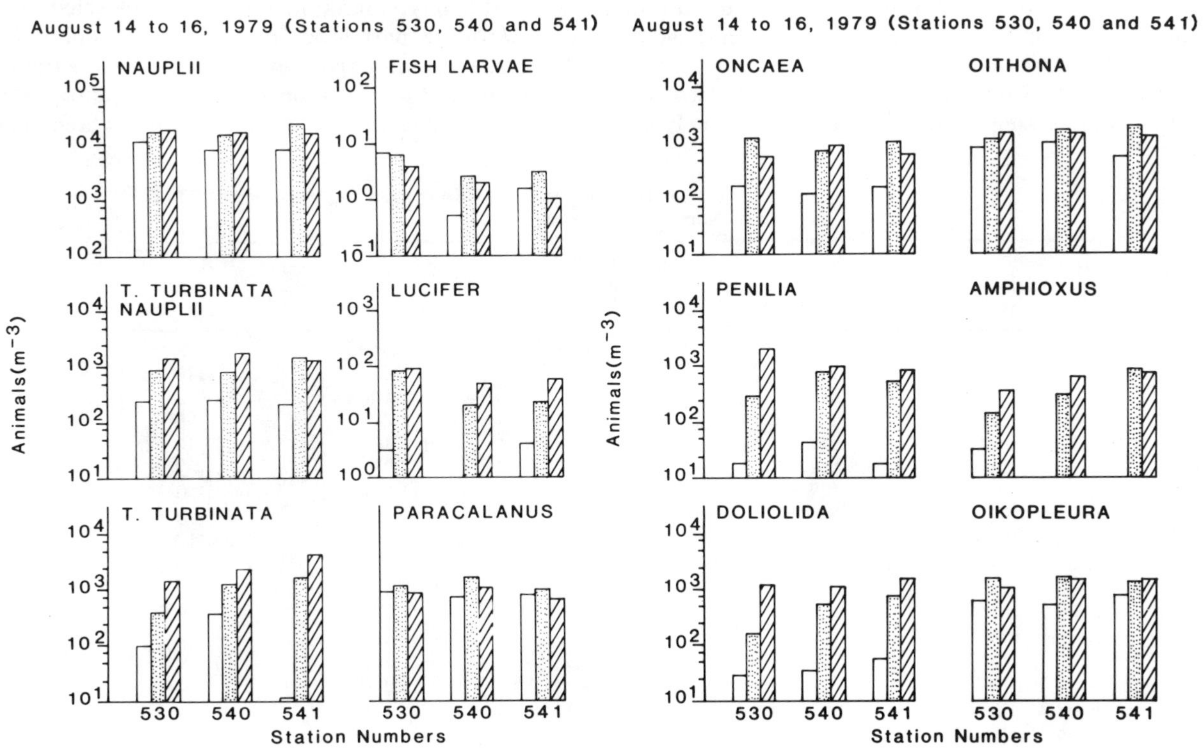

Fig. 11. Same as Figure 9.

The Abundance of Fish Larvae

High concentrations of fish larvae were often encountered on the SAB shelf, reaching abundances of up to 77 larvae m^{-3} in the UML of Onslow Bay during August 1976. During summer 1979, average concentrations were 10.5 m^{-3} in the thermocline and about half of that in the UML and in bottom waters (Table 3). The coefficient of concordance W was significant ($p < 0.05$), indicating that ranks of abundance agreed between stations in the sequence thermocline > UML > intrusion [*Tate and Clelland*, 1957]. Average concentrations were usually somewhat higher in the intrusions than in the UML, but at some stations the absolute concentration of larvae was higher in the latter.

Discussion

Upwelling at the shelf break of the SAB occurs at a frequency of 2–14 days [*Lee et al.*, 1981]. Stranding of upwelled water on the shelf proper, here called intrusions, occurs between late April and September when the density of the shelf water is lower than that of the upwelled water [*Blanton et al.*, 1981]. Stranded water masses are usually displaced in a northerly or alongshore direction. Zooplankton which develop in these water masses feed on phytoplankton, whose development depends on the original nitrate content of the upwelled water. Contrary to the situation in persistent upwelling off Oregon [*Peterson et al.*, 1979] zooplankton in stranded intrusions develop from episodic (one time) inputs of nutrients. How do zooplankton abundances in these episodic upwellings compare with those in areas with more persistent upwelling?

Comparative data from the Ivory Coast and Peru are presented in Table 2. Zooplankton composition off the Ivory Coast [*Binet*, 1979] closely resembles that on the SAB shelf during summer. Temperatures in the euphotic zone off the Ivory Coast during the main upwelling season [*Binet*, 1975] and off Peru during intensive upwelling [*Stuart*, 1981] are similar to those of summer upwelling on the SAB shelf (16° to 21°C). Concentrations of zooplankton off the Ivory Coast (*T. turbinata*, *C. furcatus*, *Eucalanus* spp., and *P. avirostris*) are close to those on the SAB shelf, excluding doliolids, which in 1979 on the SAB reached average concentrations of 818 zooids m^{-3} (Table 2). The mesh size of nets used by Binet was 200 μm, which probably resulted in an underestimate of the abundance of these copepods. Concentrations of *Oncaea* spp. and *Paracalanus* spp. off Peru were close to, while those of *Oikopleura* spp. and *Oithona* spp. were lower than, concentrations found in summer upwellings on the SAB shelf (Table 2). *Oikopleura* spp. and *Oithona* spp. are not typical indicators of upwelled waters. Presumably, *Oikopleura* spp. growth and reproduction are limited in upwellings due to the initially low abundance of small detritus particles and microflagellates on which they feed [*Paffenhöfer*, 1976]. Overall, concentrations of abundant taxa in relatively young episodic upwellings in the SAB are close to those of the same taxa persistent upwellings (Table 2).

The limited vertical migration of zooplankton in the SAB [*Paffenhöfer*, 1980] does not appear to enhance the nutrient supply in surface waters. After nitrate is depleted, usually within about 10 days after an intrusion strands [*Yoder et al.*, 1983], phytoplankton depend on recycled nutrients, originating in part from the excreta of zooplankton. The decrease in nutrient concentrations over time implies that the biomass of zooplankton which can be supported by phytoplankton also decreases. Zooplankton which survive in

TABLE 3. Concentration of Fish Larvae (Larvae per Cubic Meter) During Summer 1979 off Northeastern Florida

Date	Station	Upper Mixed Layer	Thermocline	Intrusion Near Bottom
July 12, 1979	44	5	no sample	22
July 17, 1979	164	1	2	240
	168	0.5	1.5	2
July 18, 1979	176	3	2	8
July 26, 1979	292	4.5	no sample	13
July 26, 1979	295	9	26	2
July 26, 1979	298	2	4.5	0
Aug. 1, 1979	370	6.5	11.5	3
Aug. 2, 1979	372	5	27	12
Aug. 2, 1979	374	4.5	5	14
Aug. 7/8, 1979	437	17	14	0
Aug. 9, 1979	457	3.5	5.5	0.5
Aug. 9, 1979	458	2	18	2
Aug. 14/15, 1979	530 (night)	7	6.5	1.5
Aug. 15, 1979	538 (night)	9	29	7
Aug. 16, 1979	540 (night)	0.5	2.5	2
Aug. 16, 1979	541 (night)	1.5	3	1
x		4.8	10.5	5.3
Range		0.5–17	1.5–29	0–22

nutrient-depleted environments are those that can gather sufficient energy at low particle concentrations. Examples are *Paracalanus* spp., omnivores like *Oithona* spp., *Oncaea* spp., and *T. turbinata*, or fine particle feeders like *Oikopleura* spp. which usually find sufficient food even in surface waters (Figure 10). Large particles feeders like *Eucalanus pileatus* or those which do not store energy like *P. avirostris* and doliolids may survive for a few days only after particulate concentrations decrease to about 0.5 mm^3 l^{-1}. Thus phytoplankton and zooplankton in stranded intrusions develop as short-term blooms. Will all forms which do not store energy die because of starvation unless they are eaten by predators, or is there a chance that they can reencounter phytoplankton-rich environments?

Prevailing winds during summer are from the southwest to southeast [*Weber and Blanton*, 1980]. The resulting offshore Ekman flow of surface waters generally coincides with subsurface onshore flow of offshore water which often contains high amounts of nutrients. Zooplankton which are displaced or have migrated into the UML thus are advected further offshore and, if able to survive for several days, may encounter a new phytoplankton-rich intrusion through vertical migration (Figure 4). To accomplish this, intrusions would have to occur at frequencies of less than one per week, as observed during summer 1979 [*O'Malley, 1981*; G.-A. Paffenhöfer, unpublished data, 1979]. Through repeated cycles, one intrusion could "seed" future stranding intrusions.

Concentrations of fish larvae encountered on the SAB shelf in the summer of 1979 are among the highest reported. *Houde and Chitty* [1976] observed a peak abundance of fish larvae in the eastern Gulf of Mexico extending from May through August, indicative of prolonged spawning. Average concentrations (about 1.2 larvae m^{-3}) and maximum concentrations (about 4 to 5 larvae m^{-3}) were clearly below those observed by us in 1979 (Table 3). Yet the abundances we report are probably low because the pumping system we employed was not specifically designed to catch these larvae.

A question we have been attempting to answer is whether zooplankton densities in and above summer upwellings can support the growth of fish larvae. Data on larvae of the bay anchovy *Anchoa mitchilli*, the lined sole *Achirus lineatus*, and the sea bream *Archosargus rhomboidalis* [*Houde*, 1978] indicate that survival should be 1 to 10% of the initial concentration at the abundances of zooplankton nauplii (Table 2) we observed in 1978. Survival should be higher than 1–10% in this case because (1) early copepodid stages are eaten by early larval stages of fish and (2) naupliar concentrations are underestimated with the 30-μm mesh which we employed. The maximum ration of 2-week-old larvae ranges from 160 to 420 nauplii $larva^{-1}$ d^{-1} [*Houde*, 1978]. At a maximum average concentration of 10.5 larvae m^{-3} (Table 3), fish larvae should consume from 1680 to 4410 nauplii m^{-3} d^{-1}, or 4.0 to 10.6% of the standing stock of nauplii in the thermocline (Figure 2). As adult copepod females constitute, on the average, approximately 2% of the copepodids encountered (*Paffenhöfer* [1980], Table 4), a daily egg production of 40 eggs $female^{-1}$ d^{-1} [*Checkley*, 1980] in *Paracalanus* spp. and *T. turbinata* combined in 1978 at environmental food concentrations would produce 880 and 1780 nauplii m^{-3} d^{-1} in the thermocline and intrusion, respectively. This is not sufficient to meet the daily demand of fish larvae observed. However, the abundant *Oncaea* spp. and *Oithona* spp., for which we do not have reproduction data, could make up the difference. During 1979 the situation should have been critical for at least *Achirus lineatus* larvae as the average nauplii concentration was below 18,300 m^{-3}, resulting in low survival. Overall, copepod concentrations in the thermocline were low in 1979 with combined reproductive rate of *Paracalanus* spp. and *T. turbinata* of 1530 nauplii m^{-3} d^{-1}.

Despite the high abundances of fish larvae during summer and adequate food supplies for early larvae, the stocks of adult fish found on the SAB shelf, compared with Peru or part of the Benguela Current, are low. Explanations may be:

1. The episodic nature of upwellings on the SAB shelf

TABLE 4. Estimated Percent Survival at 17 to 18 Days After Hatching [From *Houde*, 1978] and Approximate Daily Ration of Larvae of Three Subtropical Marine Fish Species [*Houde and Schekter*, 1978] at 26° to 28°C

	Food Concentration, l^{-1}	Estimated Survival, %	Daily Ration at 50 Nauplii l^{-1} of Larvae (Number of Nauplii Eaten) per Day	
			7 Days Old	14 Days Old
Anchoa mitchilli	27	1	62	420
	59	5		
	107	10		
Achirus lineatus	52	1		
	87	5	27	160
	130	10		
Archosargus rhomboidalis	9	1		
	23	5	89	200
	34	10		

results in a sporadic supply of zooplankton on which fish larvae feed. Heavy mortality occurs if the yolk-sac larvae do not encounter an upwelling environment and if they do not locate a second area of upwelling when nutrients in the first upwelling are exhausted.

2. Water displacements on the shelf can be advantageous or can cause high mortality. If displaced toward the shore, the probability of survival is high as the nearshore environment, which serves as a nursery for many fish species [*Thayer et al.*, 1974; *Shenker and Dean*, 1979], has high food concentrations (copepod juveniles). If displaced offshore, the fish larvae could come under the advective influence of the Gulf Stream. There low nauplii concentrations (usually below 10 l^{-1} (G.-A. Paffenhöfer, unpublished data, 1978) and temperatures near 28°C cause death of early larvae.

3. Finally, the predatory behavior of most abundant copepod genera and of juvenile and adult fish should result in a relatively high mortality rate of fish larvae. The high concentrations of fish larvae we encountered are temporary as their abundance is rapidly reduced by one or more of the factors cited above.

Acknowledgments. This research was supported by Department of Energy contracts E(38-1)-936, EY-76-S-09-0936, and DE-AS09-76EV00936. The author acknowledges the support of the crews of the R/V's *Blue Fin* and *Columbus Iselin*, of Bryan Wester, William Nicholas, and Dane Herring for data and sample processing, and of R. B. Hanson, D. W. Menzel, L. R. Pomeroy, S. L. Smith, and J. T. Turner for reviewing the manuscript.

References

Binet, D., Notes sur l'écologie de quelques taxons du zooplancton de Côte-d'Ivoire, 1, Ostracodes, cladoceres et cirripèdes, *Doc. Sci. Centre Rech. Oceanogr. Abidjan, IV*, 19–39, 1975.

Binet, D., Contribution à l'écologie de quelques taxons du zooplancton de Côte-d'Ivoire, 2, Dolioles, salpes, appendicularies, *Doc. Sci. Centre Rech. Oceanogr. Abidjan, VII*, 45–61, 1976.

Binet, D., Cycles biologiques et migrations ontogénétiques chez quelques copepodes pélagiques des eauz ivoiriennes, *Cah. ORSTOM. Ser. Oceanogr., 15*, 111–138, 1977.

Binet, D., Le zooplancton du plateau continental ivoirien, essai de synthèse écologique, *Oceanol. Acta, 2*, 397–410, 1979.

Blanton, J. O., L. P. Atkinson, L. J. Pietrafesa, and T. N. Lee, The intrusion of Gulf Stream water across the continental shelf due to topographically induced upwelling, *Deep Sea Res. Part A, 28*, 393–405, 1981.

Checkley, D. M., The egg production of a marine planktonic copepod in relation to its food supply: Laboratory studies, *Limnol. Oceanogr., 25*, 430–446, 1980.

Dagg, M., T. Cowles, T. Whitledge, S. Smith, S. Howe, and D. Judkins, Grazing and excretion by zooplankton in the Peru upwelling system during April 1977, *Deep Sea Res. Part A, 27*, 43–59, 1980.

Frankenberg, D., The dynamics of benthic communities off Georgia, U.S.A. *Thalassia, Jugosl., 7*, 49–55, 1971.

Gosselck, F., and E. Kuehner, Investigations on the biology of *Branchiostoma senegalense* larvae off the northwest African coast, *Mar. Biol. Berlin, 2*, 67–73, 1973.

Houde, E. D., Critical food concentrations for larvae of three species of subtropical marine fishes, *Bull. Mar. Sci., 28*, 395–411, 1978.

Houde, E. D., and N. Chitty, Seasonal abundance and distribution of zooplankton, fish eggs, and fish larvae in the eastern Gulf of Mexico, 1972–1974, *Tech. Rep. NMFS SSRF-701*, 18 pp., Natl. Oceanic and Atmos. Admin., Boulder, Colo., 1976.

Houde, E. D., and R. C. Schekter, Functional and developmental responses of fish larvae to varying prey concentrations *ICES/ELH Symp. PE 2*, 7 pp., Int. Counc. for the Explor. of the Sea, Copenhagen, 1978.

Jorgensen, C. B., *Biology of Suspension Feeding*, 357 pp., Pergamon, New York, 1966.

Judkins, D. C., Vertical distribution of zooplankton in relation to the oxygen minimum off Peru, *Deep Sea Res. Part A, 27*, 475–487, 1980.

Le Borgne, R., Etude de la production pélagique de la zone equatoriale de l'Atlantique à 4°W, *Cah. ORSTOM. Ser. Oceanogr., 15*, 333–348, 1977.

Lee, T. N., L. P. Atkinson, and R. Legeckis, Observations of a Gulf Stream frontal eddy on the Georgia continental shelf, April 1977, *Deep Sea Res., 28*, 347–378, 1981.

Lund, J. W. G., C. Kipling, and E. D. LeCren, The inverted microscope method of estimating algal numbers and the statistical basis of estimations by counting, *Hydrobiologia, 11*, 143–170, 1958.

O'Malley, P. G., The effect of summertime shelf break upwelling on nutrient flux in southeastern U.S. continental shelf waters, M.S. thesis, 52 pp., Ga. Inst. of Technol., Atlanta, 1981.

Paffenhöfer, G.-A., On the biology of appendicularia of the southeastern North Sea, in *10th European Symposium on Marine Biology*, vol. 2, edited by G. Persoone and E. Jaspers, pp. 432–455, Universa Press, Wetteren, Belgium, 1976.

Paffenhöfer, G.-A., Zooplankton distribution as related to summer hydrographic conditions in Onslow Bay, North Carolina, *Bull. Mar. Sci., 30*, 819–832, 1980.

Paffenhöfer, G.-A., Vertical zooplankton distribution on the northeastern Florida shelf and its relation to temperature and food abundance, *J. Plankton Res., 5*, 15–33, 1983.

Paffenhöfer, G.-A., and S. C. Knowles, Feeding of marine planktonic copepods on mixed phytoplankton, *Mar. Biol. Berlin, 48*, 143–152, 1978.

Paffenhöfer, G.-A., D. Deibel, L. P. Atkinson, and W. M. Dunstan, The relation of concentration and size distribution of suspended particulate matter to hydrography in Onslow Bay, North Carolina, *Deep Sea Res. Part A, 27*, 435–447, 1980.

Paffenhöfer, G.-A., B. T. Wester, and W. D. Nicholas, Zooplankton abundance in relation to state and type of intrusion on the southeastern United States during summer, *J. Mar. Res., 42*, 995–1017, 1984.

Peterson, W. T., C. B. Miller, and A. Hutchinson, Zonation and maintenance of copepod populations in the Oregon upwelling zone, Deep Sea Res. Part A, 26, 467–494, 1979.

Petipa, T. S., E. V. Pavlova, and G. N. Mironov, The food web structure, utilization and transport of energy by tropic levels in the planktonic communities, in *Marine Food Chains*, edited by J. H. Steele, pp. 142–167, Oliver and Boyd, Edinburgh, 1970.

Shenker, J. M., and J. M. Dean, The utilization of an intertidal marsh creek by larval and juvenile fishes: Abundance, diversity and temporal variation, *Estuaries, 2*, 154–163, 1979.

Smith, S. L., C. M. Boyd, and P. V. Z. Lane, Short-term variation in the vertical distribution of small copepods off the coast of southern Peru, in *Coastal Upwelling, Coastal Estuarine Sci.*, vol. 1, edited by F. A. Richards, pp. 417–426, AGU, Washington, D. C., 1981.

Steedman, H. F., Laboratory methods in the study of marine zooplankton, *J. Cons. Cons. Int. Explor. Mer, 35*, 351–358, 1974.

Stuart, D. W., Sea-surface temperatures and winds during JOINT II, 2, Temporal fluctuations, in *Coastal Upwelling, Coastal Estuarine Sci.*, vol. 1, edited by F. A. Richards, pp. 32–38, AGU, Washington, D. C., 1981.

Tate, M. W., and R. C. Clelland, *Non-Parametric and Short-cut Statistics*, 171 pp., Interstate, Danville, Ill., 1957.

Thayer, G. W., D. E. Hoss, M. A. Kjelson, W. F. Hettler, Jr., and M. W. Lacroix, Biomass of zooplankton in the Newport River estuary and the influence of postlarval fishes, *Chesapeake Sci., 15*, 9–16, 1974.

Weber, A. H., and J. O. Blanton, Monthly mean wind field for the South Atlantic Bight, *J. Phys. Oceanogr., 10*, 1256–1263, 1980.

Wiktor, K., La composition de la nourriture du maquerau (*Scomber colias*) des eaux côtiers de l'Afrique de Nord-Oest, *CM 170/J 7*, 4 pp., Cons. Int. de l'Explor. de la Mer, Copenhagen, 1970.

Yoder, J. A., L. P. Atkinson, S. S. Bishop, E. E. Hofmann, and T. N. Lee, Effect of upwelling on phytoplankton productivity on the outer southeastern U.S. continental shelf, *Cont. Shelf Res., 1*, 385–404, 1983.

The Microbial Food Web of the Southeastern U.S. Continental Shelf

L. R. POMEROY

Institute of Ecology, University of Georgia, Athens

Bacterial numbers in continental shelf waters of the southeastern United States vary between 10^5 and 10^6 ml^{-1}. While the variations generally reflect temporal and spatial variations in primary production, the narrow limits of bacterial abundance suggest a tight coupling between production and utilization of bacteria by consumers. Microbial food chains appear to be in parallel with phytoplankton-grazing food chains, utilizing soluble photosynthate released by phytoplankton and both soluble and particulate products of zooplankton. This is reflected in spatial separation of blooms of phytoplankton and bacteria associated with intrusions of nutrient-rich water on the outer shelf. Microbial numbers and respiratory rates indicate that microbial food chains are responsible for a substantial fraction of the total energy flux. Microbial production may be linked with zooplankton grazing, but the relation is not yet well described.

Introduction

Currently, a number of investigators suggest that marine food chains are more complex than earlier believed. Major components are bacteria having diameters of 0.1–0.3 μm, with an assimilation efficiency exceeding 50%. In spite of their small size, free-floating bacteria constitute most of the living surface area in seawater and sometimes constitute most of the living biomass [*Azam and Fuhrman,* 1984]. Marine heterotrophic bacteria in coastal upwellings may utilize one half to two thirds of net phytoplankton production [*Watson,* 1978]. Bacteria are estimated to use 25% of primary production off southern California [*Fuhrman et al.,* 1980]. Utilization of only the soluble products of photosynthesis by bacteria may account for 45% of primary production in the Baltic Sea [*Larsson and Hagström,* 1979]. Although these estimates vary widely and are based on different kinds of observations, they lead to the general postulate that microheterotrophs are a significant element in marine food webs [*Williams,* 1981*a*].

A key question was posed when this research was undertaken: does the microbial food web transfer significant energy and materials to higher organisms, or do protozoan consumers dissipate most of the energy sequestered by bacteria [*Pomeroy,* 1974]? Measurements of microbial respiration on the southeastern continental shelf suggest that microorganisms are indeed responsible for very substantial losses of energy [*Turner,* 1974]. That this may be the general case in the ocean is suggested by other investigations of both respiration and ammonium production by the <10-μm fraction of the plankton [*Pomeroy and Johannes,* 1966, 1968; *Williams,* 1981*b*; *Glibert,* 1982]. On the other hand, modeling of the food web of the continental shelf suggests that the flux of energy may at times pass through microbial pathways to higher consumers [*Pomeroy,* 1979; *Pace et al.,* 1984]. This question is significant for the present study, as it is for marine waters everywhere. The continental shelf of the southeastern United States has unique physicochemical regimes, but the microbial responses to them may help us understand the roles of microorganisms on continental shelves elsewhere.

Microorganisms in seawater include bacteria, fungi, microalgae, and protozoans. Except for cyanobacteria, autotrophs are considered in the paper by *Yoder* [this volume]. The distribution of fungi in the ocean has been described by a number of investigators using culture techniques [*Johnson and Sparrow,* 1966; *Fell,* 1976]. Their absolute abundance, rate of production, and significance in food webs in the sea, however, is virtually unknown. Recently, it has become possible to enumerate protozoa in seawater, though the precision of counts is still questioned. While it is too early to evaluate the significance of protozoans in marine food webs, present indications, including the observations reported here, suggest that they may be an important and underinvestigated group [*Azam et al.,* 1983]. Unfortunately, methods for enumerating protozoans were not yet developed when this work was done, and methods for measuring their production are still being developed [*Heinbokel,* 1982].

Microbial Populations

In terms of absolute numbers, free bacteria are the most abundant organisms in ocean water, varying in number from 10^4 ml^{-1} in subsurface open ocean water to 10^6 ml^{-1} in highly productive regions [*Ferguson and Rublee,* 1976; *Azam et al.,* 1983]. The most common morphological types are cocci, 0.1–0.3 μm in diameter, and less common are sigmoid forms, sometimes with a cross section of less than 0.1 μm (Figure 1). Too small to be resolved by ordinary phase microscopy, these small bacteria were discovered by scanning electron microscopy (SEM) [*Watson et al.*, 1977].

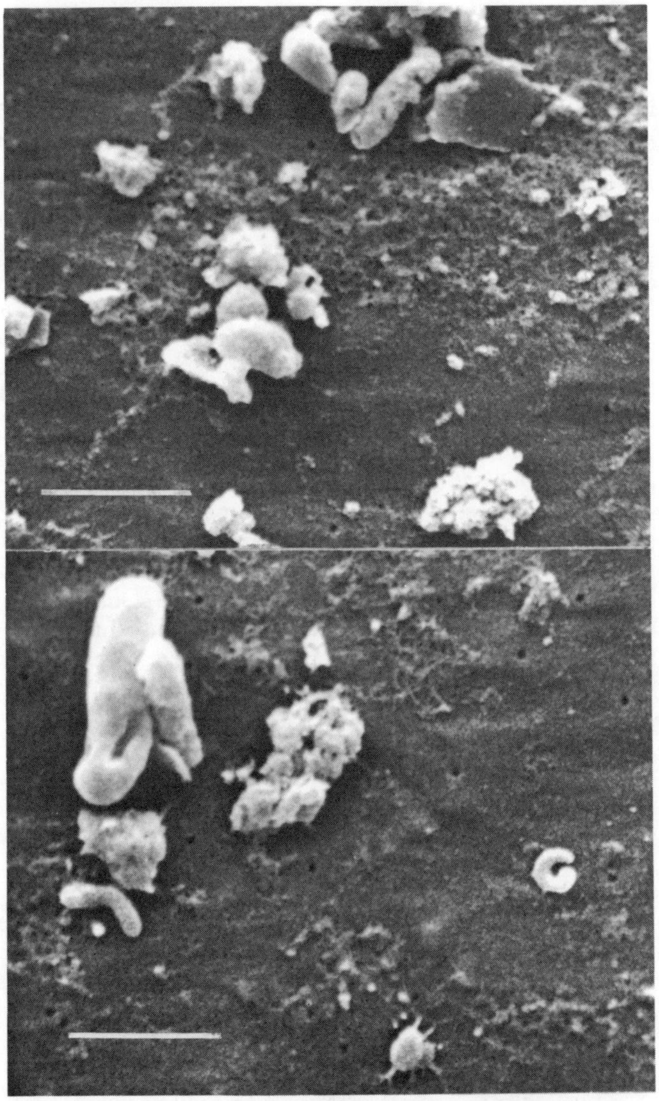

Fig. 1. Scanning electron micrographs of bacteria and particulate matter from a seawater sample from the inner continental shelf, 31°N, 81°W, August 16, 1981, 3-m depth. The scale bars are 2 μm, and the holes in the filters are 0.1 μm. The upper portion shows a flocculent particle covering most of the field and containing several 1- to 2-μm rods and some angular inorganic fragments. The matrix has dried to a thin film on the filter surface. The lower portion shows several small flocculent particles, largely devoid of bacteria except for one large rod (upper left). Lodged on the filter are a variety of forms of free bacteria: typical small cocci (lower left and upper right), a small bent rod (lower left), a comma form (lower right), and a dividing coccus with extracellular filaments (lower center).

A satisfactory counting procedure using fluorescent stains which bind to RNA or DNA is now available [*Hobbie et al.*, 1977; *Porter and Feig*, 1980].

On the southeastern shelf, bacterial numbers vary from 10^4 ml^{-1} in subsurface Gulf Stream water at the continental margin to 10^6 ml^{-1} in water adjacent to intrusions of subsurface Gulf Stream water and in water inside the coastal front [*Pomeroy et al.*, 1983]. Figure 2 shows the distribution of bacteria across the shelf during an intrusion of nutrient-rich water. Considering the potentially short generation time of bacteria and their responsiveness to environmental changes, the constancy of numbers in shelf water and in the ocean in general is striking. It suggests that some sort of negative feedback maintains a constant population in spite of rapid turnover [*Fenchel*, 1982*a*].

A morphologically distinct population of bacteria occurs on suspended particulate organic matter. Usually these bacteria are rods 1–2 μm in length, but cocci of similar size are also seen (Figure 1). Both motile forms and colonies may be present. Most organic particles in seawater do not harbor recognizable bacteria [*Wiebe and Pomeroy*, 1972]. The minority of particles which contain active bacteria are known as flocculent organic aggregates [*Gordon*, 1970], which probably have varied origins. One line of evidence suggests that they are fragments of the feces of zooplankton. The feces of larger microcrustacea are encased in a chitinous membrane and, because of their size and encasement, most probably sink to the bottom on continental shelves. However, the feces of the smaller microcrustaceans, and early life history stages of larger forms, are less compact and smaller. These may not sink out of the water as readily as those of larger microcrustacea and may contribute to the assemblage of suspended particles [*Hofmann*

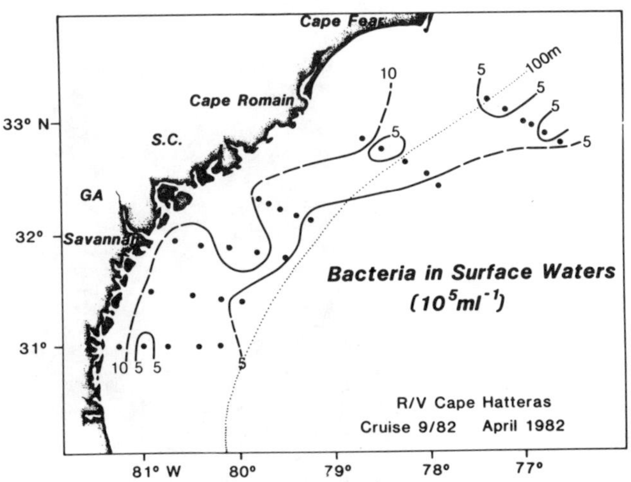

Fig. 2. Distribution of bacteria in surface water of the continental shelf of the southeastern United States. This data set is from R/V *Cape Hatteras* cruises September 1982 and April 1982. Values are 10^5 ml^{-1}. Dotted line is 100-m isobath.

et al., 1980; *Krause,* 1981]. The feces of the large pelagic tunicates in the open ocean also sink and are collected in sediment traps at great depths [*Honjo,* 1980]. Those of smaller individuals, which make up the bulk of most natural populations, fragment readily into 50- to 100-μm pieces which remain in suspension [*Pomeroy and Deibel,* 1980]. By following the sequence of events in fecal fragments in the laboratory, Pomeroy and Deibel found that bacteria, originally rare, multiply rapidly. Their numbers in individual particles are highly variable, however, because the bacteria are eaten by protozoans which tend to graze through a particle and remove virtually all of the bacteria before moving on to another particle. Particles produced in this way in the laboratory are indistinguishable microscopically from those collected from seawater (Figure 3).

Flocculent aggregates are also formed in seawater by adhesion of bacteria to each other, to nonliving organic particles, and to inorganic particles. These processes are easily duplicated in the laboratory, and the sequential development of bacterial and protozoan populations, like those associated with the degradation of fecal particles, is observed [*Robertson et al.,* 1982]. Aggregation of this kind appears to be promoted by water movement. At present it is impossible to distinguish fecal particles from aggregates formed by bubbles or bacteria by morphological or chemical tests. The identity of particles is further confused by the incorporation of fecal particles into larger aggregates and the reingestion of many particles by zooplankton. Particles regularly observed in continental shelf water are too large to be zooplankton feces but appear under the microscope to contain all of the living and nonliving elements of them. Even quite large particles of this kind appear to remain suspended and, by scanning electron microscopy, appear to have little mass other than living bacteria (Figure 1). Therefore the particulate matter in continental shelf water appears to be a mixture of fragments of fecal matter, aggregates produced by adhesion of bacteria and their extracellular products to each other and to other particles, and larger aggregates of all of the above. These have been described by *Alldredge* [1979], *Alldredge and Cox* [1982], and others cited therein. Flocculent particles of all sizes are very fragile. They may be broken by slight water movement but reaggregate in time. Certainly, water samples collected in conventional samplers and drawn through a fine orifice contain only small particles, which have, in some cases, fragmented from larger ones. All particles look similar and appear to undergo continuing aggregation and breakup because conventional sampling destroys the natural spectrum of particle sizes. However, change in size probably does not alter the composition of the particles or biological processes within them [*Pomeroy,* 1984].

Attached bacteria constitute about 1% of the total bacteria in shelf water but, because of their larger size, account for about half of the total bacterial biomass [*Hodson et al.,* 1981]. While their numbers are rather constant, as are those of the free bacteria, microscopic observations of par-

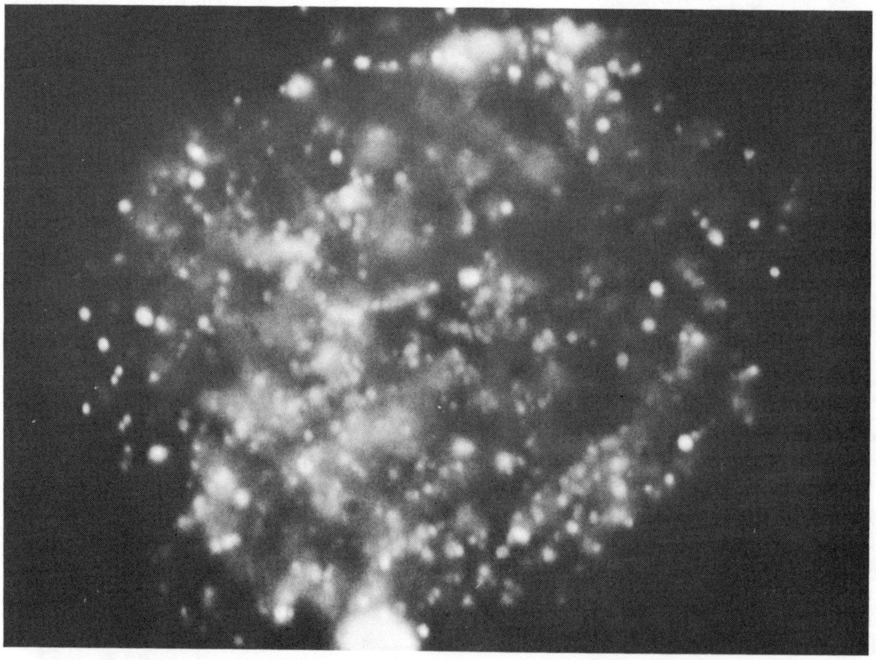

Fig. 3. Epifluorescence photomicrograph of a fecal particle in a water sample from the continental shelf. Smaller objects are bacteria stained with acridine orange. Larger objects are autofluorescent cyanobacteria and fragments of phytoplankton. Width of the field is 70 μm.

ticulate matter indicate that the attached bacteria grow actively and are constantly grazed by protozoans. As soon as particles contain a sufficiently large number of bacteria to attract protozoans, ciliates and flagellates invade and denude them.

Although, under SEM, most particles appear to have very little substance, something about the particulate environment stimulates the production of relatively large growth forms of bacteria, usually >1 μm. These large forms compete well with smaller free bacteria for dilute dissolved substrates [*Hodson et al.,* 1981], and it is possible that, in spite of their affinity for particles, the large attached bacteria obtain the greater part of their energy from dissolved materials. At this time we certainly cannot dismiss either the free or attached forms as relatively minor contributors to the flux of energy and materials in the sea. Limited information suggests that there is considerable variation in the metabolic dominance of one or the other morphological group over time and space [*Azam and Hodson,* 1977; *Bell and Abright,* 1981; *Fenchel and Jørgensen,* 1976; *Fuhrman,* 1981; *Hanson and Wiebe,* 1977; *Hoppe,* 1976, 1978; *Williams,* 1970; *Wright and Burnison,* 1979].

The chroococcoid cyanobacteria have been recognized recently as a ubiquitous and often abundant component of marine plankton [*Waterbury et al.,* 1979; *Johnson and Sieburth,* 1979]. Chroococcoids show some indication of seasonality in abundance on the southeastern shelf. Comparatively large populations have been found on the inner shelf in winter [*Pomeroy et al.,* 1983]. However, their occurrence is sporadic, and it is doubtful that they ever dominate the autotrophic biomass in this region. We know nothing about their rate of production on the southeastern shelf.

Distribution of Bacteria

The distribution of bacteria on the shelf is predictable. In the absence of intrusions of subsurface nutrient-rich Gulf Stream water, bacteria are found in abundances of $>10^5$ ml^{-1} over most of the shelf outside of the coastal front and $>10^6$ ml^{-1} inside the coastal front (Figure 2). These abundances are slightly less than summer values in continental shelf waters south of Long Island [*Ferguson and Palumbo,* 1979]. Numbers exceeding 10^6 ml^{-1} in surface water of the outer continental shelf are indicative of the presence of a subsurface intrusion of Gulf Stream water. These bodies of water can be detected by increases in the ATP content of suspended particulate matter or, better, by a reduction in the ratio of ATP to chlorophyll or total adenylates to chlorophyll [*Campbell et al.,* 1979].

In coastal waters, bacteria are not only more abundant but are larger than those on the outer shelf (0.4 versus 0.2 μm). Inside the coastal front water is more estuarine in character. Salinity is 34‰ or less, and the water column is vertically mixed. Because of the rapid rates of regeneration of nutrients from the bottom, phytoplankton production is high, providing a continuing supply of substrates for heterotrophic bacterial growth. At times, usually in the late winter or spring, inputs from rivers may be so great that coastal water of reduced salinity extends nearly to the shelf break, so the "front" becomes a halocline in a vertically stratified system. At these times, bacterial numbers and adenylate concentrations may not remain as high as in a narrow, fully mixed belt of coastal water. This depends on how long the water on the middle or outer shelf has remained stratified. Because coastal water is a regular and stable, albeit dynamic, feature [*Blanton,* 1981], there is consistently high phytoplankton production, especially along the Georgia coast, providing a continuing supply of substrates to sustain the large bacterial population. In contrast, blooms of phytoplankton associated with intrusions of Gulf Stream water on the outer shelf are ephemeral, and the water itself probably has a short residence time on the outer shelf. Therefore bacterial populations in water of the outer continental shelf are oceanic in character, 10^5 ml^{-1}, consisting mostly of 0.2-μm cocci.

Microbial Food Chains

So far as we know, free bacteria are totally dependent on dissolved substrates for energy, and attached bacteria utilize dissolved as well as particulate substrates. We also know that most of the standing stock of dissolved organic matter is probably relatively refractor [*Williams et al.,* 1969]. Although there is continuing controversy over the magnitude of the flux of dissolved organic matter released by phytoplankton, most marine ecologists tend to believe that this is the major source of heterotrophic bacterial nutrition. Phytoplankton in culture produce dissolved materials, such as glycolate, monosaccharides, amino acids, and larger molecules such as proteins and polysaccharides, which are good substrates for bacteria. Production of these substances seems to be greater when phytoplankton are stressed and, in the absence of Gulf Stream intrusions, phytoplankton on the southeastern shelf are probably under considerable nutrient stress much of the time. Compounds of the types phytoplankton release have been found in seawater in microgram to nanogram per liter quantities [*Brockman et al.,* 1979; *Mopper et al.,* 1980; *Ittekott,* 1982]. Bacteria readily utilize most of the dissolved materials released by phytoplankton [*Larsson and Hagström,* 1979], and naturally produced compounds, whatever they may be, are utilized more readily than any of the usual suite of defined substances used to estimate heterotrophic uptake rates [*Wiebe and Smith,* 1977]. However, bacterial numbers and growth rates correlate with phytoplankton biomass, not with photosynthetic rate [*Fuhrman et al.,* 1980]. Bacterial utilization of photosynthate may not be merely a direct transfer of dissolved materials between phytoplankton and bacteria [*Larsson and Hagström,* 1979] but may involve mediation by protozoans and net zooplankton.

Phytoplankton in subsurface intrusions of Gulf Stream water on the outer continental shelf release 0–15% of total photosynthate as dissolved materials [*Jacobsen,* 1981]. The fraction released varies greatly from day to day and shows no obvious relationship to the physical or biological sequence of events. Jacobsen also found that heterotrophic assimilation of the natural, undefined dissolved release products from cultures of the diatom, *Cylindrotheca fusiformis,* was much more rapid than assimilation of glycine or glyoxylate. We can conclude that phytoplankton release dissolved organic materials and that heterotrophic bacteria have a high affinity for these compounds. This is not an artifact of higher concentrations of the substrates, but it might relate to the high diversity of the naturally produced compounds. If many sets of permeases could be operated simultaneously, they would obviously assimilate more material than a single set of permeases devoted to taking up a single, defined substance.

Responses to changing supplies of dissolved materials are at least as large and as rapid as changes in population size. For example, the uptake of dissolved ATP (DATP) varies from 0.1 to nearly 5 ng l^{-1} h^{-1} over the southeastern continental shelf and slope [*Hodson et al.,* 1981]. However, DATP did not vary linearly with cell numbers or biomass. Rather, uptake of DATP on a per-cell basis was highly correlated with the absolute concentration of DATP, suggesting that permeases are mobilized by the existing population.

In the region of consistently high primary production inside the coastal front, the rate of heterotrophic utilization, in this case evaluated by uptake of ^{14}C glucose, is much more uniform. The numbers of bacteria also vary little inside of the coastal front, but if the principles learned offshore apply in the nearshore zone, lack of variation reflects the constancy of population size and surface area.

Our observations on the outer continental shelf suggest that zooplankton may also be a source of dissolved and particulate substrates for bacteria, perhaps a major one as suggested by *Eppley et al.* [1981]. The numbers of free and attached bacteria are highly correlated. Where bacteria-containing particles are more abundant, so are free bacteria. Population maxima of bacteria are not centered in the phytoplankton blooms associated with cold, nutrient-rich water which intrude onto the outer shelf. Rather, maximum bacterial populations are in the water above such intrusions (Figure 4). They are exactly congruent in time with phytoplankton blooms in the bottom water but are above them and also usually extend over a wider area. This leads us to postulate that zooplankton are intermediaries between the production of phytoplankton (as particulate food) and bacteria. Zooplankton feed along the top of the phytoplankton bloom during the day and migrate into the upper water at night. Their feces provide a source of particulate substrates for attached bacteria and dissolved substrates for both free and attached bacteria. While the evidence is circumstantial, that which is available suggests that a significant fraction of the substrates supporting bacterial production are secondarily derived from phytoplankton after ingestion by zooplankton. This conclusion was

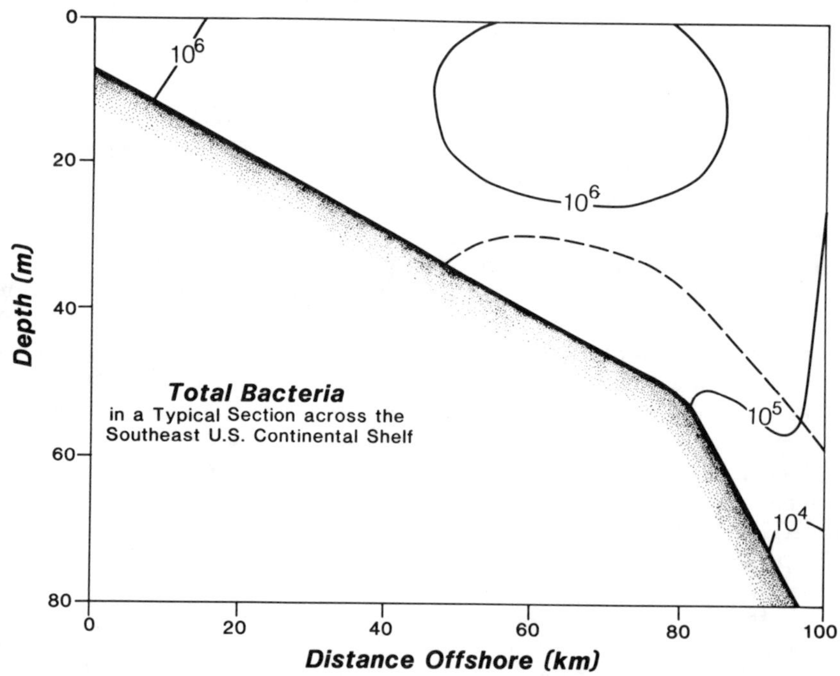

Fig. 4. Section across the continental shelf of the southeastern United States, showing the typical distribution of and total bacteria. Dashed line indicates location of an intrusion of subsurface Gulf Stream water.

also reached by *Eppley et al.* [1981], who followed the production of bacteria in a large container of seawater with and without grazing zooplankton. In their experiment, bacterial production was stimulated by the presence of microcrustacean zooplankton. Further evidence for the role of zooplankton in the production of dissolved organic matter is that peaks of dissolved material coincide with peaks of fecal pellet production and with phytoplankton blooms [*Ittekott,* 1982].

A simplified conceptual model (Figure 5) shows some of the trophic relationships which link phytoplankton, metazooplankton, microorganisms, and other consumer populations. The principal consumers of bacteria are probably protozoa, with mucus net feeders taking a lesser part of the production [*Deibel,* 1982]. However, direct measures of the rate of consumption of bacteria by consumers are not available from any part of the ocean. Moreover, the connection between protozoans and higher consumers is virtually unknown. Therefore we still do not know to what extent bacteria are a link to higher trophic levels, or if they are a sink for energy captured by phytoplankton [*Pomeroy,* 1974; *Azam and Fuhrman,* 1984].

While the absolute magnitude of flux of energy and materials through microbial food webs in not known from direct measurements, we can infer upper and lower limits from the rate of production of the substrates and knowledge of the assimilation efficiency of bacteria in nature. Production of dissolved organic matter by phytoplankton in blooms associated with intrusions of Gulf Stream water is highly variable [*Jacobsen,* 1981], suggesting that periods of favorable conditions for photosynthesis alternate with periods of stress. For this variable situation a release of

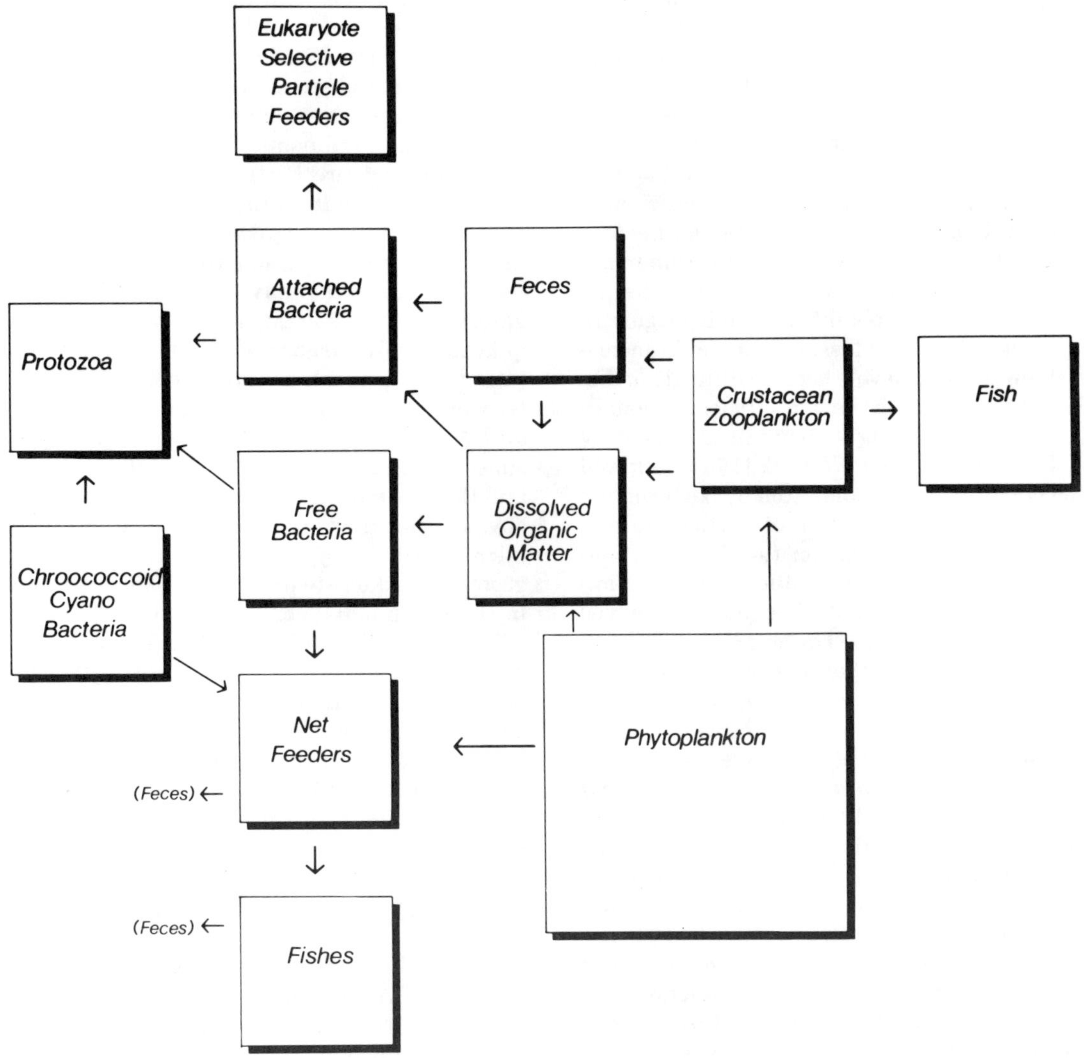

Fig. 5. Simplified conceptual model of the flux of energy through the food web of the water of a continental shelf.

10% of photosynthate may be a reasonable estimate. This material is utilized by bacteria with a net efficiency of about 60% [*Hobbie and Crawford,* 1969]. Some of the remaining 90% of photosynthate is consumed by zooplankton with an assimilation efficiency of about 15% (but highly variable). However, in addition to the fraction assimilated, about 30% of ingested food is released by zooplankton as undigested, or partially digested, fecal and excretory matter. Some fraction of phytoplankton also is damaged, or simply not ingested, by feeding zooplankton. Nearly all of these nonliving organic materials are immediately available to bacteria. While they might appear to be mostly particulate material, in fact there is a continuing production of dissolved substrates resulting from zooplankton feeding, ingestion, digestion, excretion, defecation, followed by degradation of the feces by colonizing bacteria and protozoans. This qualitative analysis of zooplankton feeding reveals why, in our observations, zooplankton appear to influence the abundance of bacteria in water. Significantly more substrates used by bacteria are produced by the activities of zooplankton than by those of phytoplankton, although, of course, all of the substrates are derived ultimately from phytoplankton production.

Analyses of microbial food webs have been inhibited by ecological dogma which states that "ecological efficiency" between trophic levels is about 10% and therefore extensions of the classical marine food web would become vanishingly small because of the rate of dissipation of energy. We now perceive that the microbial food web is largely in parallel, not in series, with the classic phytoplankton-zooplankton food chain. Trophic levels become difficult to distinguish. Questions about assimilation efficiency remain troublesome, and it is increasingly apparent that no easy answers exist. For example, *Newell et al.* [1981] followed the degradation of cultured phytoplankton by bacteria and protozoans and found the assimilation of particulate material much less efficient than that of the dissolved material. They suggest that the overall efficiency of the microbial community in using particulate detritus and dissolved materials as a whole is about 10%. However, Newell et al. excluded metazooplankton from their experiment. In other environments, invertebrate grazers and detritivores have a marked influence on both the rate and the assimilation efficiency of microbial degradation of relatively refractory particulate organic matter [*Hargrave,* 1970; *Sieburth and Dietz,* 1974; *Morrison and White,* 1980; *Findlay and Tenore,* 1982; *Tenore et al.,* 1982]. A definitive experiment to determine the assimilation efficiency of the total microbial community (including protozoans), with respect to its usual wide range of dissolved and particulate substrates, has not been performed. In designing such an experiment it will be important to evaluate how much of the production during the phytoplankton bloom is not consumed by invertebrate browsers and must be converted to detritus directly by microbial action before it is suitable food for higher organisms. If that fraction is large, calculated efficiencies may indeed prove to be low. Better knowledge of assimilation efficiency of marine heterotrophs of all kinds is clearly needed to permit quantitative evaluation of food webs.

Directly or indirectly, a highly significant fraction of the total energy fixed by phytoplankton must find its way into microbial food webs. Where the energy goes from there remains an important and largely unanswered question [*Pomeroy,* 1974; *Azam and Fuhrman,* 1984]. However, modeling exercises demonstrate that in principle a significant portion of this energy could find its way to terminal consumers, such as fishes [*Pomeroy,* 1979; *Pace et al.,* 1984]. We lack observational and experimental proof that this happens, but there are many published anecdotal accounts of larval and juvenile fishes feeding on sea snow, salps, appendicularians, and other key components of the microbial portion of the marine food web.

An even less well understood portion of the food web is that which depends on allochthonous inputs and organic matter to the sea. That there is an input from the land to the sea is accepted, and we recognize that most input comes from the atmosphere. Atmospheric dust samples taken over the ocean contain numerous organic particles including recognizable organisms [*Delaney et al.,* 1967]. Fecal particles collected directly from salps and those taken in water sampling bottles from the outer continental shelf, 100 km offshore, contain recognizable terrestrial components (Figure 6). The most common are carbon particles and pollen, but other materials have been reported by other investigators [*Ketseridis et al.,* 1976]. Less readily apparent, but quantitatively large, are inputs of volatile organic compounds and aerosols of natural origin, such as isoprene and terpenes, which precipitate in rainfall over the ocean [*Cronn and Nutmagul,* 1982]. At present the inputs of organic substrates from the land have not been well quantified, and their fate in the ocean is unknown. However, it seems obvious that these materials will be utilized primarily by microbes [*Pomeroy,* 1984]. Several investigators have suggested that the rate of heterotrophic microbial respiration in the ocean equals or exceeds the rate of photosynthesis [*Sieburth,* 1977; *Joiris,* 1977]. These investigators also questioned the accuracy of current methods of measuring photosynthesis in the sea. If allochthonous inputs of organic matter are significant in relation to photosynthesis and if microorganisms use these materials, the controversy would be resolved by identifying a source of energy to fuel respiratory metabolism not derived from marine primary production.

Control of Bacterial Populations

It is striking that bacteria vary so little in number in seawater, even in the conditions of highly varied productivity that we find in the southeast. Certainly a variation of 2 orders of magnitude in the numbers of free and attached bacteria does not reflect the ability of bacterial populations to respond to available substrates. Some control

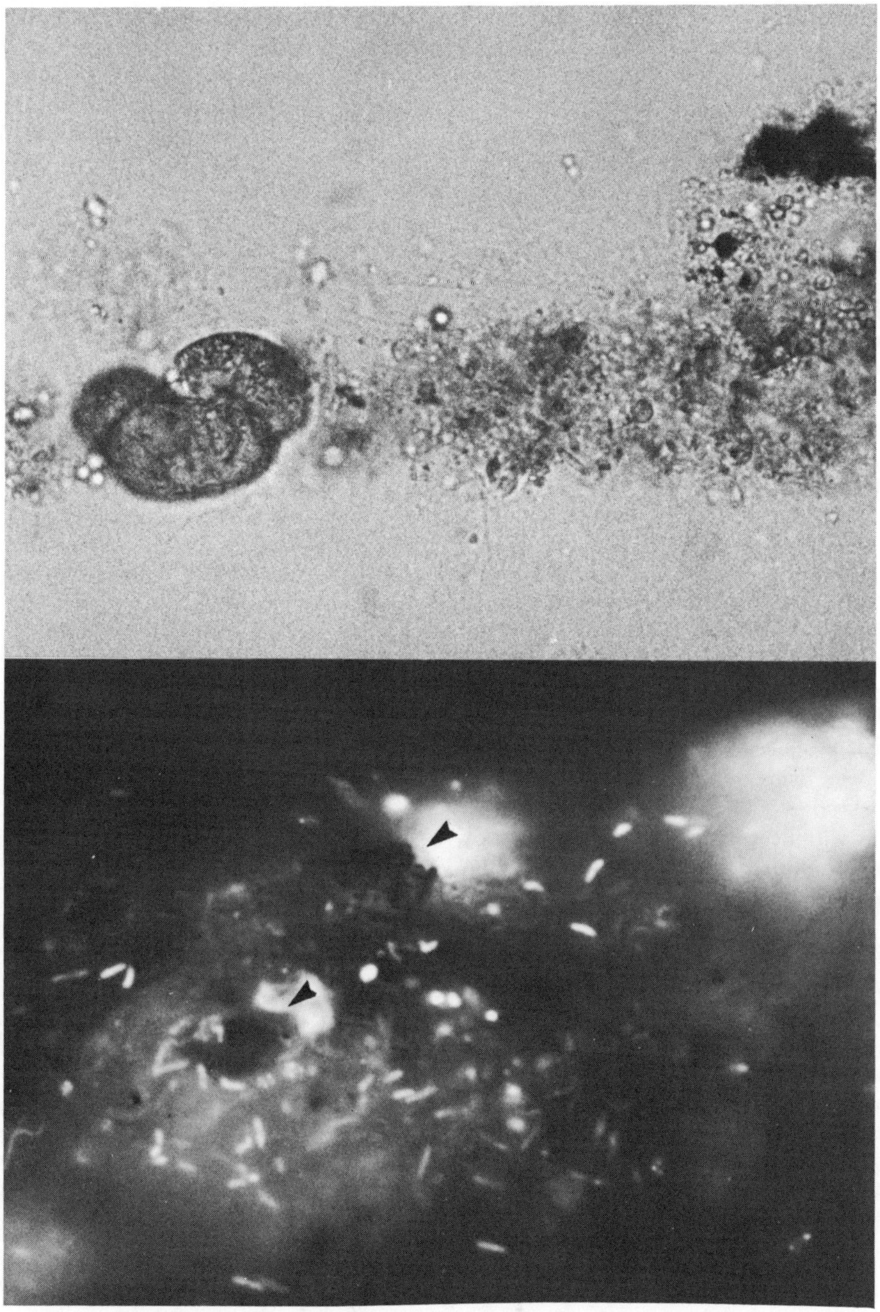

Fig. 6. Photomicrographs of feces of salps collected on the outer continental shelf showing the presence of pine pollen (upper) and carbon particles (at arrows) from smoke (lower). The fluorescent bacteria in the lower photomicrograph are 2 μm in length.

must operate, and at present we can only speculate on its nature.

The lower limit of bacterial abundance on the outer shelf, about 10^5 ml^{-1}, is a general lower limit for all surface ocean waters. This may reflect substrate limitation in which the rate of bacterial production varies with substrate supply. It may also indicate that there is a refuge level of abundance below which microflagellates cannot further reduce the population. From a series of laboratory studies with cultures of marine and freshwater microflagellates, *Fen-*

chel [1982*b*] estimated that, in natural waters, a population of bacteria of about 10^6 ml^{-1} would support the active growth of flagellates. Since some extrapolation was necessary to arrive at this estimate, it is at best the right order of magnitude. It fits well, however, with our observations, and with those from other regions, that bacterial abundances are rarely less than 10^4 ml^{-1} and rarely much more than 10^6 ml^{-1}. *Sieburth* [1984] argues against the refuge explanation largely on the basis of laboratory experiments with bacteria and protozoans. The issue is worth further consideration.

In coastal water, inside the coastal front, the situation appears to be different, but the same principles may apply. Coastal waters always contain sufficient nutrients for photosynthesis and the measured daily rate is about 1 g C m^{-2} during most of the year [*Thomas,* 1966]. Because coastal waters are vertically mixed, there is no separation of layers dominated by phytoplankton and by bacteria that we see during intrusions of the Gulf Stream on the outer shelf [*Jacobsen et al.,* 1983]. The abundance of free bacteria in the coastal water is 10^6 ml^{-1} at all depths throughout the year [*Pomeroy et al.,* 1983]. Because they are probably never substrate limited, as previously suggested, 10^6 ml^{-1} may represent a call density at which grazing by flagellates becomes very efficient. Experimental verification is needed.

Because bacteria have very short generation times and high potential rates of production, we might expect that bacteria are the first organisms to respond to environmental changes. Therefore studies of bacterial populations could serve as early warning indicators of later responses by phytoplankton and zooplankton. However, in the microbial food web of the continental shelf, bacteria depend on phytoplankton and probably zooplankton for their principal source of growth substrates. Moreover, their numbers may also be regulated by grazing protozoans and some metazoans. Consequently, changes in bacterial populations are synchronized with those of phytoplankton and zooplankton. Increases in bacterioplankton appear on the same day as increases in phytoplankton in both the lower layer of Gulf Stream water and the upper layer of shelf water [*Pomeroy et al.,* 1983]. Bacteria and other microorganisms are thus an integral part of the marine food web and are probably significant in terms of the amount of energy and materials they mediate.

Conclusions

Bacterial populations in continental shelf waters of the southeastern United States are similar to those reported in shelf waters elsewhere. Their numbers range from 10^5 to 10^6 ml^{-1}, the size of the free bacteria is 0.1–0.4 μm, and most of the free bacteria are coccoid forms with lesser numbers of curved or sigmoid cells. Attached bacteria are characteristically 1 to 2-μm rods. Variations in bacterial numbers generally follow variations in primary productivity on both a day-to-day and a seasonal basis. However, there seem to be definite upper and lower limits to population size which may be controlled by organisms feeding on the bacteria. The upper limit of abundance, about 5×10^6 ml^{-1} may represent the density at which flagellate bacteriovores grow exponentially at the expense of the bacteria, and since the microflagellates can double almost as rapidly as can the bacteria, they are an effective check on further population growth [*Fenchel,* 1982*b*]. Because of this, bacterial numbers probably do not indicate bacterial production rates. The lower limit of abundance (5×10^4 ml^{-1}) may reflect a refuge density below which bacteriovores can no longer efficiently remove bacteria from the water.

The above statements implicate the protozoans as probable key organisms in the food chain leading from dissolved and nonliving particulate organic materials to heterotrophic bacteria. We do not know what controls flagellate populations, but because their abundance appears to be comparable to that of the autotrophic microflagellates of similar size, their predators are potentially the same filter-feeding zooplankton and benthos. Thus the link between microbial and metazoan marine food webs remains uncertain. We know that microbial biomass is sometimes dominant on the continental shelf and that energy utilization by the <10-μm organisms is substantial.

Acknowledgments. This work was supported by contract DE-AS09-00639 with the U.S. Department of Energy. The scanning electron micrographs were prepared by Shirley Nishino.

References

Alldredge, A. L., The chemical composition of macroscopic aggregates in two neritic seas, *Limnol. Oceanogr., 24,* 855–866, 1979.

Alldredge, A. L., and J. L. Cox, Primary production and chemical composition of marine snow in surface waters of the Southern California Bight, *J. Mar. Res., 40,* 517–527, 1982.

Azam, F., and J. A. Fuhrman, Measurement of bacterioplankton growth in the sea and its regulation by environmental conditions, in *Heterotrophic Activity in the Sea,* edited by J. E. Hobbie and P. J. L. Williams, pp. 179–196, Plenum, New York, 1984.

Azam, F., and R. E. Hodson, Size distribution and activity of marine microheterotrophs, *Limnol. Oceanogr., 22,* 492–501, 1977.

Azam, F., T. Fenchel, J. G. Field, J. S. Gray, L.-A. Meyer-Reil, and F. Thingstad, The ecological role of water-column microbes in the sea, *Mar. Ecol. Prog. Ser., 10,* 257–263, 1983.

Bell, C. R., and L. J. Abright, Attached and free-floating bacteria in the Fraser River estuary, British Columbia, Canada, *Mar. Ecol. Prog. Ser., 6,* 317–327, 1981.

Blanton, J. O., Ocean currents along a nearshore frontal zone on the continental shelf of the southeastern United States, *J. Phys. Oceanogr., 11,* 1627–1637, 1981.

Brockman, V. H., K. Eberlein, H. D. Junge, E. Maier-Reimer, and D. Siebers, The development of a natural plankton in an outdoor tank with nutrient-poor seawater, 2. Changes in dissolved carbohydrates and amino acids, *Mar. Ecol. Prog. Ser., 1,* 283–291, 1979.

Campbell, W. B., T. R. Jacobsen, and L. R. Pomeroy, Heterotrophic-photoautotrophic index: A qualitative parameter of microbial interactions applied to a Gulf Stream intrusion, *Mar. Sci. Commun., 5,* 383–398, 1979.

Cronn, D. R., and W. Nutmagul, Analysis of atmospheric hydrocarbons during winter MONEX, *Tellus, 34,* 159–165, 1982.

Deibel, D., Laboratory measured grazing and ingestion rates of the salp, *Thalia democratica* Forskal, and the doliolid, *Dolioletta gegenbauri* Uljanin (Tunicata, Thalicea), *J. Plankton Res., 4,* 189–210, 1982.

Delaney, A. C., A. D. Delaney, D. W. Parkins, J. J. Griffin, E. D. Goldberg, and B. E. F. Reimann, Airborne dust collected at Barbados, *Geochim. Cosmochim. Acta, 31,* 885–909, 1967.

Eppley, R. W., S. G. Horrigan, J. A. Fuhrman, E. R. Brooks, C. C. Price, and K. Sellner, Origins of dissolved organic matter in southern California coastal waters: Experiments on the role of zooplankton, *Mar. Ecol. Prog. Ser., 6,* 149–159, 1981.

Fell, J. W., Yeasts in oceanic regions, in *Recent Advances in Aquatic Mycology,* edited by E. B. G. Jones, pp. 93–124, Elek Science, London, 1976.

Fenchel, T. M., Ecology of heterotrophic microflagellates, 2, Bioenergetics and growth, *Mar. Ecol. Prog. Ser., 8,* 225–231, 1982*a*.

Fenchel, T. M., Ecology of heterotrophic microflagellates, 4, Quantitative occurrence and importance as bacterial consumers, *Mar. Ecol. Prog. Ser., 9,* 35–42, 1982*b*.

Fenchel, T. M., and B. B. Jørgensen, Detritus food chains of aquatic ecosystems and the role of bacteria, *Adv. Microb. Ecol., 1,* 1–49, 1976.

Ferguson, R. L., and A. V. Palumbo, Distribution of suspended bacteria in neritic waters south of Long Island during stratified conditions, *Limnol. Oceanogr., 24,* 697–705, 1979.

Ferguson, R. L., and P. Rublee, Contribution of bacteria to standing crop of coastal plankton, *Limnol. Oceanogr., 21,* 141–145, 1976.

Findlay, S. E. G., and K. R. Tenore, Effect of a free-living marine nematode (*Diplolaimella chitwoodi*) on detrital carbon mineralization, *Mar. Ecol. Prog. Ser., 8,* 161–166, 1982.

Fuhrman, J. A., Influence of method on the apparent size distribution of bacterioplankton cells: Epifluorescence microscopy compared to scanning electron microscopy, *Mar. Ecol. Prog. Ser., 5,* 103–106, 1981.

Fuhrman, J. A., J. W. Ammerman, and F. Azam, Bacterioplankton in the coastal euphotic zone: Distribution, activity and possible relationships with phytoplankton, *Mar. Biol., 60,* 201–207, 1980.

Glibert, P. M., Regional studies of daily, seasonal, and size fraction variability in ammonium remineralization, *Mar. Biol., 70,* 209–222, 1982.

Gordon, D. C., A microscopic study of particles in the North Atlantic Ocean, *Deep Sea Res., 17,* 175–185, 1970.

Hanson, R. B., and W. J. Wiebe, Heterotrophic activity associated with particulate size fractions in a *Spartina alterniflora* salt-marsh estuary, *Mar. Biol., 42,* 321–330, 1977.

Hargrave, B. T., The utilization of benthic microflora by *Hyalella azteca* (Amphipoda), *J. Anim. Ecol., 39,* 427–437, 1970.

Heinbokel, J. F., Growth, reproduction and life cycles in marine planktonic protozoa, *Ann. Inst. Oceanogr. Paris, 58,* suppl., 213–222, 1982.

Hobbie, J. E., and C. C. Crawford, Respiration corrections for bacterial uptake of dissolved compounds in natural waters, *Limnol. Oceanogr., 14,* 528–532, 1969.

Hobbie, J. E., R. J. Daley, and S. Jasper, Use of Nuclepore filters for counting bacteria by fluorescence microscopy, *Appl. Environ. Microbiol., 33,* 1225–1228, 1977.

Hodson, R. E., A. E. Maccubbin, and L. R. Pomeroy, Dissolved adenosine triphosphate and its utilization by free-living and attached bacterioplankton, *Mar. Biol., 64,* 43–51, 1981.

Hofmann, E. E., L. J. Pietrafesa, J. M. Klinck, and L. P. Atkinson, A time-dependent model of nutrient distribution in continental shelf waters, *Ecol. Modell., 10,* 192–214, 1980.

Honjo, S., Material fluxes and modes of sedimentation in the mesopelagic and bathypelagic zones, *J. Mar. Res., 38,* 53–97, 1980.

Hoppe, H.-G., Determination and properties of actively metabolizing heterotrophic bacteria in the sea investigated by means of microautoradiography, *Mar. Biol., 36,* 291–302, 1976.

Hoppe, H.-G., Relations between active bacteria and heterotrophic potential in the sea, *Neth. J. Sea Res., 12,* 78–98, 1978.

Ittekott, K., Variations of dissolved organic matter during a plankton bloom: Qualitative aspects based on sugar and amino acid analyses, *Mar. Chem., 11,* 143–158, 1982.

Jacobsen, T. R., Autotrophic and heterotrophic measurements on the continental shelf of the southeastern United States, Ph.D. dissertation, 196 pp., Univ. of Ga., Athens, 1981.

Jacobsen, T. R., L. R. Pomeroy, and J. O. Blanton, Autotrophic and heterotrophic abundance and activity associated with a nearshore front off the Georgia coast, USA, *Estuarine Coastal Shelf Sci., 17,* 509–521, 1983.

Johnson, P. W., and J. Sieburth, Chroococcoid cyanobacteria in the sea, a ubiquitous and diverse phototrophic biomass, *Limnol. Oceanogr., 24,* 928–935, 1979.

Johnson, T. W., and F. K. Sparrow, *Fungi in Oceans and Estuaries,* 668 pp., Hafner, New York, 1966.

Joiris, C., On the role of heterotrophic bacteria in marine

ecosystems: Some problems, *Helgol. Wiss. Meeresunters., 30,* 611–621, 1977.

Ketseridis, G., J. Hahn, R. Jaeniche, and C. Junge, The organic constituents of atmospheric particulate matter, *Atmos. Environ., 10,* 603–610, 1976.

Krause, M., Vertical distribution of faecal pellets during FLEX '76, *Helgol. Wiss. Meeresunters., 34,* 313–327, 1981.

Larsson, V., and A. Hagström, Phytoplankton exudate release as an energy source for the growth of pelagic bacteria, *Mar. Biol., 52,* 199–206, 1979.

Mopper, K., R. Dawson, G. Liebezeit, and V. Ittekott, Monosaccharide spectra of natural waters, *Mar. Chem., 10,* 55–66, 1980.

Morrison, S. J., and D. C. White, Effects of grazing by estuarine gammariden amphipods on the microbiota of allochthonous detritus, *Appl. Environ. Microbiol., 40,* 659–671, 1980.

Newell, R. C., M. I. Lucar, and E. A. S. Linley, Rate of degradation and efficiency of conversion of phytoplankton debris by marine micro-organisms, *Mar. Ecol. Prog. Ser., 6,* 123–136, 1981.

Pace, J. L., J. E. Glasser, and L. R. Pomeroy, A simulation analysis of continental shelf food webs, *Mar. Biol., 82,* 47–63, 1984.

Pomeroy, L. R., The ocean's food web, a changing paradigm, *BioScience, 24,* 499–504, 1974.

Pomeroy, L. R., Secondary production mechanisms of continental shelf communities, in *Ecological Processes in Coastal and Marine Systems,* edited by J. Livingston, pp. 163–186, Plenum, New York, 1979.

Pomeroy, L. R., Microbial processes in the sea: Diversity in nature and science, in *Heterotrophic Activity in the Sea,* edited by J. E. Hobbie and P. J. L. Williams, pp. 1–24, Plenum, New York, 1984.

Pomeroy, L. R., and D. Deibel, Aggregation of organic matter by pelagic tunicates, *Limnol. Oceanogr., 25,* 643–652, 1980.

Pomeroy, L. R., and R. E. Johannes, Total plankton respiration, *Deep Sea Res., 13,* 971–973, 1966.

Pomeroy, L. R., and R. E. Johannes, Occurrence and respiration of the ultra-plankton in the upper 500 meters of the ocean, *Deep Sea Res., 15,* 381–391, 1968.

Pomeroy, L. R., L. P. Atkinson, J. O. Blanton, W. B. Campbell, T. R. Jacobsen, K. H. Kerrick, and A. M. Wood, Microbial distribution and abundance in response to physical and biological processes on the continental shelf of southeastern U.S.A., *Cont. Shelf Res., 2,* 1–20, 1983.

Porter, K. G., and Y. Feig, The use of DAPI for identifying and counting aquatic microflora, *Limnol. Oceanogr., 25,* 943–948, 1980.

Robertson, M. L., A. L. Mills, and J. C. Zieman, Microbial synthesis of detritus-like particulates from dissolved organic carbon released by tropical seagrasses, *Mar. Ecol. Prog. Ser., 7,* 279–285, 1982.

Sieburth, J. M., Convener's report on the informal session on biomass and productivity of microorganisms in planktonic ecosystems, *Helgol. Wiss. Meeresunters, 30,* 697–704, 1977.

Sieburth, J. M., Protozoan bacterivory in pelagic marine waters, in *Heterotrophic Activity in the Sea,* edited by J. E. Hobbie and P. J. L. Williams, pp. 405–444, Plenum, New York, 1984.

Sieburth, J. M., and A. S. Dietz, Biodeterioration in the sea and its inhibition, in *Effect of the Ocean Environment on Microbial Activities,* edited by R. R. Colwell and R. Y. Morita, pp. 318–326, University Park Press, Baltimore, Md., 1974.

Tenore, K. R., L. Cammen, S. E. G. Findlay, and N. Phillips, Perspectives of research on detritus: Do factors controlling the availability of detritus to macroconsumers depend on its source?, *J. Mar. Res., 40,* 473–490, 1982.

Thomas, J. P., Influence of the Altamaha River on primary production beyond the mouth of the river, M.S. thesis, 88 pp., Univ. of Ga., Athens, 1966.

Turner, R. E., Community plankton respiration in a saltmarsh tidal creek and estuary and in the continental shelf waters of Georgia, Ph.D. dissertation, 110 pp., Univ. of Ga., Athens, 1974.

Waterbury, J. B., S. W. Watson, R. R. L. Guillard, and L. E. Brand, Widespread occurrence of a unicellular, marine, planktonic cyanobacterium, *Nature, 277,* 293–294, 1979.

Watson, S. W., Role of bacteria in an upwelling ecosystem, in *Upwelling Ecosystems,* edited by R. Boje and M. Tomczak, pp. 139–154, Springer, New York, 1978.

Watson, S. W., T. J. Novitsky, H. L. Quinby, and F. W. Valois, Determination of bacterial number and biomass in the marine environment, *Appl. Environ. Microbiol., 33,* 940–946, 1977.

Wiebe, W. J., and L. R. Pomeroy, Microorganisms and their association with aggregates and detritus in the sea: A microscopic study, *Mem. Ist. Ital. Idrobiol. Dott. Marco de Marchi, 29,* suppl., 325–352, 1972.

Wiebe, W. J., and D. F. Smith, Direct measurement of dissolved organic carbon release by phytoplankton and incorporation by microheterotrophs, *Mar. Biol., 42,* 213–223, 1977.

Williams, P. J. L., Heterotrophic utilization of dissolved organic compounds in the sea, 1, Size distribution of population and relationship between respiration and incorporation of growth substance, *J. Mar. Biol. Assoc. U. K., 50,* 859–870, 1970.

Williams, P. J. L., Incorporation of microheterotrophic processes into the classical paradigm of the planktonic food web, *Kiel. Meeresforsch. Sonderh., 5,* 1–28, 1981*a*.

Williams, P. J. L., Microbial contribution to overall marine plankton metabolism: Direct measurements of respiration, *Oceanol. Acta, 4,* 359–364, 1981*b*.

Williams, P. M., H. Oeschger, and P. Kinney, Natural radiocarbon activity of the dissolved organic carbon in the north-east Pacific Ocean, *Nature, 224,* 256–258, 1969.

Wright, R. T., and B. K. Burnison, Heterotrophic activity measured with radio-labeled organic substrates, in *Native Aquatic Bacteria: Enumeration, Activity and Ecology, STP-695,* edited by J. W. Costerton and R. R. Colwell, pp. 140–155. American Society for Testing and Materials, Philadelphia, Pa., 1979.

Yoder, J. A., Environmental control of phytoplankton production on the southeastern U.S. continental shelf, in *Oceanography of the Southeastern U.S. Continental Shelf, Coastal and Estuarine Sci.,* vol. 2, edited by L. P. Atkinson, D. W. Menzel, and K. A. Bush, AGU, Washington, D. C., this volume.

Seasonal Changes in Soft Bottom Macroinfauna of the U.S. South Atlantic Bight

K. R. Tenore

Chesapeake Biological Laboratory, Solomons, Maryland

The macroinfauna of soft bottom sediments of the U.S. South Atlantic Bight has high species diversity but low densities because of unstable sediments, wide temperature fluctuation, and low nutrient inputs. The fauna is homogenous, with no persistent latitudinal gradients. Communities may be delimitated into inner coastal and outer shelf zones. The inner community is quite distinct and persists throughout the year; the spatial scales of communities of the central and outer zones charge seasonally with influx of newly settled species. The outermost slope stations were distinct and did not cluster with each other. In the autumn the central and outer shelf were divided into northern and southern zones, because of a high number of transient species in the northern area. Thus there are few persistent or resident dominants, and species composition changes seasonally, perhaps in great part moderated by current regimes.

Introduction

The U.S. South Atlantic Bight (SAB) extends from Cape Hatteras, North Carolina, to Cape Canaveral, Florida. Climatological conditions are similar over most of the continental shelf of this region. The water depth is shallow (<30 m), temperature fluctuates widely (5°–25°C), and sediments are subject to considerable wind and tidal scour [*Dörjes,* 1977]. Most sediments are coarse sand, with outcroppings of "hard bottoms" relict coral. River discharge, tidal outwelling from salt marshes, and intrusions of Gulf Stream water are the major sources of nutrients to the shelf [*Atkinson et al.,* 1978; *Odum,* 1968; *Lee et al.,* 1981]. Because of the great width of the shelf off South Carolina and Georgia (120 km) and the density barriers formed nearshore by salinity fronts, estuarine outwelling does not continuously contribute particulates to the shelf seaward of 20 km from the coast [*Haines and Dunstan,* 1975; *Brokaw and Oertel,* 1976; *Blanton and Atkinson,* 1978]. Thus most of the shelf is oligotrophic but receives intermittent intrusions of relatively nutrient rich Gulf Stream water [*Lee et al.,* 1981]. Phytoplankton blooms are frequent in areas of intrusions [*Dunstan and Atkinson,* 1976; *Yoder et al.,* 1981]. Such unique characteristics of this particular shelf region are important because the overlying primary production of the water column is the major factor affecting benthic production [*Rowe,* 1971].

Although information is available on deep-sea and nearshore benthos [*Rowe and Menzies,* 1969; *Rowe,* 1971; *Smith,* 1971; *Frankenberg,* 1971; *Frankenberg and Leiper,* 1977; *Dörjes,* 1977], there are few studies of the macroinfaunal benthos of the continental shelf proper. *Cerame-Vivas and Gray* [1966] and *Day et al.* [1971] studied distributional patterns of macrobenthos of the continental shelf off North Carolina, and *McNulty et al.* [1962] studied benthic community structure off southern Florida. These areas lie outside the shelf region of the Georgia Bight proper. Most interest has centered on the so-called "hard bottoms" that are characteristic of the southeast shelf. *Tenore et al.* [1978] and *Hanson et al.* [1981] studied benthic processes over a wide area of the Georgia continental shelf, reporting macrofaunal standing crop but not their spatial or seasonal delimitations. *Dörjes* [1977] reported densities of major taxonomic groups of infaunal benthos at 100 stations off Sapelo Island, Georgia, but noted only the dominant species and did not specify the seasons in which samples were taken. Thus there have been no comprehensive studies of the macroinfauna of the offshore continental shelf of the Georgia Bight proper, nor has there been a sampling design to describe seasonal patterns of distribution.

The South Atlantic Bight Program (SABP) funded by the Bureau of Land Management and conducted by Texas Instruments, Inc., in 1977 included seasonal sampling of the macroinfauna. In this paper we report analyses of these data that (1) delimit areas of similar community composition and their stability throughout the year, (2) characterize stability of species composition, and (3) report on patterns of standing biomass. Further, we discuss the findings in relation to several unique characteristics of the U.S. South Atlantic Bight.

Material and Methods

Sampling stations were established to delineate major spatial and seasonal trends. The stratified sampling design included 50 fixed stations spaced evenly from north to south on seven transects crossing the shelf perpendicular to the shoreline. Each transect crossed the inner, middle, and outer shelf regions. Sampling periods were winter (February 9 to March 5), spring (May 8 to May 22), summer (August 4 to August 15), and fall (November 15 to November 29). Box cores for macrofauna were obtained in winter and summer at all 50 stations and in spring and fall at 25 stations (1 BCDEF, 2 BCDEFGH, 5 BCDEFGHI, 7 BCDEF; see Figure 6). Sampling was done from the M/V *G. W. Pierce* using Raydist DRS-H or Argo navigation systems.

Bottom samples of 0.06-m^2 area were collected with a stainless steel box corer with sufficient weight to guarantee 15-cm penetration. Six cores from each station were subsampled for grain size distribution and particulate organic

TABLE 1. Average Total Macroinfaunal Density and Biomass (mg AFDW) With Percent Contribution of Polychaetes to Winter Samples

Station	Density		Biomass		No. of Taxa	Diversity (H')
	No./0.06 m^2	% Polychaetes	mg/0.06 m^2	% Polychaetes		
1A	317	28	1,042	53	121	3.81
1B	239	30	408	79	98	3.64
1C	232	29	734	68	96	4.21
1D	107	50	1,008	16(M)*	108	4.16
1E	222	82	2,279	36	186	3.88
1F	116	12	151	87	75	2.06
2A	325	25	1,878	43	136	3.56
2B	479	44	9,603	16(M)*	210	4.64
2C	554	51	1,304	50	256	5.18
2D	393	54	1,384	77	258	5.36
2E	104	66	657	61	141	4.77
2F	336	61	714	83	259	5.08
2G	94	39	534	73	110	3.98
2H	28	29	371	14(E)*	44	3.52
3A	346	60	1,631	56	138	4.52
3B	251	43	1,027	72	179	4.71
3C	212	65	556	71	171	4.89
3D	448	56	1,431	77	217	4.86
3E	143	52	864	32	168	5.16
3F	54	52	639	41	100	3.60
4A	580	43	1,036	71	120	3.16
4B	230	46	1,322	42	146	4.16
4C	1,002	61	13,583	12(M)*	180	3.99
4D	456	32	760	64	150	3.75
4E	413	48	2,500	39	203	4.80
4F	180	57	1,788	32	134	3.87
4G	42	20	474	71	48	3.29
5A	525	27	3,601	74	129	3.29
5B	535	43	1,501	51	206	4.62
5C	450	51	4,782	23(M)*	147	4.62
5D	331	45	876	62	159	4.07
5E	273	45	3,237	28(E)*	210	5.19
5F	382	41	8,686	25(E)*	182	4.20
5G	336	40	5,839	22(EM)*	216	4.74
5H	279	75	2,865	77	202	4.07
5I	6	50	18	82	14	2.09
6A	200	72	2,308	87	106	4.01
6B	494	60	3,813	50	204	4.99
6C	379	49	1,038	78	230	5.00
6D	346	33	1,432	47	238	4.92
6E	448	35	5,773	27(E)*	248	5.20
6F	377	39	3,503	40	225	5.15
6G	128	61	1,717	80	129	4.66
6H	22	40	66	42	52	3.10
7A	314	68	10,975	16(E)*	146	4.05
7B	151	44	4,825	24(EM)*	175	5.09
7C	318	54	1,267	37	228	5.12
7D	158	66	1,776	74	210	5.64
7E	145	13	943	37	75	2.35
7F	60	47	91	56	84	3.62

*Low value due to presence of more than 50% of another group (E, echinoderm; M, miscellaneous).

carbon. Subsamples for grain size were taken with stainless steel core tubes of 6-cm depth. The wet samples were sealed in jars and returned to the laboratory for analysis. Similar subcores were frozen for total organic carbon (TOC) analysis. The macrofaunal samples were washed on a 0.5-mm mesh sieve, preserved in buffered 10% formalin containing rose bengal stain, and later transferred to 70% ethanol.

The number of individuals in each taxa was counted, and biomass was estimated from wet (preserved) weight to dry and ash-free conversions. Biomass measurements of dry (DW) and ash-free dry weights (AFDW) were determined from subsamples of major taxonomic groups (decalcified molluscs, polychaetes, arthropods, echinoderms, and miscellaneous) by drying the samples at 62°C to constant weight and ashing at 500°C for 6 hours. These conversion values were used to calculate standing biomass measures from wet (preserved) weights.

Sediments for grain size analysis were homogenized and air dried, the organics were removed with 30% hydrogen peroxide, and the soluble salts were removed by washing with deionized water. The washed sample was then dried at 40°C and deflocculated with a Calgon solution. Grain size distribution was determined first by wet-sieving to separate gravel, sand, and silt/clay fractions and then by obtaining hydraulic grain size equivalents at 0.5-day intervals for the sand fraction and pipette analysis for the silt/clay composition [*Folk,* 1968]. Results of all size intervals are reported as percent of total weight of sediment. The values were used to calculate mean and median particle size, sorting coefficient, and skewness. Total organic carbon of sediments was determined by triplicate analyses with a Coleman model 33 CH analyzer.

Stepwise and multiple-regression techniques were used to correlate the density distribution of total and individual major taxa with hydrographic and sedimentary parameters. The Jaccard similarity index was used to delimit spatial and temporal clustering of stations using data for species that made up at least 0.2% of the samples. The index for each station was clustered using a complete linkage algorithm for each separate set of combined seasonal samplings.

Results

Averages of total number of individuals and biomass, the percent contribution of polychaetes, and total species diversity (H') from each station are given in Tables 1 through 4. Averages for each station were calculated from the six box core samples. Because the variance among stations is low, large fluctuations between stations reflect the patchy distribution of benthos over the shelf. Both density and biomass, especially at the outer shelf break stations, were low throughout the shelf region at all seasons (Table 5). Overall, the middle shelf stations had consistently higher

TABLE 2. Average Total Macroinfaunal Density and Biomass (mg AFDW) With Percent Contribution of Polychaetes to Spring Samples

	Density		Biomass			
Station	No./0.06 m²	% Polychaetes	mg/0.06 m²	% Polychaetes	No. of Taxa	Diversity (H')
1B	98	29	1,042	62	87	3.91
1C	332	26	598	64	126	3.85
1D	257	40	483	58	178	4.68
1E	502	28	1,109	73	207	3.25
1F	50	31	140	61	74	3.54
2B	924	48	13,115	16(M)*	249	4.84
2C	436	45	1,838	55	225	5.04
2D	412	57	2,012	90	220	4.65
2E	902	49	3,241	68	325	4.96
2F	824	18	2,091	37	249	3.61
2G	171	29	714	44	134	4.01
2H	40	28	834	19(E)*	68	3.24
5B	571	32	2,389	18(M)*	177	4.56
5C	298	53	8,171	29(M)*	148	4.02
5D	353	38	1,113	43	136	4.04
5E	455	42	3,778	24(EA)*	234	5.24
5F	633	48	5,236	57	279	4.86
5G	436	31	2,970	19(EA)*	137	3.70
5H	256	64	2,391	72	218	5.11
5I	91	46	535	36	131	4.41
7B	234	34	3,316	91	199	5.09
7C	350	57	2,468	55	242	5.37
7D	272	54	2,804	79	260	5.22
7E	106	33	727	37	72	2.71
7F	21	40	182	40	63	2.90

*Low value due to presence of more than 50% of another group (E, echinoderm; M, miscellaneous; A, arthropod).

TABLE 3. Average Total Macroinfaunal Density and Biomass (mg AFDW) With Percent Contribution of Polychaetes to Summer Samples

Station	Density		Biomass		No. of Taxa	Diversity (H')
	No./0.06 m^2	% Polychaetes	mg/0.06 m^2	% Polychaetes		
1A	50	38	851	50	86	4.21
1B	129	27	962	36	110	4.06
1C	258	24	642	34	126	4.14
1D	206	40	1,401	24(M)*	206	4.95
1E	262	28	482	52	186	3.86
1F	28	41	168	62	47	2.50
2A	40	28	797	83	50	2.54
2B	377	60	5,459	29(M)*	232	4.82
2C	283	34	1,205	66	213	5.02
2D	314	54	1,844	85	238	5.26
2E	117	41	828	67	169	4.41
2F	352	36	2,136	36	203	4.46
2G	75	30	349	43	83	3.39
2H	11	26	470	22(E)*	30	1.74
3A	134	30	1,135	30(EA)*	95	4.15
3B	122	42	1,825	21(M)*	160	5.02
3C	204	41	609	61	207	5.17
3D	250	46	794	79	182	5.08
3E	91	34	781	62	88	3.75
3F	22	29	578	6.0(E)*	37	2.36
4A	101	61	1,466	59	108	3.91
4B	1,275	27	1,477	39	116	2.35
4C	1,416	32	7,370	12(M)*	233	4.28
4D	247	32	790	48	187	4.80
4E	295	56	1,672	55	245	5.63
4F	36	62	137	47	72	4.01
4G	42	59	517	48	25	2.26
5A	66	56	1,070	43	92	4.17
5B	128	33	669	53	123	3.89
5C	459	44	6,873	35	161	4.26
5D	260	42	2,744	40	162	4.40
5E	218	30	1,098	38	202	5.28
5F	352	33	3,330	17(E)*	202	3.86
5G	382	35	3,246	54	166	4.12
5H	84	66	798	64	155	4.90
5I	18	22	112	34	29	2.14
6A	82	48	334	41	116	4.27
6B	628	28	2,903	17(EAM)*	300	4.74
6C	170	65	674	82	191	4.88
6D	220	46	1,831	74	223	5.06
6E	155	44	4,698	10(M)*	183	4.61
6F	154	45	1,664	72	191	5.13
6G	139	52	821	52	117	4.19
6H	18	37	58	77	50	2.39
7A	365	63	20,801	4.4(E)*	129	4.09
7B	546	49	2,015	27(E)*	297	5.31
7C	783	62	3,335	31	262	5.76
7D	200	63	2,082	79	208	5.09
7E	128	31	288	0.0(M_O)*	70	3.48
7F	9	56	84	61	11	2.46

*Low value due to presence of more than 50% of another group (E, echinoderm; m, miscellaneous; A, arthropod; M_O, mollusc).

standing biomass. Thus the results of the stepwise multiple regressions of total polychaete macrofauna density with sedimentary and hydrographic parameters consistently resulted in water depth being the first significant predictor variable (Table 6). Species density was high with most stations averaging greater than 100 total taxa, with the exception of the outermost slope stations where H' generally was greater than 3, and greater than 5 at 22% of the stations.

Polychaetes usually dominated the macrofauna, accounting for more than 50% of total density in both numbers and biomass. The few stations where polychaete con-

TABLE 4. Average Total Macroinfaunal Density and Biomass (mg AFDW) With Percent Contribution of Polychaetes to Fall Samples

Station	Density		Biomass		No. of Taxa	Diversity (H')
	No./0.06 m²	% Polychaetes	mg/0.06 m²	% Polychaetes		
1B	224	35	729	44	98	4.16
1C	336	42	3,208	27(E)*	182	4.84
1D	431	30	456	71	142	3.48
1E	140	62	887	39	143	4.95
1F	61	30	150	41	84	3.74
2B	1,035	32	3,767	27(AM_OM)*	221	4.43
2C	518	37	1,002	38	192	4.84
2D	669	27	2,951	34	226	4.69
2E	147	45	1,166	40	189	5.38
2F	413	36	1,247	49	148	3.93
2G	182	23	597	35	97	3.27
5B	555	38	6,250	7(M)*	159	4.47
5C	615	48	6,584	17(M)*	165	4.26
5D	213	42	1,415	17(A)*	133	4.23
5E	569	28	2,141	36	224	5.00
5F	555	15	5,590	17(E)*	183	3.43
5G	564	19	4,200	14(E)*	140	3.27
5H	121	48	1,426	39	157	5.06
5I	131	31	221	36	108	3.62
7B	133	32	4,142	11(EA)*	166	4.98
7C	197	44	6,048	11(E)*	197	5.23
7D	263	49	1,990	19(EA)*	217	5.28
7E	23	29	582	67	48	2.91
7F	16	45	138	40	41	2.97

*Low value due to presence of more than 50% of another group (E, echinoderm; M, miscellaneous; A, arthropod; M_O, mollusc).

tribution to total biomass was low had large megabenthos in the samples. At these stations the percent contribution of polychaetes was still high.

There was no clear dominance by one or several species that persisted seasonally, either throughout or over part of the shelf. Of the most abundant species, only 18 comprised more than 0.2% of the total faunal density at all stations in at least one season (Table 7). Of these species, none ever comprised more than 8% of the total animal density. The polychaete *Spiophanes bombyx* and the polychaete *Unciola* sp. comprised more than 5% relative abundance in winter and spring, respectively, but the contribution of these species was reduced in other seasons. These percentages are low in part because they represent averages from all stations. At a few stations there were sporadic high densities of a given species. For example, in spring *S. bombyx* comprised up to 22% of faunal abundance at five stations. However, there was no geographic pattern or delineation, and dense distributions were spatially patchy. Most species were rare, comprising less than 0.2% mean total density.

Sediments are fairly homogenous throughout the bight, consisting of predominantly fine to medium, moderate to well-sorted (s_1 from 0.3 to 1.0) sand. The outer slope stations have poorly sorted coarse sediments ($s_1 > 1.0$). The sediments of this region have been extensively studied [*Gorsline,* 1963; *Emery,* 1968; *Kingery,* 1973; *Tenore et al.,* 1978; *Hanson et al.,* 1981]. The innermost stations varied most in sand particle size because of inputs from major river systems. The percent clay-silt fraction of the sediments was consistently high (~10%) at the inner shelf and low (<1%) at the midshelf. These differences in sediment composition with water depth (i.e., inshore versus middle shelf stations) might have contributed to the results of the stepwise multiple regressions (Table 6). Depth was the first and median or mean particle size was the second predictor variable of macrobenthic density. These factors, however, are confounded and sediment composition appears to be the major environmental factor causing inshore and midshelf differences in benthic densities.

A Q mode cluster analysis was performed by computing the Jaccard similarity index for all sample pairs based only on presence/absence of species and by constructing a dendrogram with the complete linkage (i.e., farthest

TABLE 5. Mean Biomass (g AFDW/0.06 m²) for Total Macroinfauna for All Seasons

Season	Shelf Regions		
	Inner	Middle	Outer
Winter	1.52	2.77	0.50
Spring	0.82	2.72	0.52
Summer	1.48	2.25	0.40
Fall	. . .	3.04	0.34

TABLE 6. Results of Stepwise Multiple Regressions of Macrofaunal Density (Total and Individual Components) to Sedimentary and Hydrographic Parameters

Season	Total		Echinoderm		Arthropod		Mollusc		Polychaete	
			Macroinfaunal Group							
Winter										
r^2 = 0.45*	D	0.32	ME	0.18	C	0.17	D	0.13	D	0.24
	T	0.34	MD	0.24	D	0.34	T	0.19	ME	0.28
Spring										
r^2 = 0.76	D	0.38	MD	0.18	NS†		D	0.29	D	0.32
	ME	0.47	MD	0.24	NS		L	0.39	ME	0.45
Summer										
r^2 = 0.28	D	0.13	NS		NS		NS		D	0.14
	ME	0.23	NS		NS		NS		ME	0.21
Autumn										
r^2 = 0.57	D	0.32	SK	0.23	NS		D	0.27	D	0.36
	MD	0.42	L	0.34	NS		C	0.28	MD	0.40

D, water depth; ME, sediment rates; MD, median particle size; C, silt/clay fraction; SK, skewness; L, latitude; T, bottom water temperature.

*The correlation r^2 is given for the complete regression for the total macroinfaunal biomass for each season. The first and second significant (p = 0.05) contributor, if any, to the stepwise regression is given with the cumulative correlation coefficient.

†NS, no significant regression.

neighbor) method. The analysis was performed for each set of seasonal data and for all four seasons combined (Figures 1 through 5). Graphic representations of the major spatial delimitations are illustrated in Figures 6 through 11. In spring and autumn, fewer stations were sampled, but the graphic display areally connects clustering stations and assumes similarity of fauna at unsampled stations. With this technique the stations could be divided into bands of inner, middle, and outer (to ~200 m) shelf station clusters. The outermost slope stations did not cluster, even with each other. The clustering of inner shelf stations (A and B) seemed to persist throughout the year. However, as only the extreme north-south transects of inner stations were sampled at all four seasons, the spatial continuity of this clustering is speculative. The middle and outer shelf clusters running from north to south persisted, with slight changes in spatial size, except in autumn. At this time, two north and south clusters containing both middle and outer shelf stations were delineated. Thus depth, not latitude, was the main factor in community delimitation. This agreed with results of the stepwise multiple regression (Table 6) that showed depth and sediment particle size (also correlated with depth) to be the major predictor variables of macrofaunal biomass. There are few

TABLE 7. Seasonal Changes in Percent Relative Abundance of Macroinfaunal Species

Species	Winter	Spring	Summer	Fall
Spiophanes bombyx	5.3	7.3	1.5	0.3
Parapionosyllia longicirata	3.3	0.9	1.6	1.9
Spio pettiboneae	2.0	1.4	2.4	0.2
Exogone lourei	2.0	0.4	0.4	1.2
Oxyurostylis smithi	1.4	. . .	. . .	. . .
Prionospio cristata	1.3	0.4	2.0	0.2
Synelmis albini	1.3	0.9	1.3	1.1
Protodorvillea kefersteini	1.2	0.9	0.6	1.0
Unciola sp. b	. . .*	6.0	1.5	. . .
Ericthonius brasiliensis	0.5	1.6	. . .	. . .
Leptochela serratobita	. . .	1.6	. . .	. . .
Paleanotus sp. a	0.8	1.4	0.4	0.5
Prionospio cirrifera	0.6	0.3	1.0	. . .
Typosyllis regulata carolinae	0.5	0.8	0.7	1.2
Pseudovermilia occidentallis	. . .	. . .	0.4	1.2
Haploops sp. a	0.2	0.5	0.2	1.2
Goniadides carolinae	0.7	0.5	0.6	1.0
Spaerosyllis bulbosa	0.7	0.3	0.4	1.0

*Less than 0.2% relative abundance.

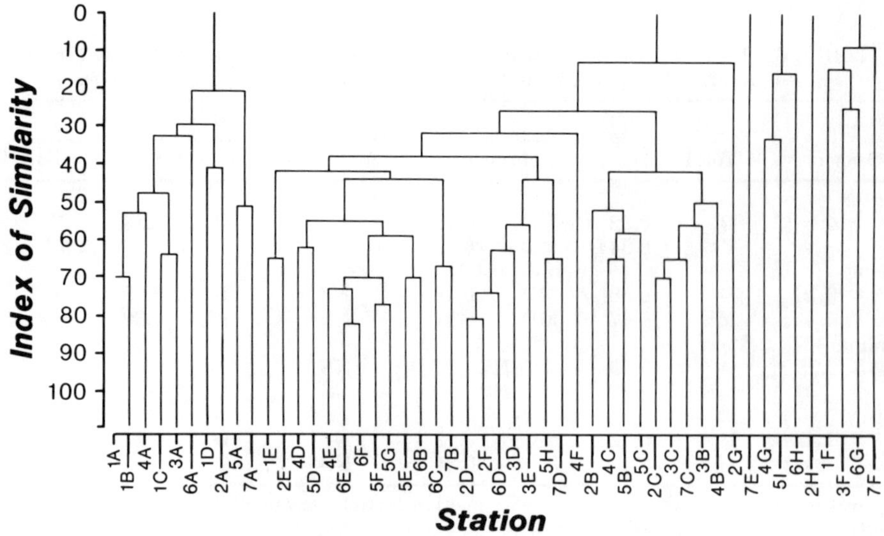

Fig. 1. Station clustering of winter macroinfauna data for the SAB using Jaccard similarity index based on species absence or presence.

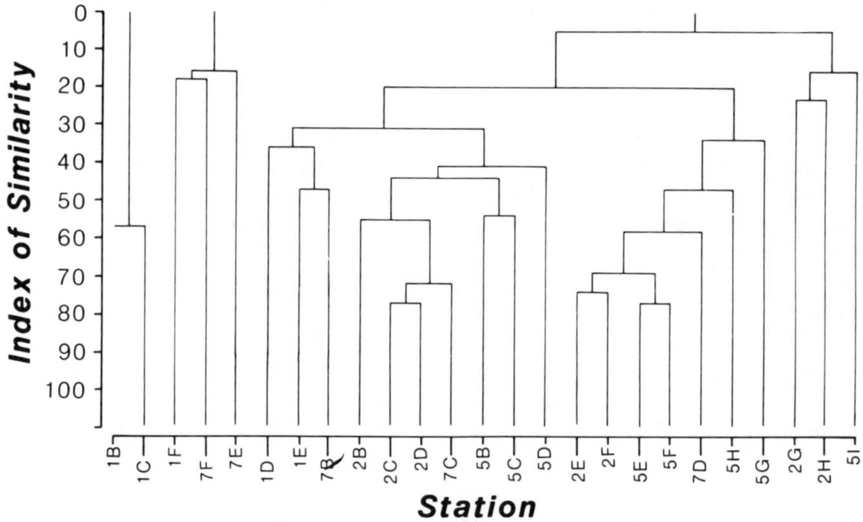

Fig. 2. Station clustering of spring macroinfauna data for the SAB using Jaccard similarity index based on species absence or presence.

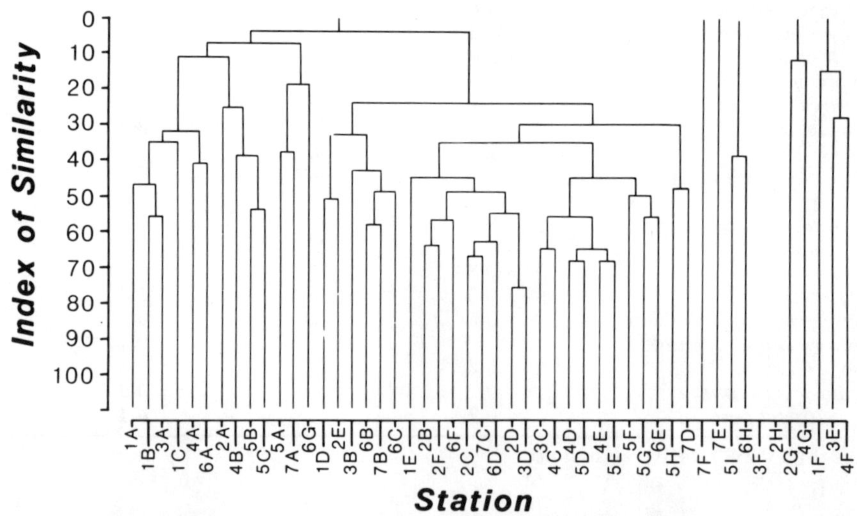

Fig. 3. Station clustering of summer macroinfauna data for the SAB using Jaccard similarity index based on species absence or presence.

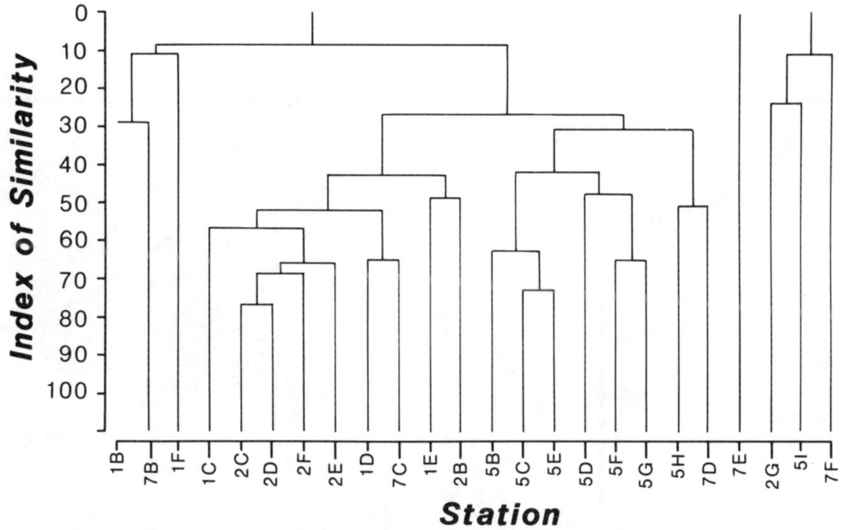

Fig. 4. Station clustering of autumn macroinfauna data for the SAB using Jaccard similarity index based on the species absence or presence.

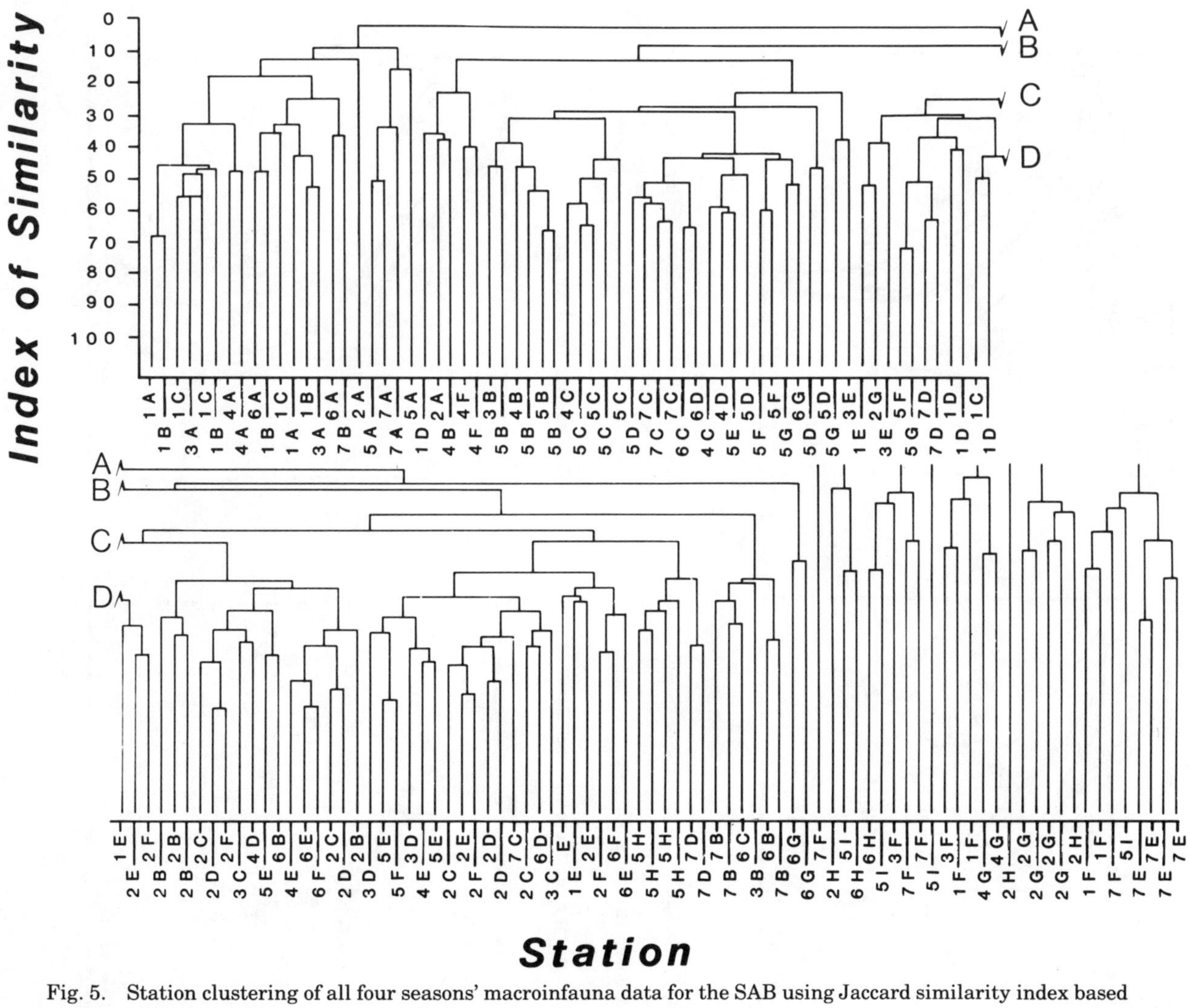

Fig. 5. Station clustering of all four seasons' macroinfauna data for the SAB using Jaccard similarity index based on species absence or presence.

Clusters for Winter Macroinfaunal Data

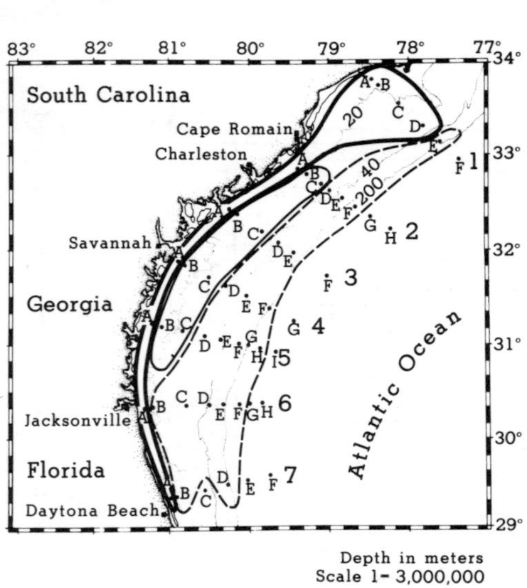

Fig. 6. Graphic display of major clusters from the winter sampling of the SAB using Jaccard similarity index based on species absence or presence.

Clusters for Spring Macroinfaunal Data

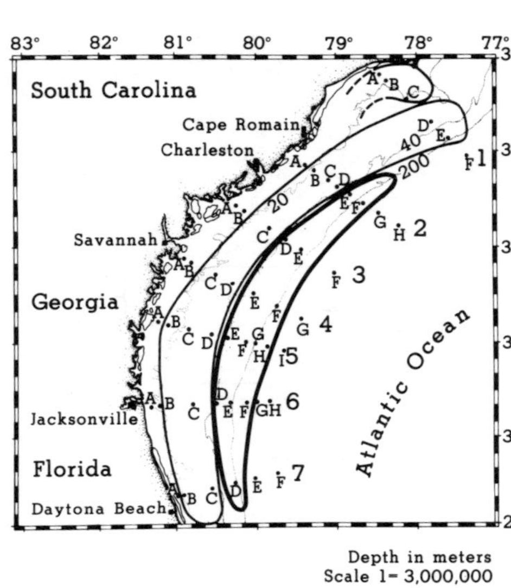

Fig. 7. Graphic display of major clusters from the spring sampling of the SAB using Jaccard similarity index based on species absence or presence.

Clusters for Summer Macroinfaunal Data

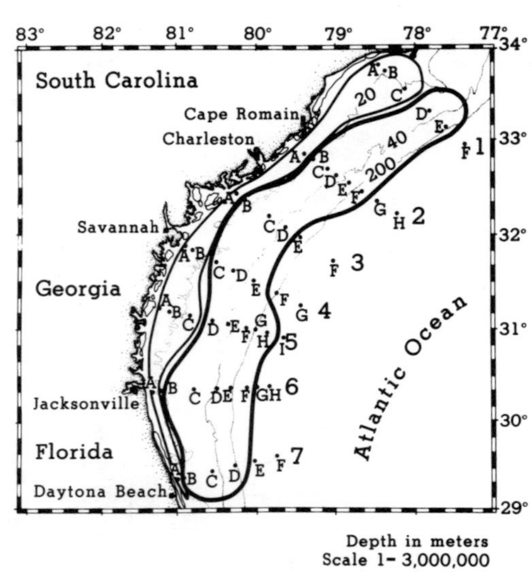

Fig. 8. Graphic display of major clusters from the summer sampling of the SAB using Jaccard similarity index based on species absence or presence.

Clusters of Fall Macroinfaunal Data

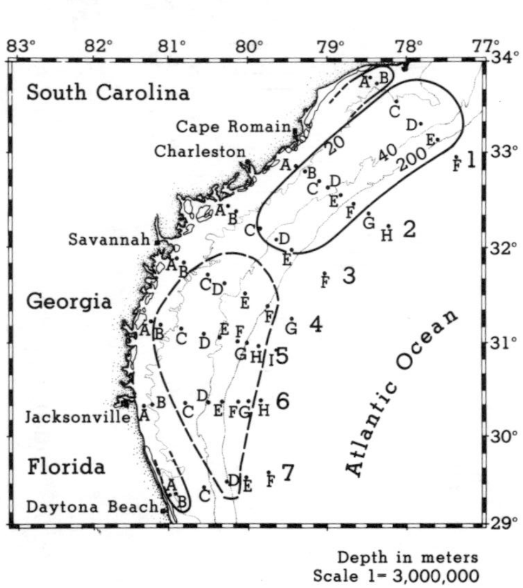

Fig. 9. Graphic display of major clusters from the autumn sampling of the SAB Jaccard similarity index based on species absence or presence.

Clusters for Total Annual Macroinfaunal Data

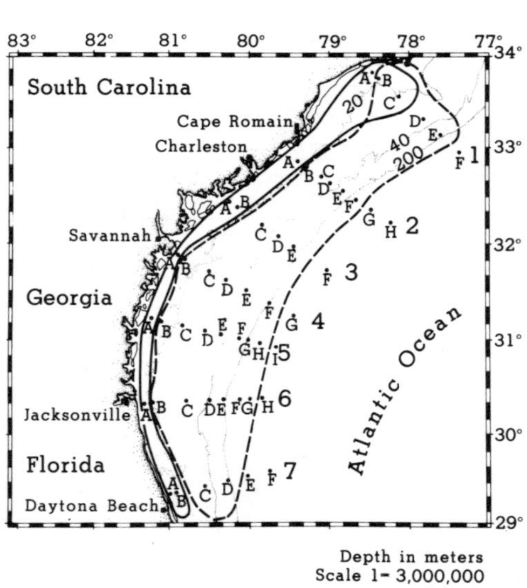

Fig. 10. Graphic display of major clusters from all four seasons' sampling of the SAB using Jaccard similarity index based on species absence or presence.

Subclustering within Broad Annual Macroinfaunal Clustering

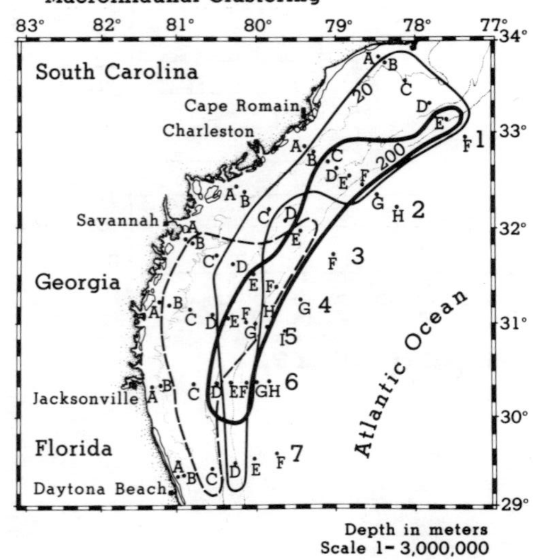

Fig. 11. Graphic display of subclustering from all four seasons' sampling of the SAB using Jaccard similarity index based on species absence or presence.

persistent dominants that characterize the faunal community delimitations, and seasonal changes are great.

Discussion

Many of the characteristics of the macrobenthos reflect the rather harsh environmental regime of the U.S. South Atlantic Bight. Because of the shallow depth of the wide shelf, water temperatures vary widely over the year, and the sediments are subject to strong tidal and wind-driven scour. Thus most species are mobile and/or surface dwellers that can withstand unstable sediment conditions. The shelf area, generally low in nutrients and primary production, depends in great part on intermittent intrusions of Gulf Stream waters that result in short-lived bursts of phytoplankton, mainly in the summer. *Hanson et al.* [1981] suggested that the uniformly low benthic densities in winter [*Tenore et al.*, 1978] give way in late spring and summer to occasional areas of elevated biomass due to enrichment from production associated with the intruded water that eventually sediments to the bottom. Many of the fauna are small opportunists with quick generation times that could respond quickly to intermittent and patchy nutrient inputs. Much of the coastal area is marshland, high in primary production, but whose contribution to outer shelf nutrients is in doubt [*Bishop et al.*, 1980; *Yoder et al.*, 1981].

Sediments are low in organic carbon, and most species are surface deposit or filter feeders, taking advantage of recently sedimented organics or microalgae. Thus standing biomass of the macrobenthos is low. Patchiness occurs in part because of areas of enrichment associated with outcroppings of "hard bottoms" that are characteristic of the bight [*Henry and Giles*, 1979].

Mostly mobile, non-tube-dwelling polychaetes dominate the fauna, which is typical of other shelf regions [*Maurer and Leathem*, 1980; *Flint and Holland*, 1980; *Boesch*, 1979]. This contrasts with the Georgia inshore communities where tube-dwelling polychaetes and amphipods are dominant [*Dörjes*, 1977].

Bottom fauna is quite homogenous throughout the bight, with no evidence of persistent latitudinal gradients. Inner, central, and outer shelf zones are continuous from north to south. These zones are also evident in the Middle Atlantic Bight [*Boesch*, 1979]. The inner shelf area was undersampled in this study, but limited data indicate faunal assemblages quite distinct from the rest of the shelf, including dominant magelonid, prionospid, and nereid polychaetes, and both burrowing and surface brittle stars. The central and outer shelf zones were more similar. Though distinct differences in dominants occurred in winter and spring, the fauna was quite homogenous in the summer. The only latitudinal delimitation that could be related to seasonal changes in shelf currents in the bight was observed in autumn. There is a pronounced southward flow of water on the central shelf [*Bumpus*, 1955] in autumn that may introduce benthic larvae from more northern populations. Much of the fauna in the northern zone is unique or persists only in the winter in the bight. This may represent a transient fauna whose presence is closely linked to changes in water movement.

Of the persistent dominants, the most characteristic is the polychaete *Spiophanes bombyx*, a discretely mobile, surface deposit feeder [*Jumars and Fauchald*, 1977; *Whitlatch*, 1980] in Gulf [*Flint and Holland*, 1980] and shelf benthic environments. *Frankenberg and Leiper* [1977] also found that *S. bombyx* peaked in late spring along the Georgia coast.

Acknowledgments. Data were obtained from the benchmark study of the Southeast Atlantic Bight sponsored by the Bureau of Land Management (now Minerals Management Service). Funding was provided by the Minerals Management Service.

References

Atkinson, L. P., J. O. Blanton, and E. B. Haines, Shelf flushing rates based on the distribution of salinity and freshwater in the Georgia Bight, *Estuarine Coastal Mar. Sci., 7,* 465–472, 1978.

Bishop, S. S., J. A. Yoder, and G.-A. Paffenhöfer, Phytoplankton and nutrient variability along a cross shelf transect off Savannah, Georgia, U.S.A., *Estuarine Coastal Mar. Sci., 11,* 359–368, 1980.

Blanton, J. O., and L. P. Atkinson, Physical transfer processes between Georgia tidal inlets and nearshore waters, in *Estuarine Interactions,* edited by M. L. Wiley, pp. 515–532, Academic, New York, 1978.

Boesch, D. F., Benthic ecological studies: Macrobenthos, *Spec. Rep. 194,* 301 pp., Appl. Mar. Sci. and Ocean Eng., Va. Inst. of Mar. Sci., Gloucester Point, 1979.

Brokaw, R. S., and G. F. Oertel, Suspended sediment data from nearshore waters of Georgia, *Tech. Rep., 76–3,* 42 pp., Ga. Mar. Sci. Center, Savannah, Georgia, 1976.

Bumpus, D. F., The circulation over the continental shelf south of Cape Hatteras, *Eos Trans. AGU, 36,* 601–611, 1955.

Cerame-Vivas, M. J., and I. E. Gray, The distributional patterns of benthic invertebrates of the continental shelf off North Carolina, *Ecology, 47,* 260–270, 1966.

Day, J. H., J. G. Field, and M. P. Montgomery, The use of numerical methods to determine the distribution of the benthic fauna across the continental shelf of North Carolina, *J. Anim. Ecol., 440,* 93–125, 1971.

Dörjes, J., Marine macrobenthic communities of the Sapelo Island, Georgia region, in *Ecology of Marine Benthos,* edited by B. Coull, pp. 399–422, University of South Carolina Press, Columbia, 1977.

Dunstan, W. M., and L. P. Atkinson, Sources of new nitrogen for the South Atlantic Bight, in *Estuarine Processes,* vol. 1, edited by M. L. Wiley, pp. 69–78, Academic, New York, 1976.

Emery, K. O., Relict sediments on the continental shelves

of the world, *Am. Assoc. Pet. Geol. Bull., 52,* 445–464, 1968.

Flint, R. W., and J. S. Holland, Benthic infaunal variability on a transect in the Gulf of Mexico, *Estuarine Coastal Mar. Sci., 10,* 1–14, 1980.

Folk, R. L., *Petrology of Sedimentary Rocks,* 192 pp., Hemphill, Austin, Tex., 1968.

Frankenberg, D., The dynamics of benthic communities off Georgia, USA, *Thalassia Jugosl., 7,* 49–55, 1971.

Frankenberg, D., and A. S. Leiper, Seasonal cycles in benthic communities of the Georgia continental shelf, in *Ecology of Marine Benthos,* edited by B. Coull, pp. 383–397, University of South Carolina Press, Columbia, 1977.

Gorsline, D. S., Bottom sediments of the Atlantic shelf and slope off the southern United States, *J. Geol., 71,* 422–440, 1963.

Haines, E. B., and W. M. Dunstan, The distribution and relation of particulate or organic materials and primary productivity in the Georgia Bight, *Estuarine Coastal Mar. Sci., 3,* 431–441, 1975.

Hanson, R. B., K. R. Tenore, S. Bishop, C. Chamberlain, M. M. Pamatmat, and J. Tietjen, Benthic enrichment in the Georgia Bight related to Gulf Stream intrusions and estuarine outwelling, *J. Mar. Res., 39,* 417–441, 1981.

Henry, V. J., and R. T. Giles, Distribution and occurrence of reefs and hardgrounds in the Georgia Bight, *U.S. Geol. Surv. Open File Rep., 80–146,* 8.1–8.36, 1979.

Jumars, P. A., and K. Fauchald, Between-community contrasts in successful polychaete feeding strategies, in *Ecology of Marine Benthos,* edited by B. Coull, pp. 1–20, University of South Carolina Press, Columbia, 1977.

Kingery, F. A., Textural analysis of shelf sands off Georgia coast, M.S. thesis, 75 pp., Calif. State Univ., San Diego, 1973.

Lee, T. N., L. P. Atkinson, and R. Legeckis, Detailed observations of a Gulf Stream spin-off eddy on the Georgia continental shelf, April 1977, *Deep Sea Res., 28,* 347–378, 1981.

Maurer, D., and W. Leathem, Dominant species of polychaetes annelids of Georges Bank, *Mar. Ecol. Prog. Ser., 3,* 135–144, 1980.

McNulty, J. K., R. C. Work, and H. B. Moore, Some relationships between the infauna of the level bottom and the sediment in north Florida, *Bull. Mar. Sci., 12,* 322–332, 1962.

Odum, E. P., A research challenge: Evaluating the productivity of coastal and estuarine water, in *Proceedings of the 2nd Sea Grant Conference,* pp. 63–64, Graduate School of Oceanography, University of Rhode Island, Newport, 1968.

Rowe, G. T., Benthic biomass and surface productivity, in *Fertility of the Sea,* vol. 2, edited by J. D. Costlow, Jr., pp. 441–454, Gordon and Breach, New York, 1971.

Rowe, G. T., and R. J. Menzies, Zonation of large benthic invertebrates in the deep sea off the Carolinas, *Deep Sea Res., 16,* 531–537, 1969.

Smith, K. L., Jr., Structural and functional aspects of a sublittoral community, Ph.D. thesis, 160 pp., Univ. of Ga., Athens, 1971.

Tenore, K. R., C. F. Chamberlain, W. M. Dunstan, R. B. Hanson, G. Sherr, and J. H. Tietjen, Possible effects of Gulf Stream intrusions and coastal runoff on the benthos of the continental shelf of the Georgia Bight, in *Estuarine Interactions,* edited by M. L. Wiley, pp. 577–598, Academic, New York, 1978.

Whitlatch, R. B., Patterns of resource utilization and coexistence in marine intertidal deposit-feeding communities, *J. Mar. Res., 38,* 743–765, 1980.

Yoder, J. A., L. P. Atkinson, T. N. Lee, H. H. Kim, and C. R. McClain, Role of Gulf Stream frontal eddies in forming phytoplankton patches on the outer southeastern shelf, *Limnol. Oceanogr., 26,* 1103–1110, 1981.

Factors Influencing the Concentration and Distribution of Trace Metals in the South Atlantic Bight

H. L. Windom and R. G. Smith, Jr.

Skidaway Institute of Oceanography, Savannah, Georgia

The concentrations and distributions of iron, manganese, cadmium, copper, nickel, and zinc in South Atlantic Bight shelf waters appear to result from the mixing of river and oceanic waters, the latter with a chemistry similar to Western North Atlantic Water from a depth of 500–600 m. Superimposed on the mixture of the two end-members are other inputs and removal. Other significant inputs of cadmium, copper, nickel, and zinc are atmospheric transport and release from nearshore suspended and bottom sediments. Release from nearshore sediments (suspended and/or bottom) appears to be significant for all metals studied. First-order removal with respect to salinity appears to control iron, zinc, and, perhaps, manganese concentrations. Accumulation in shelf sediments may account for iron and zinc removal, but manganese lost from solution may be advected off the shelf in suspended particles.

Introduction

South Atlantic Bight (SAB) shelf water consists of intruded Gulf Stream waters diluted to varying degrees by river runoff. Trace element concentrations in shelf waters are, therefore, largely determined by concentrations in the oceanic and freshwater end-members but are also influenced by atmospheric inputs and removal to, or release from, suspended and bottom sediments. In the SAB there are no significant direct anthropogenic inputs other than those associated with continental runoff and atmospheric fluxes.

Earlier studies of the trace element composition of SAB waters were based on general surveys of concentration distributions [*Windom and Smith,* 1972, 1979] or considered only some of the influences mentioned above. The relative importance of riverine and atmospheric transport of a number of trace metals was estimated by *Windom* [1981]. *Windom and Taylor* [1979] considered the importance of these two pathways specifically for mercury, and *Waslenchuk* [1978] evaluated arsenic transport to shelf water from rivers, the atmosphere, and Gulf Stream intrusions. The release of copper from nearshore sediments into SAB waters was discussed by *Windom et al.* [1983].

This paper provides a more general view of factors controlling trace element concentrations in SAB shelf waters. Recent studies of iron, manganese, cadmium, copper, nickel, and zinc in southeastern rivers, shelf waters, and atmosphere are used for this purpose.

Methods

A pumping system was used to collect water samples from middepth or 5 m, whichever was shallower. The system consisted of silicone tubing (precleaned by circulating 10% HCl through it for ~1 hour), the intake end of which was attached to a teflon vane. The intake vane was lowered to the desired depth on Kevlar line (using a teflon-coated weight) with the intake orifice pointed upstream. A peristaltic pump delivered water directly to precleaned polyethylene reservoirs. After rinsing twice with sample water, the reservoir was filled and pressurized with filtered nitrogen gas. The water was then passed through a 0.4-μm precleaned Nuclepore filter in a teflon filter holder. Finally, the samples were collected in precleaned 250-ml teflon bottles and acidified with 250 μl of National Bureau of Standards subboiling redistilled HNO_3. All filtering and sample handling was within a laminar flow clean bench.

Water samples were analyzed for cadmium, copper, nickel, and zinc using a modification of procedures reported by *Smith and Windom* [1980]. This involved 100:1 concentration by extraction with dithizone and hydroxyquinoline in chloroform and back extracting into quartz-distilled nitric acid. Iron and manganese were similarly concentrated, except that the organic phase consisted of only 0.1% hydroxyquinoline in chloroform, from a separate aliquot at *p*H 9. All analyses were by flameless atomic absorption spectrophotometry.

Most of the data on atmospheric trace metal concentrations discussed below have been described earlier [*Mullins,* 1978]. A few additional samples of air and dry fallout were collected and analyzed.

As in previous studies [*Mullins,* 1978], atmospheric samples were collected on precleaned Whatman 41 filters using a large-volume vacuum system. The filters were digested and analyzed for cadmium, copper, iron, manganese, nickel, zinc, and aluminum by flameless atomic absorption spectrophotometry. Samples of atmospheric fallout were collected on acid-washed, 1-m-diameter polyethylene funnels. These were mounted on a mast atop the R/V *Blue Fin*. Sampling was limited to times when the vessel was anchored at a middle shelf station. All exhausts from the *Blue Fin* are near the waterline, minimizing potential contamination from engine exhausts. Periodically, the funnels were washed with quartz-distilled 1 *N* HNO_3, and the wash was collected in a 1-1 polyethylene bottle. Samples were evaporated to a few milliliters in a clean bench under a heat lamp and were then digested in quartz-distilled HNO_3 and Ultrex HF.

Results

Study Area

The physiography and hydrography of the SAB have been described in earlier papers of this volume. The hydrographic climatology of the region has been described by *Atkinson et al.* [1983]. Figure 1 indicates important features of the SAB and sampling locations relevant to this discussion.

Trace Metal Concentrations in Southeastern River Waters

Eleven major rivers with a total average annual freshwater discharge of about 66 km³ drain into the South Atlantic Bight (Table 1). These rivers can be divided into two major types: Piedmont and Coastal Plain. Piedmont rivers are those with drainage basins predominantly in the Piedmont Plateau Province, which contains more crystalline rock and has greater relief than the Coastal Plain. Piedmont rivers include the Cape Fear, Pee Dee, Cooper, Savannah, and Altamaha. All have higher rates of discharge

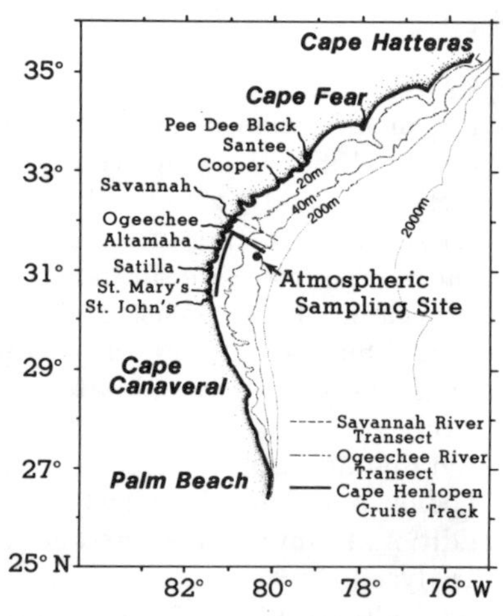

Fig. 1. Study area showing location of major rivers and sampling locations.

TABLE 1. Mean Annual Discharge of Southeastern Rivers

River	Annual Discharge, km³
Cape Fear	7.1
Pee Dee	12.4
Black	0.8
Santee	1.9
Cooper	13.3
Savannah	10.8
Ogeechee	2.1
Altamaha	12.3
Satilla	2.0
St. Mary's	0.6
St. John's	2.8

Based on 7 to 74 years of stream gage data (depending on river) for most downstream gaging stations (U.S. Geological Survey data).

than Coastal Plain rivers because of the larger areas of their drainage basins. The drainage basins of Coastal Plain rivers are confined to the Coastal Plain Province. The region is one of low relief dominated by highly weathered sedimentary rocks and many swampy areas. The Black, Satilla, St. Marys, and St. Johns are examples of this type. The Ogeechee River is a hybrid with about 95% of its drainage basin in the Coastal Plain and 5% in the Piedmont Plateau.

Waters from representative rivers of each type were analyzed for dissolved trace metals and dissolved organic carbon (DOC) (Table 2). Higher DOC levels occur in Coastal Plain than in Piedmont rivers. Iron, cadmium, and zinc concentrations are directly ($p < 0.05$) related to DOC concentration (Figure 2). Copper concentrations, however, are inversely related to DOC ($p < 0.05$). These relationships suggest that at least iron, cadmium, and zinc are transported in the rivers primarily in naturally occurring organic compounds or complexes (see, for example, *Sholkovitz and Copland* [1981]). The inverse relationship between copper and DOC suggests, alternately, that all metals and DOC concentrations may be controlled by characteristics of drainage basins, with those in the Coastal Plain being enriched in DOC, iron, cadmium, and zinc, and depleted in copper, relative to those in the Piedmont Plateau.

Metal concentrations in river water vary temporally, in some cases by as much as a factor of 5 (e.g., zinc in the Ogeechee, Table 2). Iron, arsenic, and copper concentrations vary temporally in southeastern U.S. rivers [*Waslenchuk and Windom,* 1978; *Windom et al.,* 1983] and appear to be related to discharge. The rivers studied have a fairly consistent seasonal discharge pattern, with highest flow rates in the early spring. Temporal variability in metal concentration, therefore, is probably related to weathering regimes (i.e., season) and hydrology. For example, during the early spring wet season weathering solutions within drainage basin soils that have become increasingly enriched in metals during the preceding warmer period (i.e., during the period of maximum evapotranspiration) may be more efficiently flushed into the rivers.

TABLE 2. Dissolved Trace Metal Composition of Southeastern Rivers (Mean and Standard Deviations)

River	Date	$\bar{Q}$, $m^3 s^{-1}$	DOC, μmol kg^{-1}	Fe, μmol kg^{-1}	Mn, μmol kg^{-1}	Cd, pmol kg^{-1}	Cu, nmol kg^{-1}	Ni. nmol kg^{-1}	Zn, nmol kg^{-1}
Savannah	June 1980	430	425 ± 11	0.40 ± 0.01	...	86 ± 27	12.3 ± 1.2	5.8 ± 0.3	5.5 ± 0.1
	Nov. 1980	220	175 ± 8	0.51 ± 0.12	...	87 ± 16	7.1 ± 0.8	1.9 ± 1.0	19 ± 1.0
	March 1981	230	433 ± 40	0.51 ± 0.01	0.36 ± 0.03	18 ± 4	4.4 ± 0.3	4.9 ± 0.1	5.5 ± 0.3
			(370)	(0.36)		(68)	(8.9)	(4.6)	(8.9)
Ogeechee	Aug. 1980	8	...	1.2 ± 0.1	...	94 ± 6	4.7 ± 0.2	5.8 ± 0.6	3.3 ± 0.1
	Nov. 1980	16	408 ± 33	0.9 ± 0.1	...	98 ± 10	4.4 ± 0.4	2.6 ± 0.4	16 ± 3.0
	March 1981	63	650 ± 50	2.7 ± 0.2	0.24 ± 0.10	78 ± 6	8.2 ± 0.8	4.6 ± 1.2	17.7 ± 0.5
			(600)	(2.2)		(83)	(7.1)	(4.3)	(16)
Altamaha	Nov. 1980	87	325 ± 17	0.62 ± 0.29	...	55 ± 10	9.9 ± 2.0	5.5 ± 1.5	11 ± 1.0
Satilla	Nov. 1980	4	833 ± 75	1.5 ± 0.1	...	72 ± 16	4.0 ± 0.4	2.6 ± 0.1	28 ± 3
	June 1981	3	667 ± 27	1.8 ± 0.2	...	51 ± 15	6.3 ± 2.8	7.8 ± 2.0	17 ± 1
			(760)	(1.6)		(63)	(5.0)	(4.8)	(23)
St. Mary's	Nov. 1980	5	1700 ± 110	2.1 ± 0.2	...	148 ± 14	2.1 ± 0.5	4.0 ± 0.9	31 ± 4
	June 1981	2	1260 ± 110	1.7 ± 0.4	...	129 ± 42	...	9.7 ± 3.0	17 ± 6
			(1570)	(2.0)		(142)		(5.6)	(27)
Weighted mean concentration*			390	0.55	0.33	69	8.8	4.5	9.8

Weighted mean metal concentrations ($\Sigma(Q - M)/\Sigma Q$) are given in parentheses. Mean and standard deviations are the results of triplicate trace element analyses and quadruplicate DOC analyses.

*Overall weighted mean concentrations using data from all rivers.

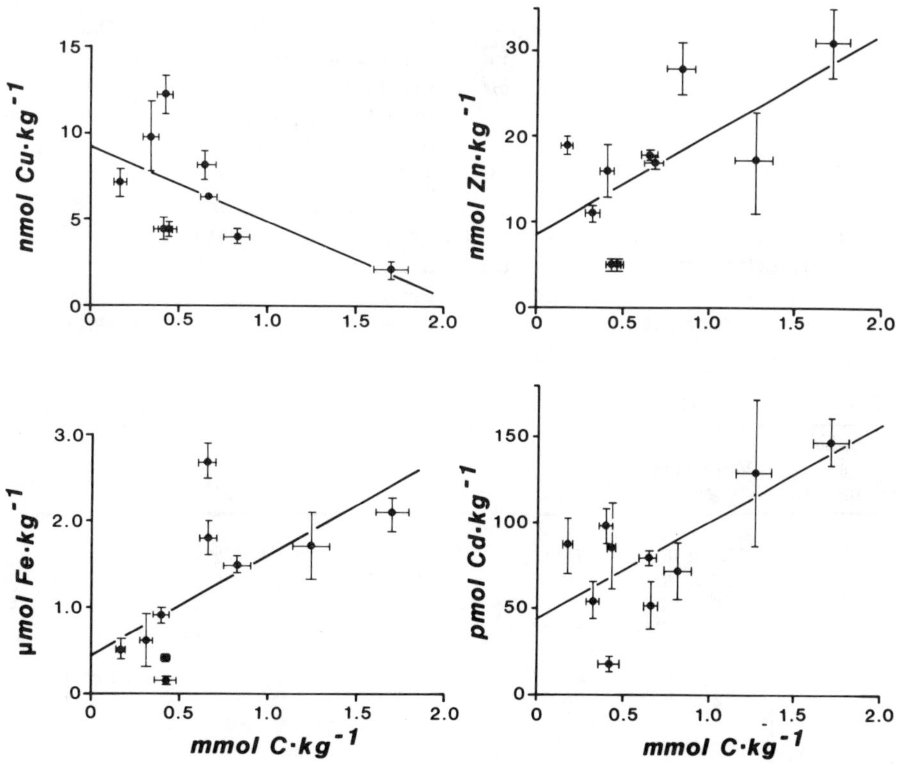

Fig. 2. Dissolved copper, iron, zinc, and cadmium concentrations in river water in relation to dissolved organic carbon concentrations.

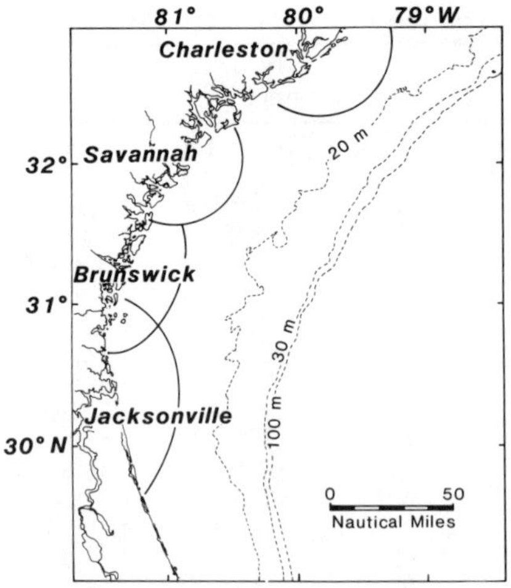

Fig. 3. Nearshore areas in which atmospheric trace metal samples were collected.

Concentrations of cadmium, copper, nickel, and zinc reported here are lower than those from other North American and European rivers [*Bewers and Yeats,* 1978; *Klinkhammer and Bender,* 1981; *Sholkovitz and Copland,* 1981; *Elderfield et al.,* 1979; *Duinker and Nolting,* 1978; *Eaton,* 1979], but iron and manganese have similar concentrations [*Wollast et al.,* 1979; *Boyle et al.,* 1974; *Graham et al.,* 1976]. Copper, nickel, and cadmium concentrations in southeastern U.S. rivers appear to be more similar to those of the Amazon, Yangtze, and Orinoco [*Boyle et al.,* 1982; *Grant et al.,* 1982].

Atmospheric Trace Metal Concentrations and Fluxes

Mullins [1978] determined the concentrations of iron, manganese, cadmium, copper, nickel, zinc, and aluminum in air samples collected over the South Atlantic Bight. He compared concentrations over the inner shelf in the vicinity of four major population and industrial areas (Figure 3) with those over the middle and outer shelf (Table 3). Although the samples were collected during periods dominated by westerly winds, there was no evidence of higher concentrations in nearshore than in offshore air, suggesting no major local anthropogenic inputs. Atmospheric metal levels over the SAB are within the range of concentrations in air samples collected at Bermuda [*Duce et al.,* 1976].

We also analyzed air and dry fallout samples from a site on the middle shelf (Figure 1). Concentrations of metals in air during the time we collected the dry fallout were similar to those reported earlier (*Mullins* [1978] and Table 3).

Rates of dry metal fallout (Table 4) at the middle shelf station were divided by atmospheric metal concentrations to calculate dry deposition velocities. The values obtained were similar in range to total deposition velocities of aerosols calculated from ^{7}Be measurements in western North Atlantic surface waters [*Young and Silker,* 1980]. Deposition velocities of manganese, iron, copper, and zinc along the Washington coast are similar to those reported here [*Crecelius,* 1981].

Concentrations and Distribution of Trace Metals in Estuarine and Shelf Waters

Water samples for trace metal analyses were collected along transects through the Savannah River Estuary and across the SAB shelf in June 1980 and March 1981. Transects through the Ogeechee Estuary, extending across the shelf, were sampled in August 1980 and March 1981 (Figure 1). Other samples were collected from the inner shelf near Savannah during November 1979 and along transects parallel and perpendicular to the Georgia coast during March 1981. The combined data show reasonably consistent trends:

Iron and Manganese. In low-salinity regions of estuaries the concentrations of dissolved iron and manganese are slightly higher than the freshwater end-member, suggesting release. At higher salinities, removal occurs (Figure 4). Concentrations of iron significantly follow ($p < 0.01$)

TABLE 3. Trace Metal Concentrations in Southeastern Continental Shelf Atmospheric Samples

Area	Number of Samples	Aluminum, ng/m³	Iron, ng/m³	Manganese, ng/m³	Cadmium, ng/m³	Copper, ng/m³	Nickel, ng/m³	Zinc, ng/m³
Nearshore*								
Charleston	5	100 ± 60	74 ± 17	4.0 ± 2.4	1.8 ± 2.1	10 ± 8	2.9 ± 15	23 ± 23
Savannah	6	160 ± 80	120 ± 50	3.2 ± 3.2	0.4 ± 0.2	4.3 ± 1.0	11 ± 11	15 ± 7
Brunswick	5	170 ± 60	80 ± 30	2.4 ± 0.8	0.4 ± 0.1	2.1 ± 0.3	0.7 ± 0.7	14 ± 3
Jacksonville	9	180 ± 60	80 ± 40	4.1 ± 1.1	0.2 ± 0.1	3.0 ± 3.3	2.2 ± 2.7	7 ± 2
Middle–outer shelf								
Nov. 1977	5	95 ± 40	190 ± 80	2.0 ± 2.2	0.3 ± 0.1	11 ± 6	4.6 ± 2.7	15 ± 2
Sept. 1979	3	32 ± 8	25 ± 15	1.6 ± 0.8	0.3 ± 0.2	8 ± 2	3.0 ± 3.0	5 ± 4
Central Atlantic†								
Mean		140	94	1.2	0.2	0.9	0.1	3.2
Range		3–2700	4–1900	0.03–27	0.005–3.2	0.07–24	0.002–1.5	0.2–32

*Data are from *Mullins* [1978].
†Data are from *Duce et al.* [1976].

TABLE 4. Dry Fluxes and Deposition Velocities of Atmospheric Trace Metals

	Fluxes, ng m^{-2} h^{-1}	Deposition Velocity, cm s^{-1}
Iron	1100 ± 1000	1.20
Manganese	60 ± 15	1.04
Cadmium	4.3 ± 1.5	0.40
Copper	190 ± 90	0.66
Nickel	270 ± 95	2.5
Zinc	390 ± 240	2.1

a first-order removal rate of -0.12×10^{-3} S^{-1} with respect to salinity. Decrease in manganese concentrations with salinity over the whole salinity range also suggests first-strongly influenced by the low- and high-salinity values. Highest deviations from the curve occur at intermediate salinities where elevated manganese concentrations suggest release from sediments. It is likely that at intermediate to low salinity manganese concentrations are controlled by estuarine-nearshore cycling similar to that described by *Evans et al.* [1977], involving removal from the water column and release from sediments.

Variability in iron concentrations in the estuarine/nearshore region (i.e., salinities less than 30×10^{-3}) appears related to the relative influence of specific river inputs. The influence of the Ogeechee and Savannah rivers on nearshore concentrations of DOC is clearly apparent (Figure 5). At salinities higher than 30×10^{-3}, DOC behaves conservatively (i.e., covaries with salinity, $p < 0.01$), with an apparent zero-salinity end-member concentration approximately that of the Savannah River, indicating the dominance of the larger Piedmont-type rivers on DOC inputs to the SAB.

Scatter in dissolved iron and manganese data at salinities higher than 30×10^{-3} may be, in part, due to offshore mixing of different river waters because some of these data were collected on a north-south transect (Figure 1). Variability in shelf waters, however, may result from differences in removal mechanisms. For example, concentrations of manganese observed in March 1981 and those observed in November 1979 follow a first-order removal model with respect to salinity ($p < 0.01$) but have removal constants (Figure 4) which differ from the overall average of -0.12×10^{-3} S. Thus two regimes appear to influence manganese concentrations in estuarine and shelf waters, an estuarine-nearshore cycling regime and an offshore mixing-removal regime. Iron concentrations at salinities above 30×10^{-3} may also vary because of a temporal variability of removal mechanisms. However, there is no significant relation between concentrations and salinity in the different data sets.

Cadmium, Copper, Nickel, and Zinc. The results of analyses of dissolved concentrations of cadmium, copper, nickel, and zinc are presented in Figure 6. Most data were derived from the mean of triplicate analyses; others are the mean of duplicate analyses. Standard deviations are not shown. Coefficients of variation were generally 10% or less and were always less than 20%.

Although the behavior of a given metal can be interpreted differently, depending on the data set used, an overall pattern emerges from the combined data. For example, trends in copper concentrations on the June 1980 Savannah transect suggest removal. The average overall trend, however, suggests release from sediments at low and intermediate salinities (also reported by *Windom et al.*, [1983]). At salinities higher than 30×10^{-3}, copper covaries with salinity ($p < 0.01$).

Highest cadmium and nickel concentrations occur at intermediate to high salinities. Cadmium maxima occur generally at salinities of approximately 30×10^{-3}. High cadmium concentrations at higher salinities were observed on the nearshore transect in March 1981 and may reflect influences of rivers enriched in this metal (Table 2). Maximum nickel concentrations were at salinities of 10 to 20×10^{-3}. Above 30×10^{-3}, concentrations of both metals covary ($p < 0.01$) with salinity.

Zinc concentrations in estuarine waters appear to reach a maximum at salinities of $\sim 10 \times 10^{-3}$ or less. Decreased concentrations at higher salinities appear to reflect removal of dissolved zinc. Dissolved zinc concentrations fit a first-order removal model (Figure 7) similar to iron and manganese. The removal constant for zinc, however, is -0.052×10^{-3} S^{-1}, or about half that of iron and manganese.

Discussion

End-Member Mixing

Two end-member mixing models (Table 5) were used to interpret observed trace metal concentrations in SAB shelf waters. The first, assuming conservative mixing, was used for DOC, cadmium, copper, and nickel. Concentrations of these metals and DOC covary with salinity ($p < 0.01$) at salinities above 30×10^{-3}. Regression curves for each parameter were extrapolated to zero and 36×10^{-3} salinity to determine apparent end-member concentrations. The second model, used for iron, manganese, and zinc, assumes first-order removal because the log of their concentrations covaried with salinity ($p < 0.01$).

These two approaches yield apparent zero-salinity end-member concentrations significantly higher than observed river concentrations, with the exception of DOC, suggesting other sources of input (discussed below). The oceanic end-member concentration of metals is assumed to be that at 36×10^{-3} salinity (Table 5).

Bruland and Franks [1983] determined manganese, nickel, copper, zinc, and cadmium concentrations in a few samples of shelf and slope water from the northeastern continental shelf. Concentrations in Sargasso Sea surface waters were used as the oceanic end-member, conservative mixing was assumed, and concentrations were extrapo-

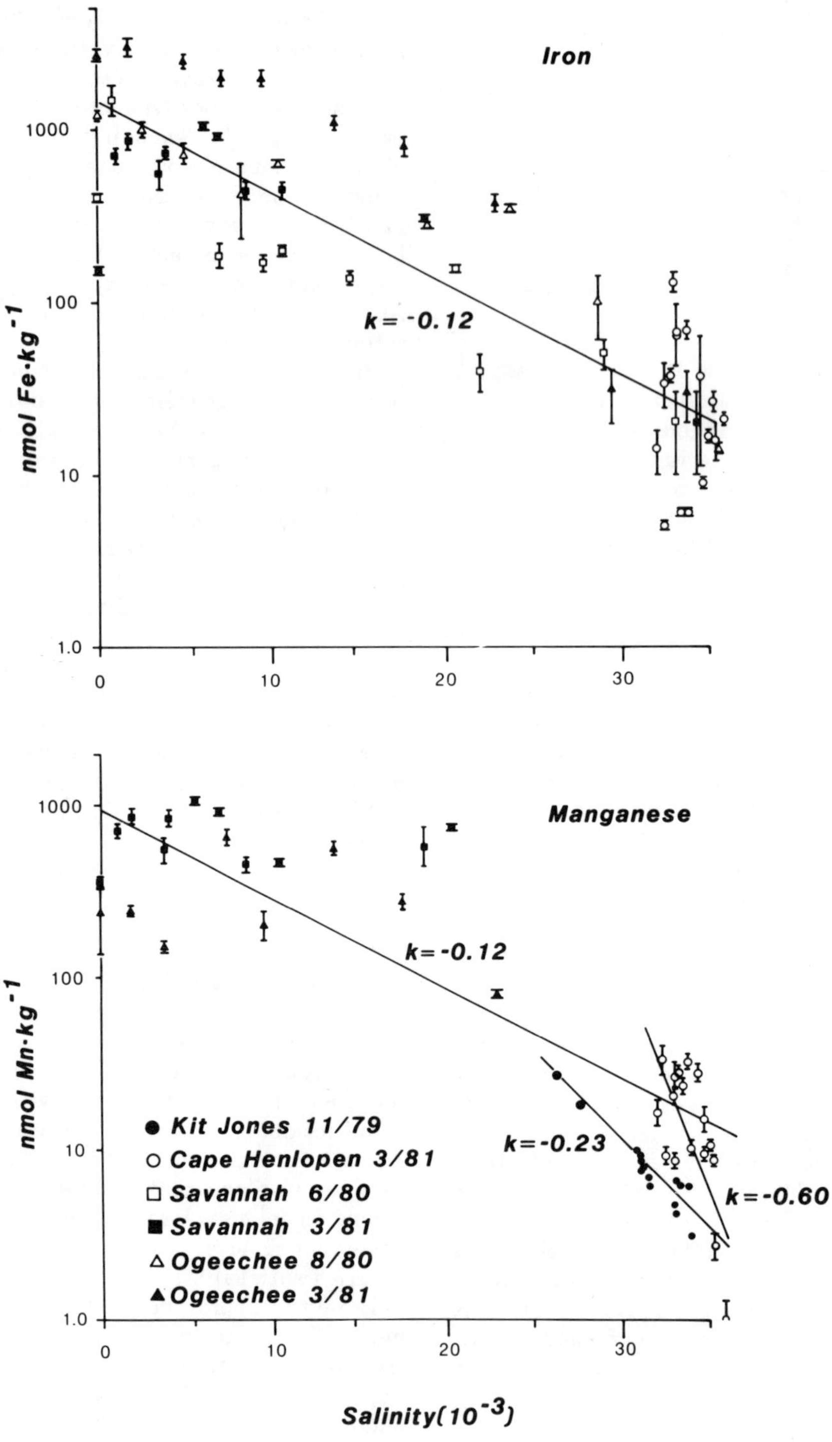

Fig. 4. Dissolved iron and manganese concentrations in relation to salinity. Values show means of triplicate analyses and standard deviations (single analyses only were performed on the *Kit Jones* samples). Values of k are the slopes of regression lines.

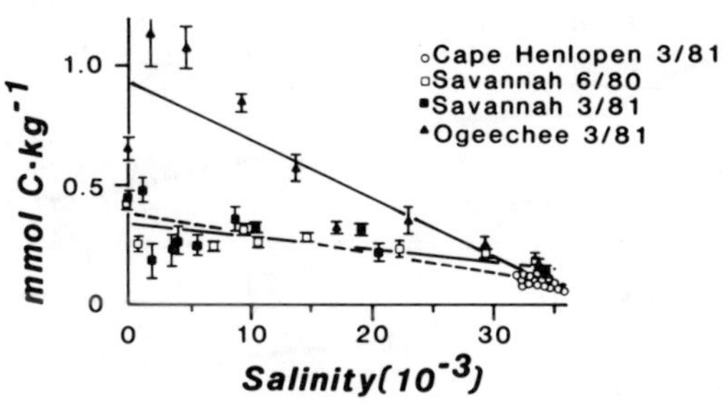

Fig. 5. Dissolved organic carbon concentrations in relation to salinity.

lated to zero salinity. Their extrapolated zero-salinity end-member concentrations of copper, nickel, cadmium, and zinc were 19, 16, 1.1, and 12 nmol kg^{-1}, respectively, similar to our results. Their calculated manganese zero-salinity end-member concentration, 114 nmol kg^{-1}, is considerably lower than what we found because they assumed manganese to behave conservatively. Trace metal concentrations in shelf waters reported by *Bruland and Franks* [1983] were virtually the same as those reported in Table 5 at salinities greater than 30×10^{-3}. *Boyle et al.* [1981] also report similar results for copper, nickel, and cadmium in northeastern U.S. shelf waters. The apparent oceanic end-member concentrations (i.e., at salinity 36×10^{-3}) we determined, however, are different from those in Sargasso Sea surface waters, the reference value used by *Bruland and Franks* [1983]. Concentrations at around 36×10^{-3} salinity in the SAB are more like those in deeper North Atlantic water [*Bruland and Franks*, 1983].

Along the western boundary of the Gulf Stream, upwelling and intrusions of Western North Atlantic Water (WNAW, also referred to as Central Atlantic Water) are, along with runoff, the major sources of materials in SAB shelf waters [*Atkinson*, 1977; *Blanton and Pietrafesa*, 1978; *Blanton et al.*, 1981]. These events force a turnover time of shelf waters once every 0.14 to 0.59 year [*Atkinson et al.*, 1983]. The WNAW, characterized by a nearly linear temperature-salinity (T-S) relationship between about 18.5°C, 36.5×10^{-3} and 11°C, 35.4×10^{-3}, spans the main thermocline below Sargasso Sea surface waters [*McCartney et al.*, 1980]. The Gulf Stream normally transports only WNAW [*Worthington*, 1976] or WNAW modified by transit through the Caribbean Sea.

Temperature and salinity profiles along a cross-shelf transect in March 1981 (Figure 1) are shown in Figure 8. An analysis of T-S relationships at stations 9 and 10 indicates an intrusion of Western North Atlantic Water, while at station 11 these relationships indicate some dilution at the surface by freshwater [*Lee and Atkinson*, 1983]. The temperatures and salinities observed at stations 9 and 10 are similar to those of WNAW at 500- to 600-m depth in

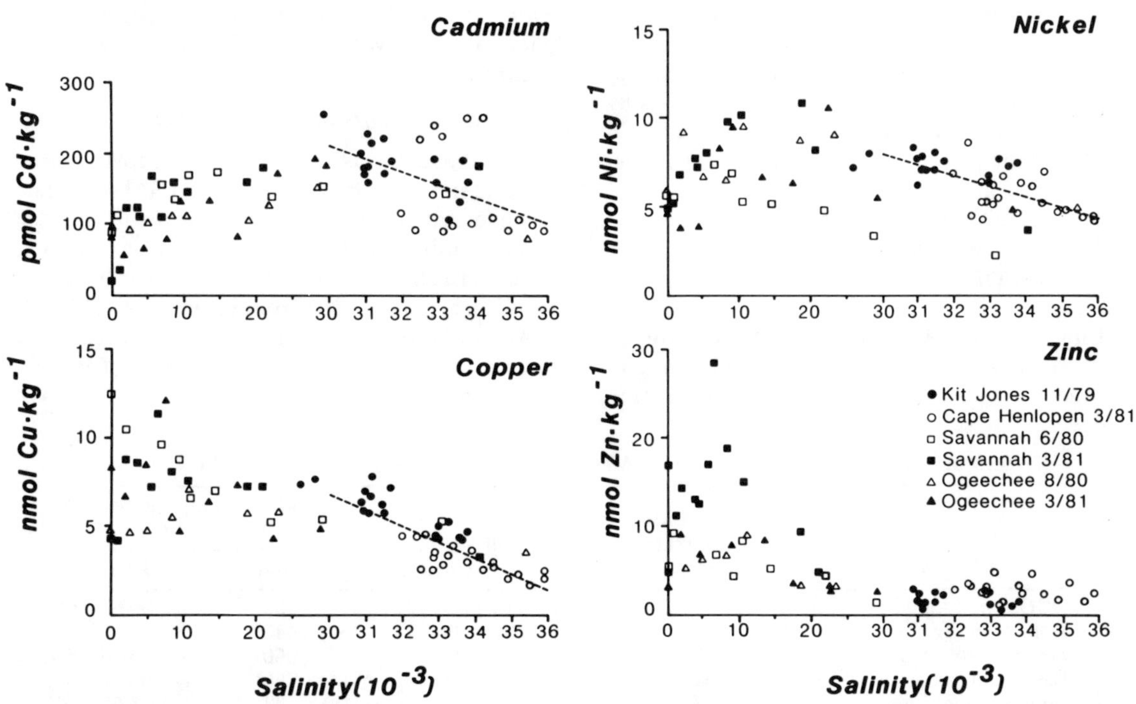

Fig. 6. Dissolved cadmium, copper, nickel, and zinc in SAB waters in relation to salinity. All values are means of replicate analyses (two or three). Dashed lines are linear regression curves for data collected for salinities of >30 ppt. Note broken salinity scale.

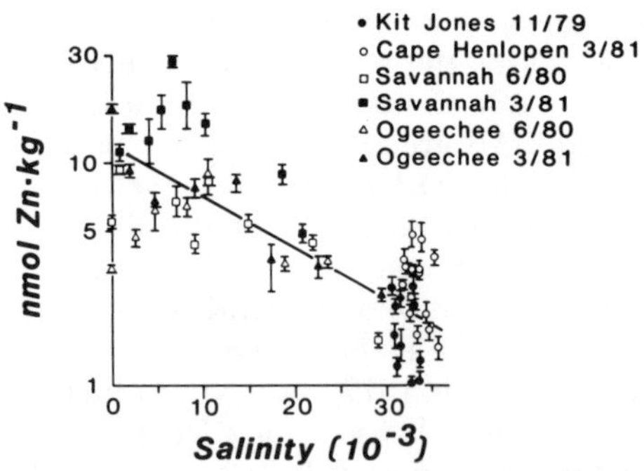

Fig. 7. Dissolved zinc in relation to salinity (same data shown in Figure 6). Line is the regression of natural log of concentration on salinity.

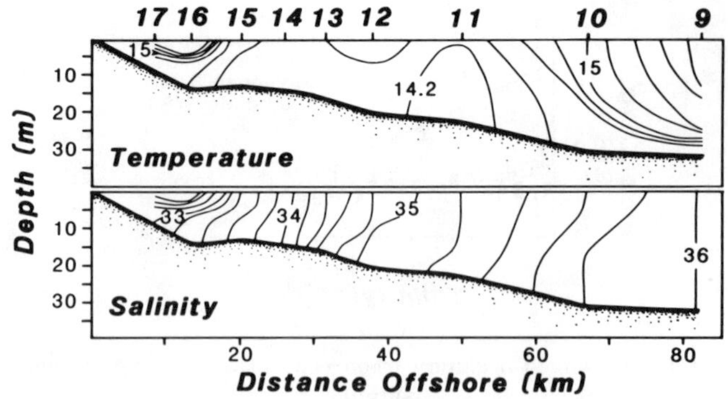

Fig. 8. Cross sections of temperature and salinity along cross shelf transect of *Cape Henlopen* cruise (Figure 1). Station numbers are given at the top of the figure.

the Sargasso Sea. This does not imply that water advects directly westward from the Sargasso Sea onto the SAB shelf. Water from the area, however, may be transported into the Gulf Stream following a circuitous route through the Caribbean Sea, Gulf of Mexico, and Straits of Florida.

A direct link between the source of SAB shelf water and that at 500 to 600 m in the Sargasso Sea cannot be established. Nevertheless, metal concentrations at the oceanic end-member of SAB shelf waters are more similar to deeper North Atlantic water than to surface Sargasso Sea water. A comparison of the 36×10^{-3} apparent end-member (Table 5) with concentrations at 500 to 600 m in the North Atlantic [*Bruland and Franks,* 1983] shows reasonable agreement for cadmium, copper, nickel, and zinc. The apparent manganese end-member concentration from our study (Table 5) is more than 10 times higher than that of 500- to 600-m North Atlantic Water. Results (Figure 4), however, indicate a steep manganese concentration gradient at high salinities in SAB waters, suggesting an oceanic end-member less than the extrapolated value and more similar to concentrations at 500 to 600 m [*Bruland and Franks,* 1983].

Assuming that 500- to 600-m North Atlantic water is the oceanic end-member for SAB waters, the dissolved metal and nutrient concentrations are controlled by regeneration processes. As this water is intruded onto the shelf, it is depleted of nutrients. For example, nitrate concentrations at stations 9 and 10 were 0.04 to 0.05 μmol/kg, depleted from expected levels of 20 mol/kg in 15°C water at depth in the Gulf Stream. Metals, however, are not significantly depleted. This suggests that metal-nutrient relationships in SAB waters are controlled by removal processes rather than the regeneration processes that clearly control these relationships in central ocean gyres. *Boyle and Huested* [1983] have suggested that metal removal from surface waters varies from region to region depending on the nutrient regime, with removal being lower in low-nutrient regimes.

Trace Metal Inputs to the SAB Shelf

The above discussion clearly indicates the importance of oceanic inputs of trace metals into SAB waters. There are, however, at least three other modes of input which may influence dissolved trace metal concentrations and distributions in shelf waters. These are river input, atmospheric transport, and release from sediments.

Assuming that the weighted mean concentrations of trace metals in rivers (Table 2) are typical of the chemistry of runoff from all rivers discharging into the SAB (Table 1),

TABLE 5. Summary of Dissolved Trace Metal Composition of South Atlantic Bight Waters

Element	Weighted Mean River Water Concentration	Shelf Water Concentration	Apparent End-Member Concentration	
			0‰	36‰
DOC, μmol kg^{-1}	390	110 ± 30	390*	90*
Iron, nmol kg^{-1}	550	32 ± 30	1440†	19†
Manganese, nmol kg^{-1}	330	12 ± 9	960†	13†
Cadmium, pmol kg^{-1}	69	160 ± 50	760*	100*
Copper, nmol kg^{-1}	8.8	4.1 ± 1.5	33*	1.5*
Nickel, nmol kg^{-1}	4.5	6.0 ± 1.5	26*	4.4*
Zinc, nmol kg^{-1}	9.8	2.3 ± 1.1	11.9†	1.8†

*Using conservative mixing model for data collected at salinity of $>30 \times 10^{-3}$
†Using first-order removal model applied to all data.

the products of these concentrations and the mean annual runoff (66 km^3) to the SAB estimates the river input. Similarly, the dry atmospheric fluxes into SAB waters (Table 4) can be integrated over the shelf area to estimate atmospheric inputs.

To compare river and atmospheric inputs, we considered the continental shelf area between 30° and 34°N latitude, the region most influenced by river runoff (Figure 1). Between these latitudes the shelf has an area of 5.9×10^4 km^2 and a volume of 1600 km^3. Calculated river and atmospheric inputs suggest that inputs from the latter significantly exceed that from the former for cadmium, copper, nickel, and zinc (Table 6). We emphasize that atmospheric inputs are calculated from dry deposition and that rainout could increase calculated fluxes even more. For example, *Duce et al.* [1976] multiplies dry fluxes by three to estimate total atmospheric inputs. Also, all atmospheric inputs do not necessarily increase the concentrations of dissolved trace metal in shelf waters. As much as 90% of the atmospheric cadmium and zinc but less than 70% of the nickel is soluble in seawater [*Mullins,* 1978].

Apparent zero-salinity end-member concentrations can be calculated from combined estimated river and atmospheric inputs (Table 6). Comparing these with apparent zero salinity end-member concentrations from mixing models (Table 5) indicates that atmospheric inputs of iron and manganese do not account for the excess of apparent overobserved zero-salinity end-member concentrations. For cadmium, approximately half of the excess can be accounted for, whereas all excess copper, nickel, and zinc of the apparent zero-salinity end-member concentration could be explained by atmospheric input.

Though the above observations are based on many assumptions, it appears that atmospheric input may account for some increases in cadmium, copper, nickel, and zinc in lower-salinity nearshore regions (Figure 6). Atmospheric inputs would be clearly more significant nearshore because of the relatively longer residence time of water there [*Atkinson et al.,* 1983] and greater surface to volume ratios compared with middle and outer shelf regions.

Release from suspended and bottom sediments appears to be another important mechanism of input to SAB waters and may account for observed increases in iron and manganese concentrations at low salinities (Figure 4). Concentrations higher than that in rivers occur within estuaries at low salinities where suspended sediment concentrations are also high. These concentrations may be influenced by release of metals from suspended particles and bottom sediments. Elevated dissolved manganese concentrations at salinities between 0 and 6×10^{-3} in the Pee Dee Estuary were attributed to desorption of Mn^{2+} from both bottom and suspended sediments [*Hess,* 1980]. *Windom et al.* [1983] suggest that copper is released at salinities less than 7×10^{-3} by desorption from suspended sediment and may also be released from bottom sediments at higher salinities. Of the metals studied, deviations in dissolved cadmium concentrations from ideal mixing suggest that this metal is probably influenced most by release from sediments.

Any release of metals from the sediments that significantly influences concentrations in water must be confined to the estuarine-nearshore region. The reason is that fine grain, relatively metal-rich bottom sediments are confined to within about 5 km of shore [*Bigham,* 1972] and trace metals associated with suspended particulates decrease exponentially in concentration with distance from shore. Metal concentrations on particulates are 10% or less of the dissolved fraction at distances more than 10 km from shore [*Wallace,* 1980]. Iron is the only metal for which particulate concentrations, which may reach 10 nmol kg^{-1}, are a significant fraction of the total in open shelf waters.

Removal

Our results suggest that iron, manganese, and zinc are removed from solution in SAB waters during mixing of freshwater and ocean water. Dissolved concentrations C of these metals can be assumed to follow first-order removal curves with respect to salinity S, such that

$$\ln C = kS + a \tag{1}$$

where k is the first-order removal constant and a is the natural log of the zero-salinity end-member concentration.

TABLE 6. Comparison of Atmospheric and Riverine Trace Metal Inputs to Shelf Waters

	Atmospheric Input,* nmol kg^{-1} yr^{-1}	River Input,* nmol kg^{-1} yr^{-1}	Atmospheric Input ÷ River Input	Resulting Zero-Salinity End-Member,† nmol kg^{-1}
Iron	6.3	23	0.27	700
Manganese	0.22	14	0.01	330
Cadmium	0.012	0.0026	4.6	0.39
Copper	0.95	0.37	2.6	32
Nickel	1.4	0.19	7.4	38
Zinc	1.9	0.40	4.8	57

*Assuming shelf area is 5.9×10^4 km^2 and shelf water volume is 1600 km^3 [*Atkinson et al.,* 1983] and total freshwater runoff of 66 km^3 (Table 1).

†Concentration is $(R + 1)\,C_r$, where R is the ratio of the atmospheric input to the river input and C_r is the observed weighted mean river water concentration (Table 5).

Li and Chan [1979], following the approach of *Boyle et al.* [1974], derived the following equation for the mass balance of a nonconservative element in an estuary:

$$C_R + I_c/Q_R = C_E - (S_E - S_R)(dC/dS)_{S=S_E} \quad (2)$$

Subscripts R and E refer to river and estuarine concentrations or salinities, respectively, Q_R is the river discharge, and I_c is either the addition or removal of the element from the estuarine box bounded by the zero-salinity surface (river end-member) and a second surface with salinity S_E.

Choosing the outer boundary of the system (i.e., box) as the 36×10^{-3} surface and substituting dC/dS of equation (1) into equation (2), data in Table 5 can be used to calculate removal of iron, manganese, and zinc. These calculations indicate that about 90% of the iron and manganese and about 55% of the zinc inputs (using the apparent zero-salinity end-member concentrations, Table 5) are removed from the estuarine-shelf waters during mixing.

Most removal of these elements is in estuaries or within 5 km of the shore, though some may occur on the shelf proper. An estimate of the importance of removal on the shelf can be derived from the metal content of shelf sediments and assumed sedimentation rates. Mean concentrations of manganese [from *Bothner et al.*, 1980], iron, cadmium, copper, nickel, and zinc [from *Windom and Betzer*, 1979] in sediments are given in Table 7 along with calculated removal rates assuming a 1 mm yr^{-1} sedimentation rate. Unfortunately, there are no estimates of sedimentation rates on the SAB shelf, but the sediments are relict, reworked materials, and sedimentation rates are probably much lower than the assumed 1 mm yr^{-1}.

Assuming that the total trace metal content of shelf sediments (Table 7) is derived from overlying shelf waters, maximum metal removal rates can be calculated for given sedimentation rates. Annual metal inputs to the shelf can be estimated as the products of the zero-salinity end-member concentrations (Table 5) and the total mean annual freshwater discharge (66 km^3). Figure 9 shows the percentages of annual inputs that can be accounted for by the calculated sediment removal for given sedimentation rates. These results suggest that a sedimentation rate of only several cm/10^3 yr is necessary to account for the observed removal of iron and zinc. This rate of sedimentation, while probably reasonable, cannot explain the removal of manganese. Such a sedimentation rate would also result in insignificant removals of cadmium, copper, and nickel, agreeing with conclusions that these three metals are conservatively mixed across the shelf.

TABLE 7. Metal Concentration in Shelf Sediment: Estimate of Potential Removal of Metal Due to Accumulation in Sediments

	Mean Concentration In Shelf Sediments,* $\mu g\ g^{-1}$	Potential Rate of Removal From Water Column,† nmol $kg^{-1}\ yr^{-1}$
Iron	1090	1775†
Manganese	47‡	81
Cadmium	0.026	0.021
Copper	0.57	0.82
Nickel	0.87	1.3
Zinc	2.4	3.3

*From *Windom and Betzer* [1979] excluding slope sediments.

†Assuming a sedimentation rate of 1 mm/yr and sediment density of 2.5 g cm^{-3}.

‡From *Bothner et al.* [1980].

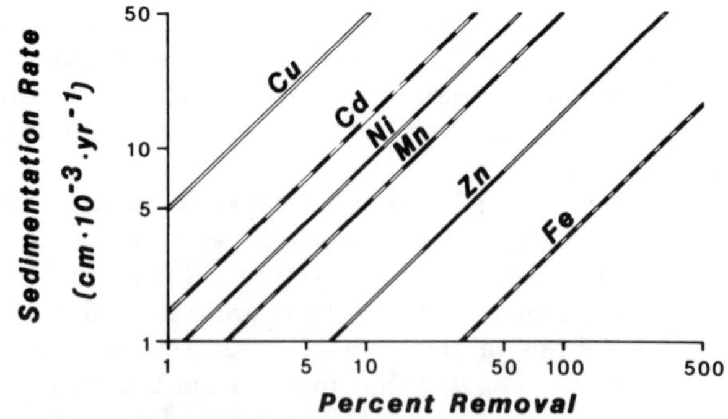

Fig. 9. Theoretical efficiency of trace metal removal due to sedimentation on SAB shelf (in percent of total input), using data from *Windom and Betzer* [1979] and *Bothner et al.* [1980].

The above discussion suggests that manganese does not accumulate on the shelf to the degree that iron does. Manganese may be adsorbed from solution onto fine particles which settle slowly. If the residence time of these particles in the water column is longer than the 0.14- to 0.59-year residence time of water, manganese removed from solution may be advected off the shelf in particulate form.

Acknowledgments. The authors wish to thank the crews of the research vessels *Blue Fin, Kit Jones,* and *Cape Henlopen*. Thanks are also due to Dannah McCauley, who typed the manuscript, and Michael Bewers and Thomas Church, who critically reviewed the manuscript. This research was sponsored by the U.S. Department of Energy (contract DE-AS09-76EV00890).

References

Atkinson, L. P., Modes of Gulf Stream intrusion into South Atlantic Bight shelf waters, *Geophys. Res. Lett., 4,* 583–586, 1977.

Atkinson, L. P., T. N. Lee, J. O. Blanton, and W. S. Chandler, Climatology of the southeastern United States continental shelf waters, *J. Geophys. Res., 88,* 4705–4718, 1983.

Bewers, J. M., and P. A. Yeats, Trace metals in the waters of a partially mixed estuary, *Estuarine Coastal Mar. Sci., 7,* 147–162, 1978.

Bigham, G. N., Clay mineral transport on the inner continental shelf of Georgia, M.S. thesis, 49 pp., Ga. Inst. of Technol., Atlanta, 1972.

Blanton, J. O., and L. J. Pietrafesa, Flushing of the continental shelf south of Cape Hatteras by the Gulf Stream, *Geophys. Res. Lett., 5,* 495–498, 1978.

Blanton, J. O., L. P. Atkinson, L. J. Pietrafesa, and T. N. Lee, The intrusion of Gulf Stream water across the continental shelf due to topographically-induced upwelling, *Deep Sea Res. Part A, 28,* 393–405, 1981.

Bothner, M. H., P. J. Aruscarage, W. M. Ferrebee, and P. A. Baedecker, Trace metal concentrations in sediment cores from the continental shelf off the southeastern United States, *Estuarine Coastal Mar. Sci., 10,* 523–541, 1980.

Boyle, E. A., and S. S. Huested, Aspects of the surface distributions of copper, nickel, cadmium and lead in the North Atlantic and North Pacific, in *Trace Metals in Seawater,* edited by C. S. Wong, E. Boyle, K. W. Bruland, J. D. Burton, and E. D. Goldberg, pp. 379–394, Plenum, New York, 1983.

Boyle, E. A., R. Collier, A. T. Dengler, J. M. Edmond, A. C. Ng, and R. F. Stallard, On the chemical mass-balance in estuaries, *Geochim. Cosmochim. Acta, 38,* 1719–1728, 1974.

Boyle, E. A., S. S. Huested, and S. P. Jones, On the distribution of copper, nickel, and cadmium in the surface waters of the North Atlantic and North Pacific Ocean, *J. Geophys. Res., 86,* 8048–8066, 1981.

Boyle, E. A., S. S. Huested, and B. Grant, The chemical mass-balance of the Amazon plume, 2, Copper, nickel, and cadmium, *Deep Sea Res., 29,* 1355–1364, 1982.

Bruland, K. W., and R. P. Franks, Mn, Ni, Cu, Zn and Cd in the western North Atlantic, in *Trace Metals in Seawater,* edited by C. S. Wong, E. Boyle, K. W. Bruland, J. D. Burton, and E. D. Goldberg, pp. 395–414, Plenum, New York, 1983.

Crecelius, E. A., Prediction of marine atmospheric deposition rates using total ^{7}Be deposition velocities, *Atmos. Environ., 15,* 579–582, 1981.

Duce, R. A., G. L. Hoffman, B. J. Ray, I. S. Fletcher, P. R. Walsh, E. J. Hoffman, J. M. Miller, J. L. Heffter, G. T. Wallace, J. L. Fasching, and S. R. Piotrowicz, Trace metals in the marine atmosphere: Sources and fluxes, in *Marine Pollutant Transfer,* edited by H. L. Windom and R. A. Duce, pp. 77–119, D. C. Heath, Lexington, Mass., 1976.

Duinker, J. C., and R. F. Nolting, Mixing, removal and mobilization of trace metals in the Rhine estuary, *Neth. J. Sea Res., 12,* 205–223, 1978.

Eaton, A., Observations on the geochemistry of soluble copper, iron, nickel and zinc in the San Francisco Bay estuary, *Environ. Sci. Technol., 13,* 425–432, 1979.

Elderfield, H., A. Hepworth, P. N. Edwards, and L. M. Holliday, Zinc in the Conwy River and estuary, *Estuarine Coastal Mar. Sci., 9,* 403–422, 1979.

Evans, D. W., N. H. Cutshall, F. A. Cross, and D. A. Wolfe, Manganese cycling in the Newport estuary, North Carolina, *Estuarine Coastal Mar. Sci., 5,* 71–80, 1977.

Graham, W. F., M. L. Bender, and G. P. Klinkhammer, Manganese in Narragansett Bay, *Limnol. Oceanogr., 21,* 665–673, 1976.

Grant, B., M.-H. Hu, E. Boyle, and J. Edmond, Comparison of the trace metal chemistry in the Amazon, Orinoco, and Yangtze plumes (abstract), *Eos Trans. AGU, 63,* 48, 1982.

Hess, J., Chemical behavior of Mn and Fe in the Pee Dee River, Winyah Bay estuary, South Carolina, M.S. thesis, 53 pp., Univ. of S. C., Columbia, 1980.

Klinkhammer, G. P., and M. L. Bender, Trace metal distributions in the Hudson River estuary, *Estuarine Coastal Shelf Sci., 12,* 629–643, 1981.

Lee, T. N., and L. P. Atkinson, Low-frequency current and temperature variability from Gulf Stream frontal eddies and atmospheric forcing along the southeast U.S. continental shelf, *J. Geophys. Res., 88,* 4541–4568, 1983.

Li, Y., and L. Chan, Desorption of Ba and ^{226}Ra from riverborne sediments in the Hudson estuary, *Earth Planet. Sci. Lett., 43,* 343–350, 1979.

McCartney, M. S., L. V. Worthington, and M. E. Raymer, Anomalous water mass distributions at 55W in the North Atlantic in 1977, *J. Mar. Res., 38,* 147–174, 1980.

Mullins, B. M., Geochemical aspects of atmospherically transported trace metals over the Georgia Bight, M.S. thesis, 66 pp., Ga. Inst. of Technol., Atlanta, 1978.

Sholkovitz, E. R., and D. Copland, The coagulation, solubility and adsorption properties of Fe, Mn, Cu, Ni, Cd, Co and humic acids in a river water, *Geochim. Cosmochim. Acta, 45,* 181–189, 1981.

Smith, R. G., Jr., and H. L. Windom, A solvent extraction technique for determining nanogram per liter concentrations of cadmium, copper, nickel and zinc in seawater, *Anal. Chim. Acta, 113,* 39–46, 1980.

Wallace, G. T., Trace element geochemistry of the South Atlantic Bight, progress report, contract DE-AS09-76EV00890, 28 pp., U.S. Dep. of Energy, Washington, D. C., 1980.

Waslenchuk, D. G., The budget and geochemistry of arsenic in a continental shelf environment, *Mar. Chem., 7,* 39–52, 1978.

Waslenchuk, D. G., and H. L. Windom, Factors controlling the estuarine chemistry of arsenic, *Estuarine Coastal Mar, Sci., 7,* 455–462, 1978.

Windom, H. L., Comparison of atmospheric and riverine transport of trace elements to the continental shelf environment, in *River Input to Ocean Systems,* pp. 360–369, United Nations Environmental Program and United Nations Education, Scientific and Cultural Organization, Geneva, 1981.

Windom, H. L., and P. R. Betzer, Trace metal chemistry, in *Results of Studies of Georgia Bight of the South Atlantic Ocean: South Atlantic Benchmark Program, Outer Continental Shelf Environmental Study,* final report, contract AA 550-CT7-2, pp. 153–196, Bureau of Land Management, Washington, D. C., 1979.

Windom, H. L., and R. G. Smith, Jr., Distribution of cadmium, cobalt, nickel and zinc in southeastern United States continental shelf waters, *Deep Sea Res., 19,* 727–730, 1972.
Windom, H. L., and R. G. Smith, Jr., Copper concentrations in surface waters off the southeastern Atlantic coast, U.S.A., *Mar. Chem., 7,* 157–163, 1979.
Windom, H. L., and F. E. Taylor, The flux of mercury in the South Atlantic Bight, *Deep Sea Res. Part A, 26,* 283–292, 1979.
Windom, H. L., G. Wallace, R. Smith, N. Dudek, M. Maeda, R. Dulmage, and F. Storti, Behavior of copper in southeastern United States estuaries, *Mar. Chem., 12,* 183–193, 1983.
Wollast, R., G. Billen, and J. C. Duinker, Behavior of manganese in the Rhine and Scheldt estuaries, 1, Physicochemical aspects, *Estuarine Coastal Mar. Sci., 9,* 161–169, 1979.
Worthington, L. V., *On the North Atlantic Circulation, Johns Hopkins Oceanogr. Stud.,* vol. 6, 110 pp., Johns Hopkins University Press, Baltimore, Md., 1976.
Young, J. A., and W. B. Silker, Aerosol deposition velocities on the Pacific and Atlantic oceans calculated from ^{7}Be measurements, *Earth Planet. Sci. Lett., 50,* 92–104, 1980.

Summary and Future Plans for South Atlantic Bight Oceanography

K. A. Bush, J. O. Blanton, and L. P. Atkinson

Skidaway Institute of Oceanography, Savannah, Georgia

Summary: South Atlantic Bight Oceanography

The continental shelf of the southeastern United States is unique among United States coastal regions because of its direct coupling with the Gulf Stream boundary current. A wide shallow shelf with relatively smooth, gently sloping bottom topography and large tidal amplitudes further characterize the South Atlantic Bight (SAB) marine environment. Latitudinal variations occur in weather-related physical processes, in Gulf Stream–shelf water interactions, and in nearshore processes such as freshwater influx and estuary–shelf water interactions. Significant cross-shelf variations occur in physical, chemical, and biological processes. Winds provide the primary forcing to circulation in regions not directly influenced by the Gulf Stream. Fluctuations in the Bermuda-Azores High, the Icelandic Low, and the Ohio Valley High influence the wind regime of the SAB. Significant forcing at periods of 3–10 days occurs from the passage of cold fronts. Infrequent major impulses result from the passage of tropical cyclones (hurricanes) during summer and early autumn.

Although historically lacking in oceanographic exploration, the SAB is now the focus of extensive ongoing research on the influence of the Gulf Stream on shelf oceanography, shelf mixing processes, and the coupling between estuarine waters and inner and middle shelf waters. Multidisciplinary research programs have been undertaken since the mid-1970's to determine the physical processes that drive the circulation and control the chemistry, biology, and geology of the SAB. These programs were designed to synoptically study Gulf Stream eddies and intrusions onto the shelf. Long-term current, temperature, and pressure records from moored instruments have provided information on temporal and spatial correlations of oceanic processes. The use of modern technology, particularly satellite remote imagery, has enhanced the observation and interpretation of these processes.

The SAB shelf is divided into three regions based on topography and dynamics. The inner shelf, with water depths less than 20 m, is dominated by tidal currents, river inflow, local winds, and interaction with coastal estuaries. This is the only region influenced by the diurnal wind regime. It is characterized by a band of low-salinity water from river runoff, which establishes a frontal zone adjacent to the coast. Tidal currents dominate inner shelf variability, accounting for approximately 80% of the total kinetic energy there. The middle shelf, with water depths of 20–40 m, is dominated by local wind forcing and pressure gradients and exhibits simple locally forced Ekman dynamics, with northward mean flow. There is some interaction with both the coastal environment and the Gulf Stream. This region is vertically mixed in fall and winter and stratified in spring and summer as a result of decreased wind mixing, cross-shelf density forcing, and intrusions of upwelled Gulf Stream water. Approximately 50% of the current variability in the middle shelf is due to tides, and 50% is due to low-frequency events. The outer shelf, with water depth of 40–75 m, is dominated by Gulf Stream events. Low-frequency flow variability on the outer shelf results from meanders and eddies in the Gulf Stream front that occur at near weekly intervals.

The northern and southern regions of the SAB are each bounded by a prominent cape. The Carolina Capes region, north of Cape Romain, North Carolina, is particularly distinguished by complex topography, prominent shoals, and cuspate bays. Physical processes here are primarily driven by synoptic scale wind and Gulf Stream events. Gulf Stream meanders and eddies can interact with bottom topography and modify middle and inner shelf water masses by means of intensified upwelling.

The distribution of nutrients and hydrographic parameters in the SAB depends most importantly on interactions of shelf waters with the Gulf Stream, although atmospheric forcing and interaction with rivers and estuaries are also important. Primary production on the inner shelf is relatively constant and sustained by nutrient recycling and by nutrient inputs from rivers and salt marshes. Upwelled Gulf Stream intrusions are the major source of nutrients to the middle and outer shelf regions, with upwelled nitrate an important source of new nitrogen on the outer shelf. Intense phytoplankton blooms occur on the middle and outer shelf within intrusions of upwelled water. The episodic nature of these blooms results in a discontinuous food supply for zooplankton. High zooplankton abundance occurs within summer intrusions where upwelled waters remain on the shelf longer than during other seasons. The SAB does not support large stocks of adult fish.

Because of the consistently high productivity of the inner shelf, bacteria there are not limited by the availability of substrates. They are larger and an order of magnitude more abundant on the inner shelf than on the middle and outer shelves. During intrusions on the outer shelf, bacteria increase in number but not in size. Soft bottom microinfauna show a distinct inner shelf community that persists throughout the year. In general, microinfauna have high species diversity and low densities.

Trace element concentrations and distributions are largely controlled by the mixing of river and oceanic waters but are also influenced by atmospheric transport and release from nearshore suspended and bottom sediments. The estuarine and inner shelf environment is a primary location for the deposition and resuspension of trace elements. The inner shelf front inhibits the direct transfer of dissolved and particulate material from the inner to middle shelf.

Direction of Future Investigations

Oceanographic studies in the SAB have provided extensive information on the physical, biological, and chemical processes and interactions that occur in the middle and outer shelf regions. However, inner shelf processes, including the transport and dispersal of materials entering the region with freshwater discharge, are not well understood. Evidence suggests that low-salinity water near the coast is carried offshore in spring and alongshore and

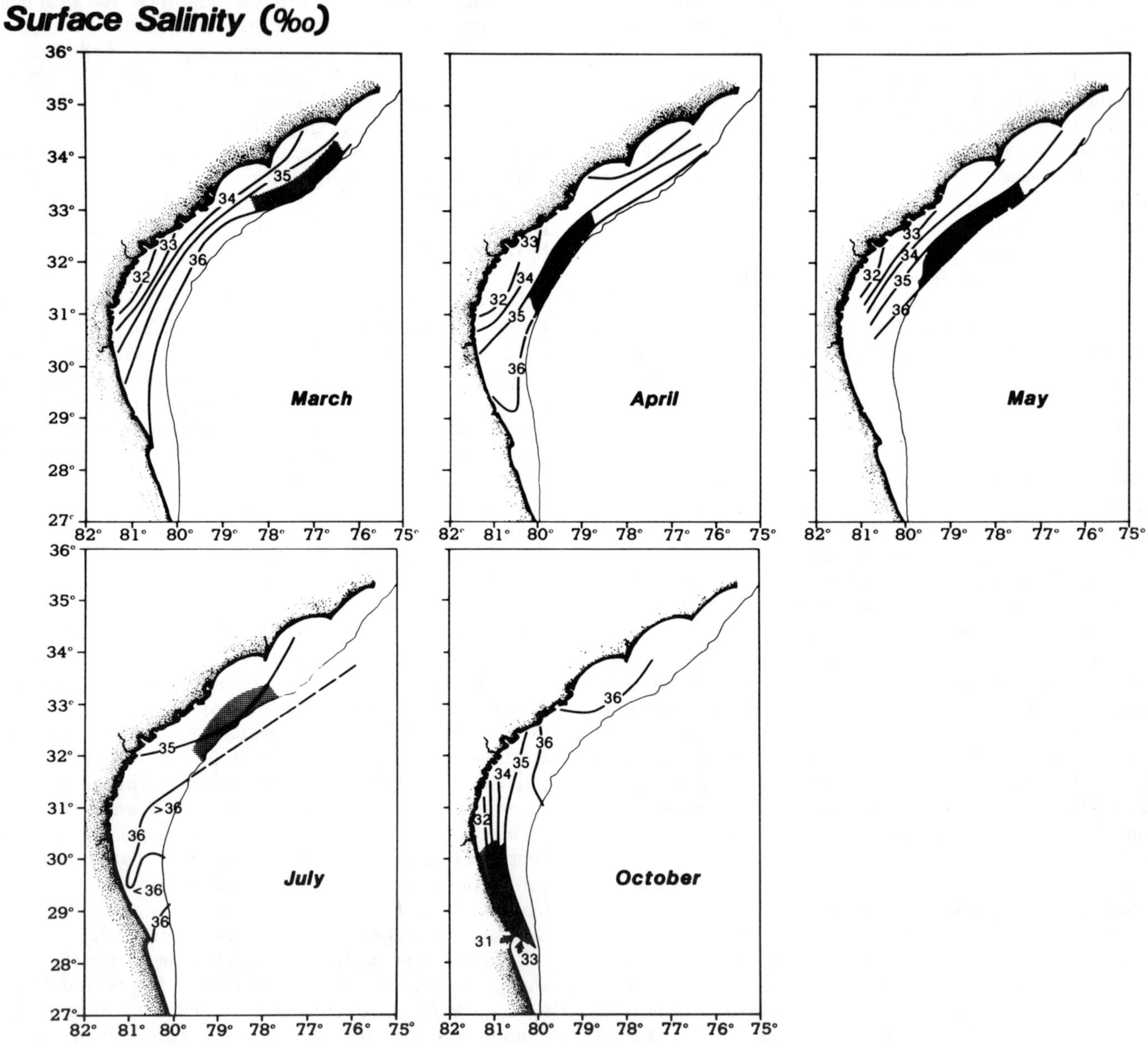

Fig. 1. Mean surface salinity for spring and fall months. Areas with diagonal lines indicate where fresher coastal waters are extending across the shelf to the shelf break. Data from *Atkinson et al.* [1983].

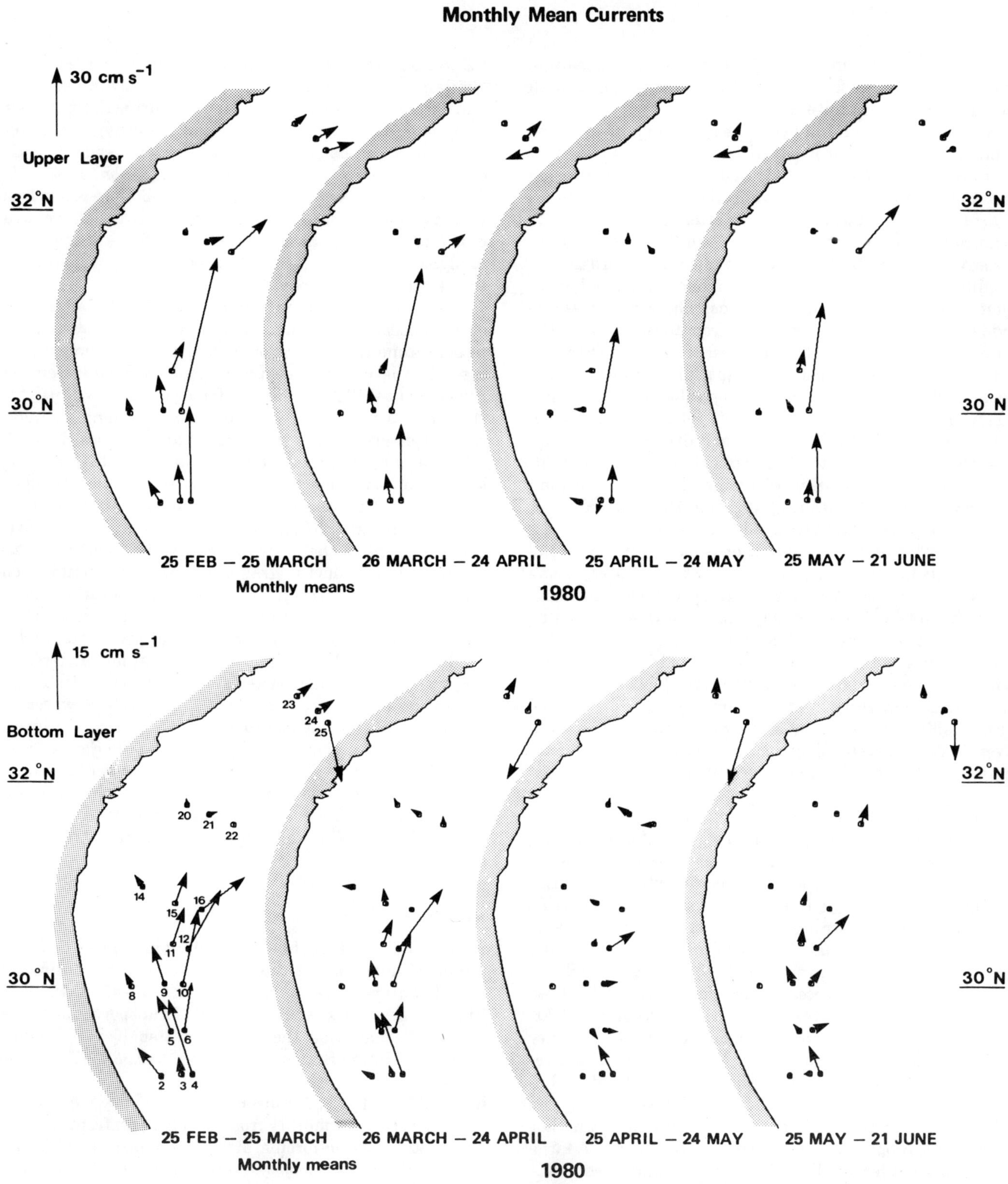

Fig. 2. Monthly mean near-surface (top panel) and near-bottom (lower panel) flows at the 28-, 40-, and 75-m isobaths during spring 1980 (February to June). A persistent southward flow in the vicinity of 32°N is evident, particularly for near-bottom currents. Current meter mooring numbers are given in the lower left panel [from *Lee et al.*, this volume].

southward in autumn. This low-salinity water appears to reach the shelf break at 32° to 34°N in the spring and off northern Florida in autumn.

Nearshore fronts inhibit cross-shelf mixing during most of the year and are important in determining the transport and fate of recycled materials and of materials entering the continental shelf waters from rivers, marshes, and estuaries. Dissolved and suspended matter occur in relatively high concentrations in frontal zones on the inner shelf off Georgia and South Carolina, as indicated by cross-shelf profiles of chlorophyll *a*, dissolved and suspended trace elements, and density distributions. There is increasing evidence that material is released from these fronts to the outer shelf where it may be entrained into the Gulf Stream. Whether the process is slow and spatially uniform, or episodic and spatially irregular, as suggested by satellite imagery and shipboard observations, is yet to be determined.

Climatological distributions of salinity (Figure 1) suggest that low-salinity water is carried northward and offshore in spring and southward against the coast in autumn [*Atkinson et al.*, 1983]. During April and May, surface water with salinity less than 36‰ is found along the shelf break between 32° and 33°N. This water is clearly a mixture of low-salinity inner shelf water and Gulf Stream water. Spring and summer winds (March through July) blow northeastward and offshore, presumably removing inner shelf surface water to offshore regions. During October, lowest-salinity water is found adjacent to the coast of Florida. Autumn winds (September through November) blow southwestward and average currents are southward. This regime presumably carries water southward toward Cape Canaveral in a narrow band against the coast, where the water may be deflected offshore by currents steered by bathymetry.

The removal of low-salinity water adjacent to the Georgia coast occurs most rapidly in spring just after maximum river discharge [*Blanton and Atkinson*, 1983], and the removal rate is related to alongshore wind stress. Water in the upper layers of the inner shelf frontal zone is carried northward, with a strong offshore component. This supports the argument that the low-salinity water originally present in surface layers adjacent to the Georgia and South Carolina coasts is advected toward the outer shelf between 32° and 33°N. Imagery from the National Oceanic and Atmospheric Administration's Coastal Zone Color Scanner (CZCS) also suggests that processes on the inner shelf release material offshore. Nearshore tonguelike distributions of chlorophyll *a* point offshore at intervals along the coast. Whether this distribution is typical and whether tongues are propagated or stationary is not known. A large pool of high chlorophyll *a* concentration water seen situated near the shelf break may result from the cyclonic Charleston Gyre north of 32°N. The circulation resulting from this gyre could cause low-salinity water that had advected into the western side of the eddy to be entrained into the Gulf Stream, where it would be removed from the shelf. Further study of chlorophyll *a* distributions from CZCS satellite-derived imagery is required to determine if similar eddies in the vicinity of Cape Canaveral could remove shelf water in a similar manner during autumn. Current meter data from GABEX I support the hypothesis that shelf water can be advected into a cyclonic eddy near the shelf break north of 32°N [*Lee and Atkinson*, 1983]. The data show northward advection of near-bottom water from regions south of 32°N and southward advection from regions north of this area (Figure 2). A finite element hydrodynamic model [*Lee et al.*, this volume] also supports the existence of a removal area near 32°N.

A key objective of further research in this area is to obtain observations of nearshore circulation processes and to determine rates of major biological and chemical processes in the nearshore environment. A Spring Removal Experiment (SPREX) planned for April 1985 will use two ships and a large array of moored instruments to investigate springtime nearshore removal processes. Low-salinity water appears near the shelf break at 32°N in spring, but neither the trajectory of this water nor the biological and chemical processes that occur during its transit are known. During SPREX, current meters will be deployed across the entire shelf. Two ships will be used over a 20-day period to conduct Lagrangian flow experiments, to obtain continuous horizontal profiles of temperature, salinity, and chlorophyll *a*, to map hydrographic properties of shelf waters, to determine trace metal gradients, and to conduct biological sampling (time series) for phytoplankton, zooplankton, and bacteria. In addition, tritium concentration will be used as a tracer of nearshore water.

During autumn, low-salinity water is carried southward along the coast and is usually found along the Florida coast in October. This water is presumed to be entrained and removed by energetic eddies influenced by bottom topography near Cape Canaveral. A future experiment is planned for the Georgia-Florida area to investigate this phenomenon.

References

Atkinson, L. P., T. N. Lee, J. O. Blanton, and W. S. Chandler, Climatology of the southeastern U.S. continental shelf waters, *J. Geophys. Res., 88*, 4705–4718, 1983.

Blanton, J. O., and L. P. Atkinson, Transport and fate of river discharge on the continental shelf of the southeastern United States, *J. Geophys. Res., 88*, 4730–4739, 1983.

Lee, T. N., and L. P. Atkinson, Low-frequency current and temperature variability from Gulf Stream frontal eddies and atmospheric forcing along the southeast U.S. continental shelf, *J. Geophys. Res., 88*, 4541–4568, 1983.

Lee, T. N., V. Kourafalou, J. D. Wang, W. J. Ho, J. O. Blanton, L. P. Atkinson, and L. J. Pietrafesa, Shelf circulation from Cape Canaveral to Cape Fear during winter, in *Oceanography of the Southeastern U.S. Continental Shelf, Coastal and Estuarine Sci.*, vol. 2, edited by L. P. Atkinson, D. W. Menzel, and K. A. Bush, this volume.